LARGE AC GENERATORS

AND THE SMART GRID

NUMERICAL SOLUTIONS

By Orlando N Acosta

Large AC Generators and the Smart Grid

Numerical Solutions

PREFACE

The electrical power grid fundamentally consists of a giant conglomerate of generating power plants, very high voltage transmission lines and electrical substations. Actually, the power grid is an irregular network of electrical wires and generating stations. The HV transmission lines interconnect most of the power generating plants through dedicated substations which receive the power from the adjunct generating station and convert it to very high voltage before sending the electrical power to the grid. This simple design was conceived early in the 20th century and was, and still is, the largest and most efficient way of supplying power to a big country like ours. In the other hand the smart-grid is a figment of the imagination of uneducated people trying to apply what they think they know. This group of well-meaning people had merged with a large number of desperate politicians looking for a lifesaving cause, and a bunch of "wannabe" entrepreneurs. In the other hand the smart grid is not a grid; because what they refer to are always low voltage radial distribution feeders. Off course, this is very convenient to them, because all the improvements so far proposed take place at the user's home and no in the power grid. The people will lose control of all the electrical devices in their homes, including smart air conditioners, smart-dish-washing machines, and smart electric sockets. However, the most important improvement for the people, the one that will free them from storms triggered blackouts has not been proposed. *Convert the local distribution feeders and all wires bringing the electrical power into the user's homes from overhead to underground.*

The main problem with the existing power grid is that in rare occasions when a generator gets unstable induces instability in other nearby or remote connected generators creating a giant blackout. The solution of this problem is collaboration among the utilities connected to the power grid. They should establish a set of rules for every generator connected to the grid must comply with. Including large AC generators design details, required generator and power system protection, and power plant operation protocols.

Chapters 1- 5 provide a review of fundamental power electrical engineering concepts, Chapters 6 and 7 deals with the design and operation of large AC synchronous generators. Chapter 8 explains the methods for grounding the generators neutral and solves a typical example. Chapters 9 – 11 deals with the numerical solution of the stability of multi machine power systems, and solve a transient stability problem in a four generators power grid. Chapter 12 provides an elementary review of the electromagnetic pulse that could be produced as a consequence of an attack against the United States using fission atomic weapons. This giant pulse could impact almost the entire country simultaneously and severely damaging all three high voltage power grids existing in the United States with the exceptions of Alaska and the Hawaiian Islands. Once the pulse is generated there is not a 100 percent effective defense against it, with the exception of burying deep underground all the grids' components. However, the possibility of this attack occurring is very remote, because the enemy surely knows that the U.S. has the largest fleet of nuclear submarines in the world, and they are everywhere, deep under all the oceans of the world. And each of them carries sixteen nuclear fusion weapons already attached to high precision missiles. So, is far easier for the enemy to attack the grid directly, in fact I believe that they are already taking the necessary data. Our power grids have become very vulnerable since almost all the protection and controlling devices were digitized. No too long ago, the power systems protection was based on induction disk relay units or plunger type relays and for low voltage applications bimetallic-bar based molded case breakers were and still are very common. *It was impossible to change the settings of these relays without actually going to the site where the relay was installed.* When I was a young engineer I did exactly that many times. Essentially, it consists in pulling *out a bolt from one hole and screwing it into another hole.* Now, on the contrary, the settings are digitized numbers assigned to pins in a register or microprocessor. Once you penetrate the programmed firewall you can change the settings in the blink of an eye. The old relays offer safer protection because they are immune to attacks launched from remote locations.

Although I have not attempted to give specific credit at each point, I have used the books listed in the bibliography as sources of data and information, and my style of writing has been influenced by them.

Orlando N Acosta, MSEE

Contents

CHAPTER 1

Power Electrical Engineering

Basic Knowledge

1.1 Magnetic Circuit Formulas and Units

In a magnetic circuit the lines of magnetic flux are always closed lines and they are continuous without ends. The magnetic flux commonly designated as ϕ is expressed in Maxwell or Weber

1 weber = 108 maxwell

1 maxwell = 1 line

The flux density usually designated as B is commonly expressed in gauss which is a maxwell per square centimeter. Sometimes the weber per square meter is also used. Symbolically:

$$B = \frac{\phi}{A} \qquad \text{gauss} \qquad\qquad (1.1)$$

B = flux density in gauss

ϕ = flux in maxwell

A = magnetic circuit net cross-sectional area in square centimeter taken perpendicular to the flux lines.

1 gauss = 1 line per square centimeter

1 gauss = 6.452 lines per square inch

1 gauss = 10-4 weber per square meter

1 gauss = 10-8 volt seconds / cm2

1 tesla = 1 weber per square meter = 104 gauss

The magnetomotive force provided by a magnetizing coil is proportional to the electric current flowing in the coil and to the number of turns, wound in the same direction, of the coil. Generalizing, the magnetomotive force is proportional to the product of the electric current and the number of linkages between the coil wire and the magnetic circuit. Actually, it is a measure of the strength of the source of magnetic flux. The unit of magnetomotive force in the cgs system is the Gilbert. Symbolically:

$$F = N \cdot I \qquad ampere - turn \qquad \text{1 amper-turn = 1.257 gilbert} \qquad (1.2)$$

Magnetic intensity is the magnetic potential drop per unit length. And it is also known as magnetic field intensity or magnetizing force. The magnetic intensity usually designated as H is expressed in amper-turn per inch or per meter. In the cgs system is expressed in oersted. Symbolically:

$$H = \frac{N \cdot I}{1} \qquad \frac{ampere - turn}{inch} \qquad \text{1 ampere-turn per inch = 0.495 oersted} \quad (1.3)$$

$$H = \frac{F}{1}$$

$$F = H \cdot 1$$

$$F = H_1 \cdot 1_1 + H_2 \cdot 1_2 + ... + H_n \cdot 1_n$$

The magnetic permeability of a material is the ratio of B to H and is represented by μ. symbolically:

$$\mu = \frac{B}{H} \qquad \text{gauss per oersted} \qquad (1.4)$$

Where B is in gausses and H is in oersted. The permeability of vacuum, μ 0, is 1 gauss per oersted in the cgs system of units. In the RMKS units μ 0 is 1.257 x 10-6 weber per ampere-turn-meter. The specific permeability, μ s, is the ration of the permeability of a material to the permeability of vacuum. In cgs units the specific permeability is equal to the permeability of the material. In general when μ is given without specifying its units, what is meant is specific permeability. In a given magnetic circuit the ratio of the mmf to the flux is called the reluctance of the magnetic circuit. Symbolically:

$$R = \frac{F}{\phi} \qquad \text{ampere-turn per weber} \qquad (1.5)$$

In considering a part of a magnetic circuit with constant cross section, A, and of length l we can write Eq. (1.5) as follows:

$$R = \frac{H \cdot l}{B \cdot A} = \frac{l}{\mu \cdot A} \qquad (1.6)$$

The reluctance of a magnetic circuit concept is analogous to the resistance concept of an electric circuit. *However, instead of dissipating electrical energy it stores magnetic energy.* The magnetic field, H, forces its magnetic flux to follow the path of least magnetic reluctance.

1.2 Electromagnetic Induction:

Lenz's Law:

$$e = -N \cdot \left(\frac{d}{dt}\phi \right) \text{volt} \qquad (1.7)$$

e = induced electromotive force (emf) in volts

N = number of turns

φ = magnetic flux in webers

t = time in seconds

If the flux is expressed in maxwells Eq. (1.7) becomes:

$$\left[e = -N \cdot \left(\frac{d}{dt}\phi \right) \cdot 10^{-8} \right] \text{volt} \qquad (1.8)$$

The electromotive force induced in a coil of wire is proportional to the rate of change of the magnetic flux linking with the electrical circuit. Strictly speaking Faraday's law is only applicable to **wires** linking with the flux of a magnetic circuit. And it could produce misleading results if apply to the interaction between magnetic fields and metal surfaces or solids. However the Maxwell-Faraday equation which is beyond the scope of this book is applicable to all situations.

The minus sign in equations (1.7 - 1.8) means that the voltage induced in the electrical circuit produces a current which creates a magnetic field which opposes the magnetic field change that generated the voltage. In fact, the voltage induced by a magnetic flux change is of such polarity that always try to keep the magnetic flux through the coil constant. The minus sign is the symbolic and succinct statement of the Lenz's law. This law is very important for understanding the operation of many electrical apparatus, such as synchronous generators. Figure 1-1 illustrates the application of the electromagnetic induction Eq. (1-7) to a simple case.

$$\phi_{max} := 1 \text{weber} \qquad f := 60\text{Hz} \qquad \omega := 377 \cdot \frac{\text{rad}}{\text{sec}}$$

$$N := 100 \qquad t := 0, 0.001 .. 0.017 \qquad \phi = 1 \cdot \sin(377 \cdot t)$$

$$\frac{d}{dt}\phi = \frac{1}{377} \cdot \cos(377 \cdot t)$$

Applying Eq. (1.7) we obtain

$$e = -\frac{100}{377} \cdot \cos(377 \cdot t)$$

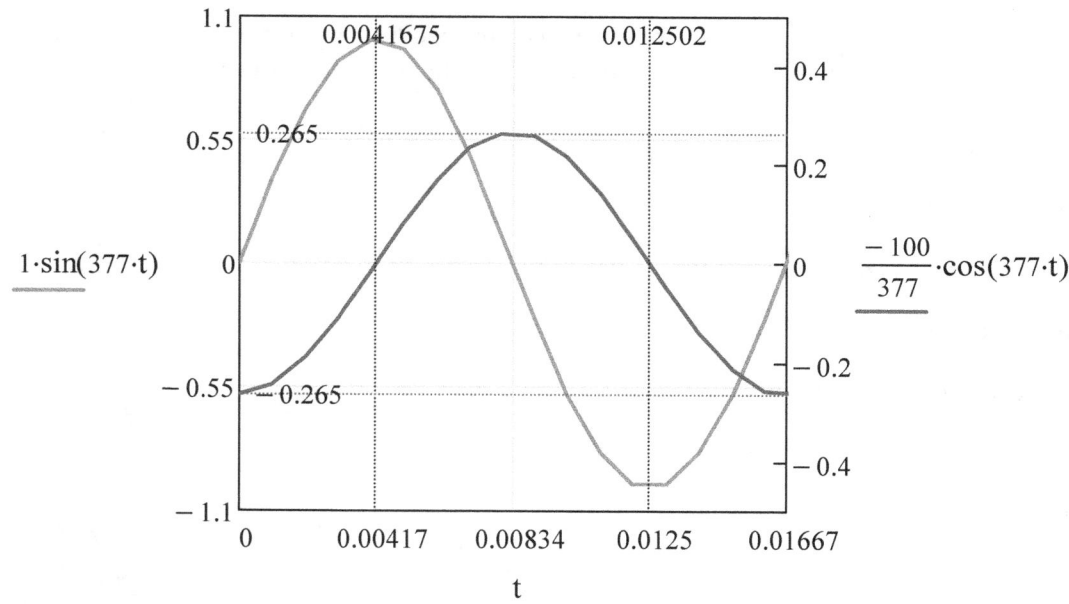

Time in seconds

Figure 1-1 Illustration of the Lenz's magnetic induction law

From Fig. (1-1) it is obvious than the electromotive force induced in the wire lags the changing flux that induced it by 90 degrees. The slope of the flux curve at the peak is horizontal (zero), and at that moment, the induced voltage is crossing the zero line at maximum rate of change.

1.3 Force on Current Carrying Conductor Situated in a Magnetic Field.

Let us consider a current carrying conductor (part of an electric circuit) that is situated perpendicular to a uniform magnetic field of flux density B. In this case there is a force acting on the conductor that tends to move it perpendicular to the magnetic field as illustrated in Fig. 1-2.

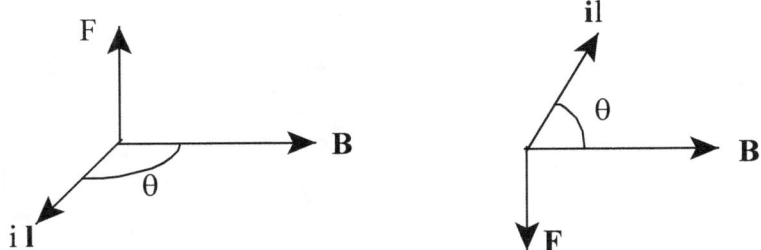

Figure 1-2 Force acting on the conductor

Vector equation (1.9) provides the force acting on the conductor en magnitude and direction and is applicable even when the conductor is not at right angle with respect to B.

$$\mathbf{F} = i\mathbf{l} \times \mathbf{B} \tag{1.9}$$

F = force acting on the conductor

l = length of the conductor linking with B

B = magnetic flux density

i = current circulating in the conductor

θ = angle between the conductor and B

Equation (1.9) could also be expressed as:

$$\mathbf{F} = il.B.\sin(\theta)\mathbf{u} \tag{1.10}$$

Where **u** is a unit vector perpendicular to the plane of il and **B** and such that **il, B, u** form a right handed system. if θ = 90 degrees then Eq. (1.10) becomes:

$$\mathbf{F} = il.B.\mathbf{u} \tag{1.11}$$

For the RMKS system of units F is in newtons, B is in webers per square meter, the current in amperes and the conductor length in meters.

1.4 Torque on Current Carrying Coil

Consider a single turn, current carrying, rectangular coil of wire coupled to a uniform and stationary magnetic field with flux density B and with sides 1 and 3 always perpendicular to B. Figures 1-3a and 1-3b illustrate the setup.

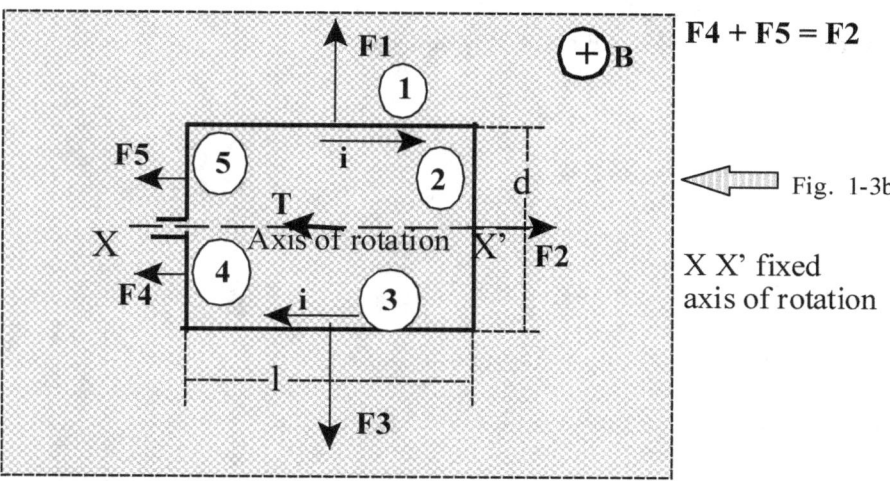

$$F4 + F5 = F2$$

Fig. 1-3b

X X' fixed
axis of rotation

Figure 1-3a Rectangular coil conducting current and placed
inside a uniform and stationary magnetic field of density B

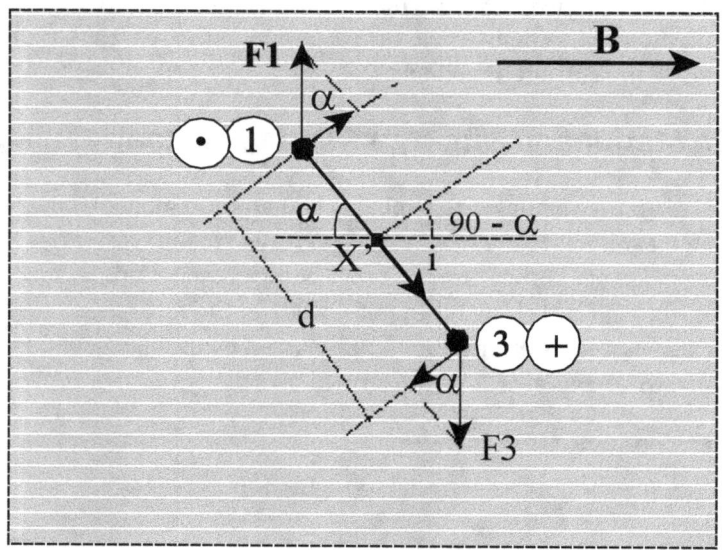

Figure 1-3b Coil side view showing the different line
of actions of the forces acting on conductors 1 and 3.

As illustrated in Fig. 3a the forces acting on the coil are: F1, F2, F3, F4, F5. However, the sum of
the forces acting on wire segments 4 and 5 is equal and with the same line of action as the
force acting on wire segment 2, but of opposite direction, and therefore they cancel each
other. So, the resultant torque produced by F2, F4 and F5 is zero. The angle θ between **il and B**
for side 1 or 3 always is 90 degrees. Although, F1 and F3 are equal in magnitude, ilB, in general
they do not have the same line of action and therefore there is a net torque which tends to

rotate the coil clockwise about the axis XX'. The torque acting on the coil is a vector pointing into Fig. 3b at X' and acting along the axis XX' as illustrated in Fib. 3a. The magnitude of the total torque acting on the coil is d.i.l.B.cos(α). Symbolically:

$$F1 = F2 = i \cdot l \cdot B$$

$$T = 2 \cdot \frac{d}{2} \cdot i \cdot l \cdot B \cdot \cos(\alpha) = d \cdot i \cdot l \cdot B \cdot \cos(\alpha)$$

(1.12)

Which can be simplified as shown in equation (1.13).

$$T = A \cdot i \cdot B \cdot \cos(\alpha)$$ Where A is the coil area.

(1.13)

If the coil has N turns then Eq. (1.13) becomes:

$$T = N \cdot A \cdot i \cdot B \cdot \cos(\alpha)$$

(1.14)

Eqs (1.12) to (1.14) are applicable to any coil shape enclosing *a plane area.* Furthermore maximum torque occurs when the plane of the coil is parallel to the vector representing the flux density, that is when α = 0 or α = 180 degrees and the torque becomes zero when the plane of the coil is perpendicular to the flux density vector. The reader must realize that the sign of the torque also depends of the current direction.

1.5 Voltage Induced in Stationary Wire by a Moving Magnetic Field.

The stationary coil illustrated in Fig. 1-4 is coupled to a moving and uniform magnetic field of flux density B. Assume that the magnitude of the moving flux density in the air gap varies cosinusoidally with the mechanical angle of the rotor, and that phasor **B** maintains a radial direction while moving. As depicted in Fig. 1-4 the mechanical angle of the rotor, **α,** is measured with respect to the plane of the stationary coil. The magnetic flux linking with the coil is maximum when α is equal to 90 degrees.

Stationary coil upper conductor

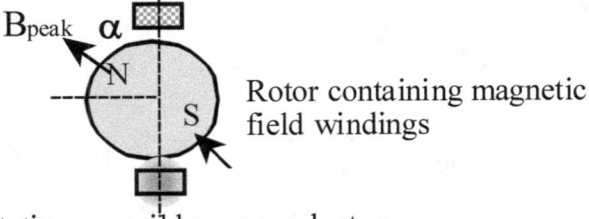

Rotor containing magnetic field windings

Stationary coil lower conductor

Figure 1-4 cross sectional view of stationary conductors and moving magnetic field.

The magnetic flux density at any point of the air gap is given by Eq. (1.15), where α is the angular displacement of the peak value of the flux density with respect to the plane defined by the stationary coil. The winding embed in the rotating rotor produces a moving magnetic field. If the rotor rotates at constant angular speed ω, then the value of the air gap flux density at any position around the stator is given by Eq. (1.16).

$$B = B_{peak} \cdot \cos(\alpha) \tag{1.15}$$

Flux density distribution in the air gap

$$B = B_{peak} \cdot \cos(\omega \cdot t) \tag{1.16}$$

$$\alpha = \omega \cdot t$$

$$\lambda = N \cdot A \cdot B_{peak} \cdot \cos(\omega \cdot t) \qquad \text{Flux linking the with coil} \tag{1.17}$$

The electromotive force induced in the coil is

$$e = -\frac{d}{dt}\lambda \tag{1.18}$$

$$e = -N \cdot A \cdot B_{peak} \cdot \frac{d}{dt}\cos(\omega \cdot t) \tag{1.19}$$

N = number of turns, 1

A = net area encircled by the stationary coil, cm2

ω = angular velocity in electrical radians or degrees per second

t = time in seconds

Bpeak = magnitude of flux density peak, gauss

Differentiating Eq. (1.19) we obtain:

$$e = A \cdot \omega \cdot B_{peak} \cdot \sin(\omega \cdot t) \tag{1.20}$$

$$A = 2 \cdot 1 \cdot r$$

Where l is the coil conductors active length and 2r the separation between the upper and lower conductors of the coil. The emf induced in the coil is

$$e = 2l \cdot r \cdot \omega \cdot B_{peak} \cdot \sin(\omega \cdot t) \cdot 10^{-8} \text{volt} \qquad \text{See Eq. (1.8)} \qquad (1.21)$$

The effective or root-mean-square value of the emf induced in the coil is

$$E = \frac{\omega}{\sqrt{2}} \cdot \phi_{peak} \cdot 10^{-8} \text{volt}$$

Where

$$\phi_{peak} = 2 \cdot l \cdot r \cdot B_{peak} \qquad (1.22)$$

The reader must realize that the rms value of the voltage vector induced in the coil is equal to vectorial addition of the rms voltage induced in each conductor. The emf induced per *conductor* is:

$$e = l \cdot r \cdot \omega \cdot B_{peak} \cdot \sin(\omega \cdot t) \cdot 10^{-8} \text{volt} = l \cdot r \cdot \omega \cdot B_{peak} \cdot \sin(\alpha) \cdot 10^{-8}$$
$$(1.23)$$

$$B = B_{peak} \cdot \cos(\alpha) \qquad \text{See Eq. (1.15)}$$

$$B_{peak} := 1$$

$$B = \cos(\alpha)$$

$$e = l \cdot r \cdot \omega \cdot \sin(\alpha) \cdot 10^{-8} \text{volt} \qquad \text{So e and B are perpendicular phasors}$$

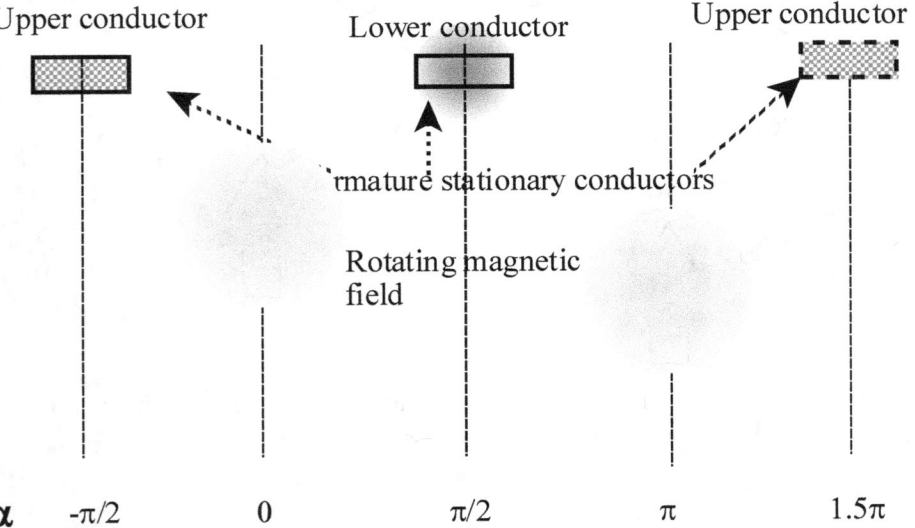

Upper conductor   Lower conductor   Upper conductor

'rmature stationary conductors

Rotating magnetic field

α   -π/2   0   π/2   π   1.5π

(a)

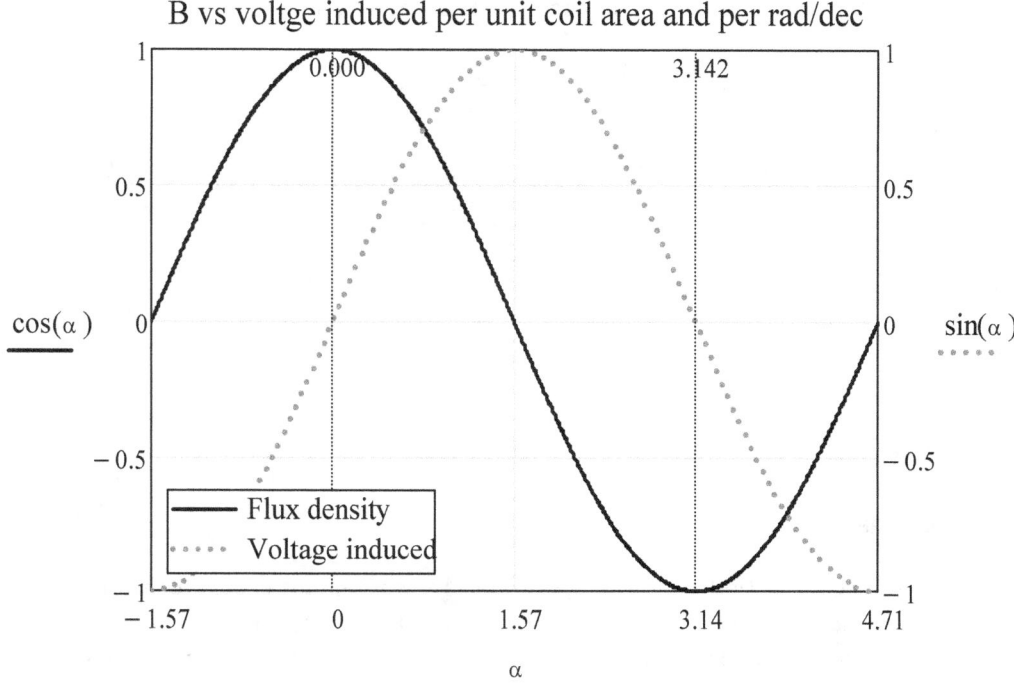

B vs voltge induced per unit coil area and per rad/dec

α

Angular displacement in radians

(b)

Figures 1-5a&b illustrate the flux density variation around the air gap and the voltage induced in the coil of a two pole generator.

In the illustration depicted in Fig. 1-5a the stationary coil in the armature is formed by two conductors π electrical radians apart, this coil links maximum flux when the N pole of the rotor points to the center of the coil. (midway between the upper and lower conductors,) however, at this instant, the two stationary conductors of the armature coil are midway between poles and are cutting no-flux and therefore the induced voltage is zero. When the magnetic poles have moved $\pi/2$ radians the flux linking with the coil becomes zero, but the poles are now in line with the armature conductors and the voltage induced in each coil conductor is at maximum value and of the same direction. The reader must realize that to find the rms value of the voltage generated in the coil a vectorial addition of the voltage induced in each conductor is required. From Fig.1-5b it is obvious that the voltage induced in a coil lags $\pi/2$ radians the creating source, the linking and changing magnetic flux. However, the voltage induced in the coil *conductors* is in synch or in phase with the flux density at each conductor. The instantaneous voltage induced in a armature conductor is given by Eqns. (1.23 and 1.25). And is equal to the instantaneous value of the magnetic flux density at the conductor location in gausses multiplied by the active conductor length in centimeters and by the component of the conductor velocity perpendicular to the flux density in cm/sec.

We know that $\quad\quad\quad r \cdot \omega = v$

Hence, the instantaneous emf induced per conductor of the coil is:

$$e = 1 \cdot v \cdot B_{peak} \cdot \sin(\omega \cdot t) \cdot 10^{-8} = 1 \cdot v \cdot B_{peak} \cdot \sin(\alpha) \cdot 10^{-8} \qquad (1.24)$$

l = conductors active length, cm

v = velocity of the moving magnetic field relative to the conductor, cm/sec

The vectorial version of Eq. (1.23) is:

$$\vec{e} = \left(\vec{v} \times \overrightarrow{B_{peak}}\right) \cdot 1 \qquad (1.25)$$

$$e = vBl \cdot \sin(\alpha)$$

1.6 Wave Analysis

If f(x) is a non-sinusoidal periodic wave, single-valued and continuous except for a finite number of finite discontinuities, and which does not have an infinite number of maxima or minima in the neighborhood of any point could be represented by the sum of a number of sine and cosine waves of different frequencies. This theorem expressed in symbolical language becomes an equation known as the Fourier Series.

$$f(x) = DC + \sum_{n=1}^{\infty} \left(A_n \cdot \sin\left(\frac{n\pi \cdot x}{L}\right) + B_n \cdot \cos\left(\frac{n \cdot \pi \cdot x}{L}\right) \right) \qquad (1.26)$$

Where 2L is the period, such that

$$f(x + 2L) = f(x)$$

Because the sine and cosine functions have a 2π period, Eq. (1.26) becomes:

$$f(x) = DC + \sum_{n=1}^{\infty} \left(A_n \cdot \sin(n\,x) + B_n \cdot \cos(n \cdot x) \right) \qquad (1.27)$$

Where,

$$DC = \frac{1}{2\pi} \cdot \int_0^{2\pi} f(x)\,dx \qquad (1.28)$$

$$A_n = \frac{1}{\pi} \cdot \int_{-\pi}^{\pi} f(x) \cdot \sin(n \cdot x) \, dx$$

$$(1.29)$$

$$B_n = \frac{1}{\pi} \cdot \int_{-\pi}^{\pi} f(x) \cdot \cos(n \cdot x) \, dx$$

$$(1.30)$$

The reader must realize that the integral of a periodic function is the same over any integration interval whose length equals the period of the function. In general, the integration limits of Eqns. (1.28 - 1.30) could also be as follow:

$$DC = \frac{1}{2\pi} \cdot \int_{\Theta}^{\Theta + 2\pi} f(x) \, dx$$

$$(1.31)$$

$$A_n = \frac{1}{\pi} \cdot \int_{\Theta}^{\Theta + 2\pi} f(x) \cdot \sin(n \cdot x) \, dx$$

$$(1.32)$$

$$B_n = \frac{1}{\pi} \cdot \int_{\Theta}^{\Theta + 2\pi} f(x) \cdot \cos(n \cdot x) \, dx$$

$$(1.33)$$

Where Θ is any real number. When $\Theta = 0$ Eqns. (1.31 - 1.33) become

$$DC = \frac{1}{2\pi} \cdot \int_{0}^{2\pi} f(x) \, dx$$

$$(1.34)$$

$$A_n = \frac{1}{\pi} \cdot \int_{0}^{2\pi} f(x) \cdot \sin(n \cdot x) \, dx$$

$$(1.35)$$

$$B_n = \frac{1}{\pi} \cdot \int_{0}^{2\pi} f(x) \cdot \cos(n \cdot x) \, dx$$

$$(1.36)$$

In simple format Eq. (1.27) becomes:

$$f(x) = DC + A_1 \cdot \sin(x) + B_1 \cdot \cos(x) + A_2 \cdot \sin(2x) + B_2 \cdot \cos(2x) + \blacksquare$$

$$A_3 \cdot \sin(3x) + B_3 \cdot \cos(3x) + \dots + A_n \cdot \sin(nx) + B_n \cdot \text{con}(nx)$$

$$(1.37)$$

If n is any integer different from zero, we know that

$$\int_0^{2\pi} \sin(nx)\, dx = 0$$

Of course, the positive area is equal to the negative area for sine and cosine functions.

$$\int_0^{2\pi} \cos(nx)\, dx = 0$$

Furthermore, in Eq. (1.37) the product of any two different functions is also zero and therefore they are pair wise orthogonal in the interval 0 - 2π.

$$\int_0^{2\pi} \sin(nx) \cdot \sin(mx)\, dx = 0 \qquad \int_0^{2\pi} \cos(nx) \cdot \cos(mx)\, dx = 0$$

Where n and m are different integers.

$$\int_0^{2\pi} \sin(nx) \cdot \cos(mx)\, dx = 0$$

Where n and m are any integers

In general the cosine coefficient B_n and the sine coefficient A_n are mutual exclusive of each other and only one of them is necessary in the Fourier expansion. In Eqns. (1.35 and 1.36), if the non-sinusoidal original function is even then A_n is zero, similarly if the original function is odd then B_n is zero. Symbolically:

If f(x) is even, then $\quad f(x) = f(-x) \quad$ and $\quad f(x) \cdot \sin(nx) \quad$ is odd

This definition of an even function implies that the plot of any even function is symmetric with respect to the y-axis. Furthermore, $f(x) \cdot \sin(nx)$ is odd, because $\sin(nx)$ is an odd function and the product of an even function times odd function is an odd function. Therefore A_n is zero for all values of n, because the area of an odd function in a 2π interval is zero. See Eq. (1.35).

If f(x) is odd, then $f(x) \neq f(-x)$ and $\quad f(x) \cdot \cos(nx) \quad$ is odd

Because cos (nx) is and even function and the product of an odd function by an even function is and odd function. Therefore B_n is zero for all values of n because the area of an odd function in a 2π interval is zero. Although, theoretically, according with Eq. (1.26) an infinite number of terms are required to duplicate a non-sinusoidal periodical wave. In engineering practice, however, only a few terms are required because of the small effect of the high frequency terms. With the exception of the DC term, all the other term of this complex wave are sinusoidal. And therefore each one of them could be treated separately and independently of the others by the regular methods of circuit analysis, including phasor representation. To express sin(nx) and cos(nx) in exponential form we proceed as follows.

$$\varepsilon^{j \cdot n \cdot x} = \cos(n \cdot x) + j \cdot \sin(n \cdot x) \qquad \varepsilon^{-j \cdot n \cdot x} = \cos(n \cdot x) - j \cdot \sin(n \cdot x)$$

$$\varepsilon^{j \cdot n \cdot x} + \varepsilon^{-j \cdot n \cdot x} = 2\cos(n \cdot x) \qquad \cos(n \cdot x) = \frac{\varepsilon^{j \cdot n \cdot x} + \varepsilon^{-j \cdot n \cdot x}}{2} \qquad (1.38)$$

$$\varepsilon^{j \cdot n \cdot x} - \varepsilon^{-j \cdot n \cdot x} = 2 \cdot j \cdot \sin(n \cdot x) \qquad \sin(n \cdot x) = \frac{\varepsilon^{j \cdot n \cdot x} - \varepsilon^{-j \cdot n \cdot x}}{2 \cdot j} \qquad (1.39)$$

Wave analysis consists in determining which particular version of Eq. (1.26) is a good substitute for the non-sinusoidal wave subject of the analysis. From a practical point of view all that is required is determining the values of the coefficients DC, An, Bn.

The following known facts reduce the computation effort considerable: any wave consisting of symmetrical positive and negative loops cannot contain even harmonics nor DC bias. Besides, if the areas of the positive and negative loops are the same the DC coefficient is zero. Mathematically: if the value of f(x) from zero to π is repeated, except for sign, between π and 2π, then the wave possess half-wave symmetry. Symbolically:

$$f(x + \pi) = -f(x) \qquad \text{or} \qquad f(x) = -f(x + \pi) \qquad (1.40)$$

Consequently, the analysis of waves possessing half-wave symmetry is done in half cycle only, or π radians. Wave possessing half-wave symmetry could also be symmetrical about their mid ordinate of its positive and negative loops, or about $\pi/2$ and $3\pi/2$. This kind of wave possess quarter-wave symmetry and the wave analysis could be performed only in a quarter of a cycle or $\pi/2$ radians. *Waves with quarter-wave symmetry contains no even harmonics and all the remaining harmonics including the fundamental pass through zero at the same time.* Two waves are of the same shape if they possess the same harmonics, if the ratio of the magnitudes of the corresponding harmonic are constant, and if when the fundamentals are in phase, all the corresponding harmonics are in phase.

1.7 Per-unit Quantities

The per-unit method of expressing electrical quantities is based in the Ohm's law. This law relates three variable with one equation, therefore only two variables can be arbitrarily selected as independent. In the power industry the most common selection is 3-phase MVA and line to line kilovolts as the two independent base quantities, and express all other quantities in function of them. It is necessary to say that the MVA quantity is a combination of two of the three original quantities related by the Ohm's law.

Per-unit value of a quantity is the ratio of the quantity to its base. The percent value is the per-unit value times 100.

For balanced 3-phase systems the per unit value of a line-to-neutral voltage with line to neutral base is the same than the per unit value, at the same location, of the line-to-line voltage with line-to-line base. Likewise, the total 3-phase MVA is three times the MVA per phase, because the 3-phase MVA base is three times the base MVA per phase, then the per-unit value of the 3-phase MVA with a 3-phase MVA base is also the same to the per-unit value of the MVA per phase with an MVA per phase base. Symbolically:

$$\frac{V_{LN}}{V_{LNb}}$$

Per-unit value of the line-to-neutral voltage

In a balanced three-phase circuit the per-unit value of the line to line voltage is equal to the per-unit value of the line to neutral voltage

$$\frac{V_{LL}}{V_{LLb}} = \frac{\sqrt{3} \cdot V_{LN}}{\sqrt{3} \cdot V_{LNb}} = \frac{V_{LN}}{V_{LNb}}$$

In the case of Mega volt-amperes the per-unit value of 3-phase MVA is the same than the per-unit value of the MVA per phase.

$$\frac{MVA}{MVA_b} = \frac{3 \cdot MVA_{ph}}{3 \cdot MVA_{bph}} = \frac{MVA_{ph}}{MVA_{bph}}$$

Usually, computations in balanced three-phase circuits are made on a per phase basis in the same way as calculations are made for single-phase circuits. However, the three-phase Mega volt-amperes of a balanced three-phase system are given in terms of the line-to-line voltage and line current, by the same equation, regardless if the system is wye-or-delta-connected. Symbolically:

$$MVA = \sqrt{3} \cdot kV_{LL} \cdot kI_L$$

From here on, unless otherwise specified, the notation for base quantities is:

MVAb = Base three-phase megavolt-amperes

kVb = Base Line-to-line kilovolts

Ib = Base line current, in kA

Zb = Base line to neutral impedance in ohms

$$MVA_b = \sqrt{3} \cdot kV_b \cdot I_b \qquad I_b = \frac{MVA_b}{\sqrt{3} \cdot kV_b} \qquad \text{Base kiloamperes} \qquad (1.41)$$

$$Z_b = \frac{kV_b}{\sqrt{3} \cdot I_b} = \frac{kV_b \cdot \sqrt{3} \cdot kV_b}{\sqrt{3} \cdot MVA_b} = \frac{\left(kV_b\right)^2}{MVA_b} \qquad \text{Base line to neutral ohms}$$

$$(1.42)$$

Conversion of per-unit impedance values from one base to another:

Per-unit impedance = actual impedance / base impedance.

Symbolically:

$$\frac{\Omega}{\Omega_b} \qquad (1.43)$$

Substituting Eq. (1.42) into Eq. (1.43) we obtain:

$$\frac{\Omega \cdot MVA_b}{kV_b{}^2} \qquad \text{per-unit impedance} \qquad (1.44)$$

If we express the per-unit impedance in two different bases, we obtain:

$$\frac{\Omega \cdot MVA_{b1}}{\left(kV_{b1}\right)^2} = Z_1 \qquad\qquad \frac{\Omega \cdot MVA_{b2}}{\left(kV_{b2}\right)^2} = Z_2$$

$$\frac{Z_2}{Z_1} = \frac{\Omega \cdot MVA_{b2} \cdot \left(kV_{b1}\right)^2}{\Omega \cdot MVA_{b1} \cdot \left(kV_{b2}\right)^2}$$

Impedance in base 2 given the impedance in base 1

$$Z_2 = Z_1 \cdot \left(\frac{MVA_{b2}}{MVA_{b1}}\right) \cdot \left(\frac{kV_{b1}}{kV_{b2}}\right)^2 \tag{1.45}$$

Given the per-unit impedance in base 1; equation (1.45) provides the per-unit impedance in base 2

1.8 Network Reduction Techniques

The simplest technique is to use wye-delta or star-mesh transformation formulas. This method works well with simple networks, but it should not be used with larger and more complicated ones. The method of node elimination by matrix algebra is presented below.

$$I = Y\,V \qquad \text{standard node equation in matrix notation}$$

Where **I** and **V** are column matrices or vectors and **Y** is a symmetrical square matrix. The column matrices must be arranged so that the elements associated with nodes to be eliminated are in the lower rows of the matrices. The admittance matrix is partitioned so that elements associated with nodes to be eliminated are separated from the other elements by horizontal and vertical lines.

$$\begin{pmatrix} I_A \\ I_x \end{pmatrix} = \begin{pmatrix} K & L \\ L^T & M \end{pmatrix} \cdot \begin{pmatrix} V_A \\ V_x \end{pmatrix} \tag{1.46}$$

Where **Ix** is the submatrix composed of the currents entering the nodes to be eliminated and **Vx** is the submatrix composed of the voltages of these nodes. Every element of **Ix** must be zero, because if not, these nodes could not be eliminated. Matrix **K** contains the self- and mutual admittances of the node to be retained. Matrix **M** is a square matrix whose order is equal to the number of nodes to be eliminated and it contains the self- and mutual admittances of the nodes to be eliminated. Matrix **L** contains the mutual admittances common to a node to be retained and to one to be eliminated. Matrix **LT** is the transpose of **L**. From Eq. (1.44) we obtain the following equations:

$$I_A = K \cdot V_A + L \cdot V_x \tag{1.47}$$

$$I_x = L^T \cdot V_A + M \cdot V_x \tag{1.48}$$

Since all the elements of **Ix** are zero, subtracting $L^T \cdot V_A$ from both sides of Eq. (1.48) and multiplying both sides by **M-1** we have

$$-L^T \cdot V_A = M \cdot V_x \qquad M \cdot M^{-1} = M^{-1} \cdot M = U \qquad\qquad \text{Unit matrix}$$

$$-M^{-1} \cdot \left(L^T \cdot V_A \right) = V_x$$

Therefore, substituting V_x in Eq. (1.47) we have

$$I_A = K \cdot V_A - L \cdot \left[M^{-1} \cdot \left(L^T \cdot V_A \right) \right] \quad I_A = K \cdot V_A - L \cdot M^{-1} L^T \cdot V_A$$

$$I_A = \left(K - L \cdot M^{-1} \cdot L^T \right) \cdot V_A$$

$$I_A = Y \cdot V_A$$

Which is a node equation where the admittance matrix is:

$$Y = K - L \cdot M^{-1} \cdot L^T \tag{1.49}$$

The admittance matrix, Eq. (1.49) allows us to assembly the circuit with the unwanted nodes eliminated.

Example 1-1

Figure 1-6 provides the per unit admittance diagram of a power system based on 230KV and 300 MVA. Find the equivalent network with nodes (busses) 2, 3 and 6 eliminated. These nodes could be eliminated because the loads connected to them are considered to be constant. Actually, we are assuming that these constant loads are not affected and do not participate in the dynamic of any disturbance occurring somewhere else in the network.

Nodes to be eliminated: 2, 3, 6

Nodes to be retained: 1, 4, 5

Equation (1.49) allow us to reassembly the circuit with nodes 2, 3 and 6 eliminated

$$Y = \begin{pmatrix} K & L \\ L^T & M \end{pmatrix}$$

To eliminate nodes, the admittance matrix is partitioned into four submatrices where **K** is a 3x3 square matrix whose order is equal to the number of nodes to be retained and is composed of the self- and mutual admittances of the nodes to be *retained.* Each element of the principal diagonal of **K** is equal to the sum of all the admittances terminating on the node identified in the matrix by repeated subscripts.

$$K = \begin{pmatrix} Y_{1,1} & Y_{1,4} & Y_{1,5} \\ Y_{4,1} & Y_{4,4} & Y_{4,5} \\ Y_{5,1} & Y_{5,4} & Y_{5,5} \end{pmatrix} \qquad\qquad j := \sqrt{-1}$$

From Fig. 1-6 we obtain all the admittances:

$$Y_{1,1} := 1.248 - 7.8j \qquad Y_{1,4} := 0 \qquad\qquad Y_{1,5} := 0$$

$$Y_{4,1} := 0 \qquad\qquad Y_{4,4} := 0.974 - 6.272j \qquad Y_{4,5} := 0$$

$$Y_{5,1} := 0 \qquad\qquad Y_{5,4} := 0 \qquad Y_{5,5} := 0.974 - 6.272j$$

$$K := \begin{pmatrix} 1.248 - 7.8j & 0 & 0 \\ 0 & 0.974 - 6.272j & 0 \\ 0 & 0 & 0.974 - 6.272j \end{pmatrix}$$

Nodes to be retained: 1, 4, 5

Matrix **M** is a square matrix whose order is equal to the number of nodes to be eliminated (3). Matrix **M** is composed of the self and mutual admittances of the nodes to be eliminated. Any element of the principal diagonal is equal to the sum of all admittances connected to the node. Each element *not* in the diagonal is equal to the negative sum of all admittances connected directly between the pair of nodes identified by the double subscript. Nodes to be eliminated: 2, 3, 6.

$$M = \begin{pmatrix} Y_{2,2} & Y_{2,3} & Y_{2,6} \\ Y_{3,2} & Y_{3,3} & Y_{3,6} \\ Y_{6,2} & Y_{6,3} & Y_{6,6} \end{pmatrix}$$

$$Y_{2,2} := 1.04 - 5.897j \qquad Y_{2,3} := 0 \qquad\qquad Y_{2,6} := 0$$

$$Y_{3,2} := 0 \qquad\qquad Y_{3,3} := 1.04 - 5.897j \qquad Y_{3,6} := 0$$

$$Y_{6,2} := 0 \qquad\qquad Y_{6,3} := 0 \qquad Y_{6,6} := 1.116 - 8.55j$$

$$M := \begin{pmatrix} 1.04 - 5.897j & 0 & 0 \\ 0 & 1.04 - 5.897j & 0 \\ 0 & 0 & 1.116 - 8.55j \end{pmatrix}$$

Nodes to be eliminated: 2, 3, 6

$$M^{-1} = \begin{pmatrix} 0.029 + 0.164j & 0 & 0 \\ 0 & 0.029 + 0.164j & 0 \\ 0 & 0 & 0.015 + 0.115j \end{pmatrix}$$

Matrix **L** is a square matrix of the same order as the number of nodes to be retained and is composed of only those **mutual** admittances common to a node to be retained and to one to be eliminated and each element is equal to the negative of the admittance between nodes.

Nodes to be retained: 1, 4, 5   Nodes to be eliminated: 2, 3, 6

$$L = \begin{pmatrix} Y_{1,2} & Y_{1,3} & Y_{1,6} \\ Y_{4,2} & L_{4,3} & L_{4,6} \\ Y_{5,2} & Y_{5,3} & Y_{5,6} \end{pmatrix}$$

$$L := \begin{pmatrix} -0.624 + 3.9j & -0.624 + 3.9j & 0 \\ -0.416 + 1.997j & 0 & -0.558 + 4.275j \\ 0 & -0.416 + 1.997j & -0.558 + 4.275j \end{pmatrix}$$

$$L^{T} = \begin{pmatrix} -0.624 + 3.9j & -0.416 + 1.997j & 0 \\ -0.624 + 3.9j & 0 & -0.416 + 1.997j \\ 0 & -0.558 + 4.275j & -0.558 + 4.275j \end{pmatrix}$$

$$Y := K - L \cdot M^{-1} \cdot L^{T} \qquad \text{See Eq. (1.49)}$$

Equation (1.49) allow us to assemble, the circuit with the unwanted nodes eliminated.

$$Y = \begin{pmatrix} 0.507 - 2.643j & -0.253 + 1.321j & -0.253 + 1.321j \\ -0.253 + 1.321j & 0.532 - 3.459j & -0.279 + 2.138j \\ -0.253 + 1.321j & -0.279 + 2.138j & 0.532 - 3.459j \end{pmatrix} \qquad (1.50)$$

This Y matrix provides the information to assemble the new simplified network. See Fig 1-7

$$Y = \begin{pmatrix} Y_{1,1} & Y_{1,4} & Y_{1,5} \\ Y_{4,1} & Y_{4,4} & Y_{4,5} \\ Y_{5,1} & Y_{5,4} & Y_{5,5} \end{pmatrix}$$

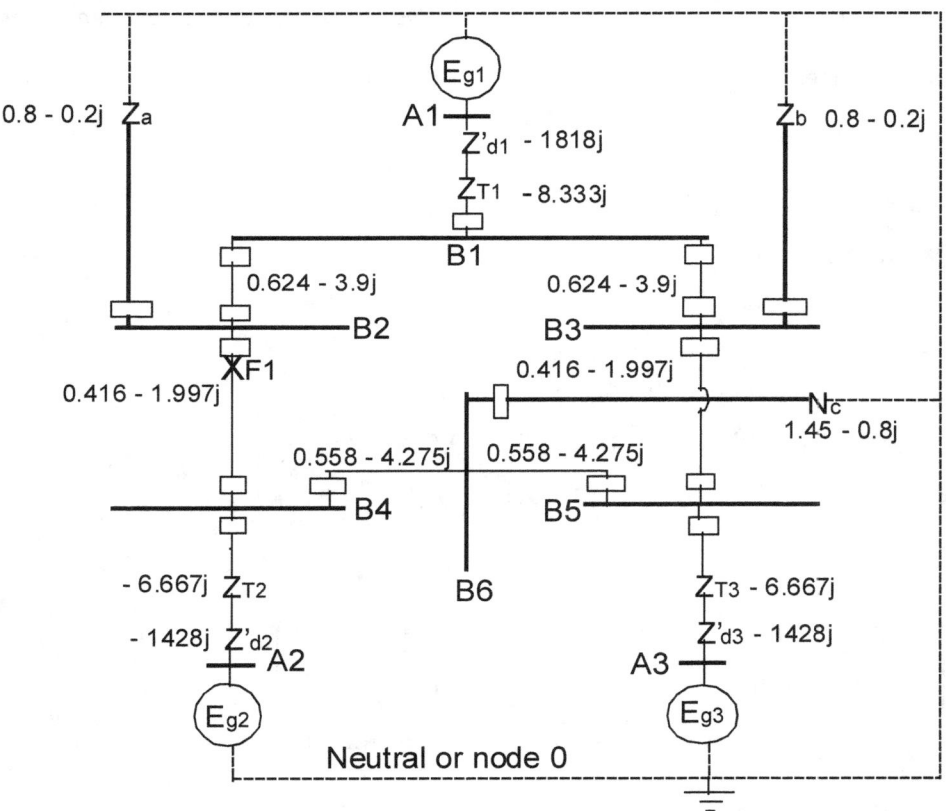

Figure 1-6 One-phase admittance diagram including loads, generators and transformers, 230 KV and 300 MVA base.

Nodes to be retained: 1, 4, 5

From the Y-matrix, Eq. (1.50) we reach the conclusion that each of the diagonal (self-admittance) elements is the negative of the sum of the off-diagonal (mutual-admittance) elements in the same row. This definition implies that there can be no admittances between any to the buses and ground. Or that the network is totally isolated with no coupling to ground or anywhere else.

$$Y_{1,1} := 0.507 - 2.643 \qquad Y_{1,4} := -0.253 + 1.321 \qquad Y_{1,5} := -0.253 + 1.321$$

$$Y_{4,1} := -0.253 + 1.321 \qquad Y_{4,4} := 0.532 - 3.459 \qquad Y_{4,5} := -0.279 + 2.138$$

$$Y_{5,1} := -0.253 + 1.321 \qquad Y_{5,4} := -0.279 + 2.138 \qquad Y_{5,5} := 0.532 - 3.459$$

The resulting network after eliminating nodes: 2, 3, 6, which are load buses, is illustrated in Figure 1-7.

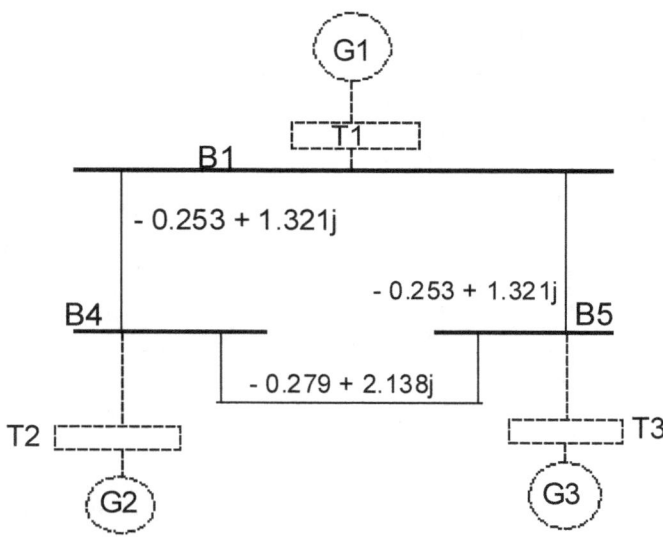

Figure 1-7 Per unit admitance diagram of the power system with 3 nodes eliminated. 230 KV and 300 MVA base.

In the original bus admittance diagram of the network, see Figure 1-6, generators and loads were considered outside the network, and therefore they were not included in the network admittance matrix. The impedances connected directly between nodes in the resulting network, are the negative of the inverse of the mutual admittances. Therefore, the impedances between nodes 1, 4, 5 are:

$$Z_{1,4} := -\left(Y_{1,4}\right)^{-1} = 0.14 + 0.73j \qquad \text{Per unit}$$

$$Z_{4,5} := -\left(Y_{4,5}\right)^{-1} = 0.06 + 0.46j \qquad \text{Per unit}$$

$$Z_{5,1} := -\left(Y_{5,1}\right)^{-1} = 0.14 + 0.73j \qquad \text{Per unit}$$

1.9 Thévenin's Theorem

Applicable to linear, two-terminal circuits which can be enunciated as follows. If a load with impedance ZL is connected between any two points of an energized circuit, then the resulting current IL through this impedance is the potential difference Ep between these points prior to the connection, divided by the sum of the connected impedance ZL and the impedance Zin, which is the impedance of the rest of the circuit looking back into the circuit from the points across which the impedance ZL is connected. To evaluate Zin, all sources of emf must be assumed to be zero and replaced by their internal impedances

1.10 Transformers Equivalent Circuit

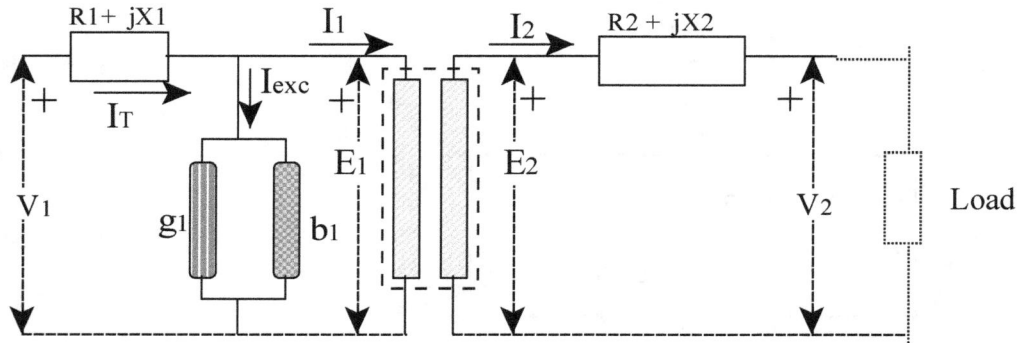

Figure 1-8 Equivalent circuit of a two-winding, iron core, transformer.

The single-phase transformer shown in Fig. 1-8 could also be consider as representing one phase of a three-phase transformer.

R1 is the resistance of the primary winding

R2 is the resistance of the secondary winding

X1 is the leakage reactance of the primary winding

X2 is the leakage reactance of the secondary winding

I2 is the secondary or load current

I1 is the load component of the total primary current

IT is the total primary current

Iexc is the current demanded by the transformer's core and consists of the magnetizing and core loss components

g1 is the apparent conductance of the iron-core shunt reactance

b1 is the apparent susceptance of the iron-core shunt reactance

E1 is the primary induced emf by the resultant mutual flux

E2 is the secondary induced emf by the resultant mutual flux

V1 is the voltage applied to the primary

V2 is the voltage received by the load

The transformer depicted in Fig. 1-8 consists of and ideal single-phase transformer plus the necessary circuit elements to mimic the magnetic characteristics of the iron-core present in a real power transformer The conductance is the real part of the shunt reactor admittance and the imaginary part is the susceptance. The following equations helps to understand the transformer description:

$$\vec{I_T} = \vec{I_1} + \overrightarrow{I_{exc}}$$ Total primary current (1.51)

$$\overrightarrow{I_{exc}} = \overrightarrow{\left(I_{mag} + I_{cL}\right)}$$ Magnetizing plus core losses currents

The load component of the primary current produce a magnetomotive force, $\overrightarrow{\left(N_1 \cdot I_1\right)}$ Which opposes and balance the magnetomotive force produced by the secondary current. Actually both magnetomotive forces are equal and of opposite direction. The resultant flux, linking both windings is produced by the magnetizing component, I_{mag}, of the primary current.

$$\overrightarrow{\left(N_1 \cdot I_1\right)} = -\overrightarrow{\left(N_2 \cdot I_2\right)}$$ (1.52)

The primary winding mmf balances the secondary winding mmf.

$$\vec{I_1} = \frac{-N_2}{N_1} \cdot \vec{I_2} \qquad a = \frac{N_1}{N_2}$$

Load component of the primary current
$$\vec{I_1} = \frac{\overrightarrow{(-I_2)}}{a}$$
(1.53)

$$\vec{I_2} = -a \cdot \vec{I_1}$$ Load current or secondary current
(1.54)

The ratio of the generated emfs in the ideal transformer, part of the model, illustrated in Fig. 1-8 is

$$\frac{E_1}{E_2} = \frac{N_1}{N_2} = a$$
(1.55)

The magnitude of the emfs induced in the transformer windings are proportional to the number of turns on the winding. The ratio of transformation is designated as **a**

The volts per turn of induced emf is the same for each winding

$$\frac{E_1}{N_1} = \frac{E_2}{N_2}$$
(1.56)

Fig. 1.9 is the transformer circuit shown on Fig. 1-8 referred to the primary side and Fig. 1.10 is the transformer circuit shown on Fig. 1-8 referred to the to the secondary side. R1 and R2 were neglected as well as the iron core shunt reactor and the excitation current which is around two percent of the total primary current. *The load component of the primary current and the secondary current have opposite directions with respect of the core flux*

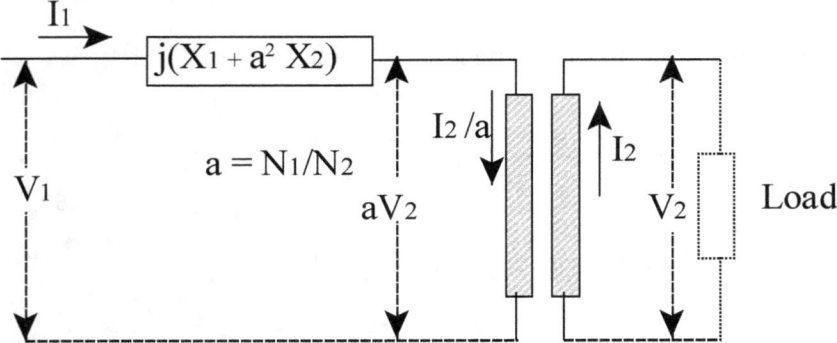

Figure 1-9 Iron core transformer with resistances and excitation shunt circuit neglected and transformer secondary reactance referred to the primary side.

Primary voltage drop referred to secondary:

$$I_1 \cdot j\left(X_1 + a^2 X_2\right) = a \cdot I_1 \cdot j\left(\frac{X_1}{a^2} + X_2\right) = I_1 \cdot j\left(\frac{X_1}{a} + a \cdot X_2\right)$$

(1.57)

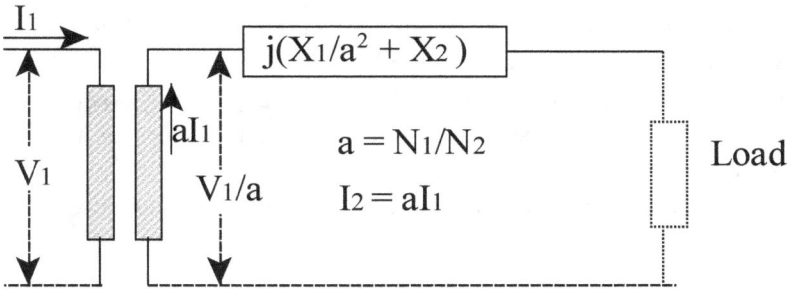

Figure 1-10 Iron core transformer with resistances and excitation shunt circuit neglected and transformer primary reactance referred to the secondary side.

Secondary voltage drop referred to primary:

$$I_2 \cdot j\left(\frac{X_1}{a^2} + X_2\right) = \frac{I_2}{a} \cdot j \cdot \left(X_1 + a^2 \cdot X_2\right) = I_2 \cdot j \cdot \left(\frac{X_1}{a} + a \cdot X_2\right)$$

(1.58)

To convert the values of transformer parameters from one side to the other do the following:

Primary to secondary:	Secondary to primary:
Divide primary voltages by **a**	Multiply secondary voltages by **a**
Multiply primary currents by **a**	Divide secondary currents by **a**
Divide primary resistances by **a2**	Multiply secondary resistances by **a2**
Divide primary reactances by **a2**	Multiply secondary reactances by **a2**

1.11 Elementary Operations with Matrices

Below the reader will find some important concepts above matrices:

- A matrix does not have a value and it does not need to be square
- A square matrix is symmetrical if the square matrix and its transpose are identical
- A matrix have and inverse, provided it is square and its determinant is different from zero
- The product of two matrices is non-commutative, meaning the order of the factors alter the product
- Division by a matrix is not defined. The objective is achieved multiplying by its inverse
- A unit matrix is a square matrix in which all elements on the principal diagonal are 1 and all off-diagonal elements are zero.
- The transpose of a matrix is the matrix obtained by interchanging the rows and columns. Meaning writing the rows as columns and in the same order. For instance, Row 1 becomes column 1.
- The product of a matrix and a column phasor produces another column phasor.

Matrix addition The sum of two matrix of the *same size,* with the same number of rows and columns, is another matrix of the same size whose elements are the sum of the corresponding elements.

$$P = \begin{pmatrix} 8 & 3 \\ 2 & 1 \end{pmatrix} \quad Q = \begin{pmatrix} 5 & -2 \\ 7 & 6 \end{pmatrix} \quad P + Q = \begin{pmatrix} 8 & 3 \\ 2 & 1 \end{pmatrix} + \begin{pmatrix} 5 & -2 \\ 7 & 6 \end{pmatrix} = \begin{pmatrix} 13 & 1 \\ 9 & 7 \end{pmatrix}$$

The subtraction is obtained as follows:

$$P - Q = P + (-Q) = \begin{pmatrix} 8 & 3 \\ 2 & 1 \end{pmatrix} + \begin{pmatrix} -5 & 2 \\ -7 & -6 \end{pmatrix} = \begin{pmatrix} 3 & 5 \\ -5 & -5 \end{pmatrix}$$

Or directly

$$P - Q = \begin{pmatrix} 8 & 3 \\ 2 & 1 \end{pmatrix} - \begin{pmatrix} 5 & -2 \\ 7 & 6 \end{pmatrix} = \begin{pmatrix} 3 & 5 \\ -5 & -5 \end{pmatrix}$$

Matrix multiplication

The product of **A** which is a (mxn) matrix by **B** which is a (nxp) matrix

Where:

m = number of rows of matrix **A**

n = number of columns of matrix **A** and the number of rows of matrix **B**

p = number of columns of matrix **B**

In general, to carry out the multiplication of two matrices the number of columns in first matrix must be equal to the number of rows in the second matrix.

Let us consider the multiplication of a 3x2 by a 2x2 matrices

$$\begin{pmatrix} 1 & 2 \\ 3 & 4 \\ 5 & 6 \end{pmatrix} \cdot \begin{pmatrix} 7 & 8 \\ 9 & 10 \end{pmatrix} = \begin{pmatrix} 1 \times 7 + 2 \times 9 & 1 \times 8 + 2 \times 10 \\ 3 \times 7 + 4 \times 9 & 3 \times 8 + 4 \times 10 \\ 5 \times 7 + 6 \times 9 & 5 \times 8 + 6 \times 10 \end{pmatrix} = \begin{pmatrix} 25 & 28 \\ 57 & 64 \\ 89 & 100 \end{pmatrix}$$

$$\begin{pmatrix} 7 & 8 \\ 9 & 10 \end{pmatrix} \cdot \begin{pmatrix} 1 & 2 \\ 3 & 4 \\ 5 & 6 \end{pmatrix} = (2 \times 2) \cdot (3x2)$$

Not compatible because the first matrix has two columns and the second matrix has three rows

Unit matrices:

$$U = \begin{pmatrix} 1 & 0 \\ 0 & 1 \end{pmatrix}_{\text{second order}} \qquad U = \begin{pmatrix} 1 & 0 & 0 \\ 0 & 1 & 0 \\ 0 & 0 & 1 \end{pmatrix}_{\text{third order}}$$

The product of a square matrix by the unit of the same order yields back the original matrix.

$$\begin{pmatrix} 1 & 0 \\ 0 & 1 \end{pmatrix} \cdot \begin{pmatrix} 7 & 8 \\ 9 & 10 \end{pmatrix} = \begin{pmatrix} 7 & 8 \\ 9 & 10 \end{pmatrix}$$

$$\begin{pmatrix} 9 & 10 & 11 \\ 12 & 13 & 14 \\ 15 & 16 & 17 \end{pmatrix} \cdot \begin{pmatrix} 1 & 0 & 0 \\ 0 & 1 & 0 \\ 0 & 0 & 1 \end{pmatrix} = \begin{pmatrix} 9 & 10 & 11 \\ 12 & 13 & 14 \\ 15 & 16 & 17 \end{pmatrix}$$

Matrix transpose

$$\begin{pmatrix} 9 & 10 & 11 \\ 12 & 13 & 14 \\ 15 & 16 & 17 \end{pmatrix}^{T} = \begin{pmatrix} 9 & 12 & 15 \\ 10 & 13 & 16 \\ 11 & 14 & 17 \end{pmatrix} \qquad (1 \quad 1 \quad 1)^{T} = \begin{pmatrix} 1 \\ 1 \\ 1 \end{pmatrix}$$

$$\begin{pmatrix} 1 \\ 1 \\ 1 \end{pmatrix}^T = (1 \quad 1 \quad 1)$$

$$(1 \quad 1 \quad 1) \cdot \begin{pmatrix} 9 & 10 & 11 \\ 12 & 13 & 14 \\ 15 & 16 & 17 \end{pmatrix} = (36 \quad 39 \quad 42)$$

Converts a three rows square matrix into a one-row matrix

$$(1 \quad 1 \quad 1 \quad 1) \cdot \begin{pmatrix} 9 & 10 & 11 \\ 12 & 13 & 14 \\ 15 & 16 & 17 \end{pmatrix}$$

This matrix operation is **not valid** because the first matrix has four columns and the second matrix only has three rows

$$\begin{pmatrix} 1 \\ 1 \\ 1 \end{pmatrix} \cdot (5 \quad 7 \quad 9) = \begin{pmatrix} 5 & 7 & 9 \\ 5 & 7 & 9 \\ 5 & 7 & 9 \end{pmatrix}$$

Valid operation. First matrix one column, second matrix one row. Converts a one-row matrix into a square matrix.

$$\begin{pmatrix} 1 \\ 1 \\ 1 \end{pmatrix} \cdot (11 \quad 12 \quad 13 \quad 14) = \begin{pmatrix} 11 & 12 & 13 & 14 \\ 11 & 12 & 13 & 14 \\ 11 & 12 & 13 & 14 \end{pmatrix}$$

Valid operation. First matrix one column, second matrix one row. Converts a one-row, four column matrix into 3x4 matrix

CHAPTER 2

Coupled Electrical and Magnetic Circuits

2.1 Background Knowledge

The interaction between the electric and magnetic circuits and in a general sense electromagnetism is the domain of electrical engineering. In fact, transformers, generators, motors and practically anything electrical are based in this interaction.

The law of electromagnetic induction is expressed symbolically by equation (2.1)

$$e = N \cdot \left(\frac{d}{dt} \phi \right) \cdot 10^{-8} \cdot \text{volts}$$

(2.1)

e = induced emf in volts

N = number of turns of wire linking with the flux

ϕ = magnetic flux in Maxwell or lines

t = time in seconds

Gauss = Maxwell per square centimeter

$$1 \text{Tesla} = 1 \frac{\text{Weber}}{\text{m}^2} = \frac{10^8 \text{lines}}{10^4 \text{cm}^2} = 10^4 \text{gauss}$$

1 Maxwell = 1 Line

1 Weber = 108 Maxwell

$$1 \text{gauss} = 1 \frac{\text{line}}{\text{cm}^2}$$

1 Gauss = 6.452 Lines per square inch

Integrating equation (2.1) we obtain:

$$\int e\, dt = N \cdot \Delta\phi \quad \cdot 10^{-8} \cdot volt - second \tag{2.2}$$

$$\Delta\phi \; = A_{Fe} \cdot \Delta B$$

If Δ B is expressed in kilogauss and knowing that one kilogauss is equal to 1000 lines per square centimeter then the integral of the volt-seconds becomes:

$$\int e\, dt = N \cdot A_{Fe} \cdot \Delta B \; \cdot 10^{-5} \cdot volt - second \tag{2.3}$$

A_{Fe} = Net cross sectional area of the magnetic circuit in square centimeters

Δ B = Magnetic flux density changes in kilogauss

Equations (2.2) and (2.3) are good for any wave shape and frequency of the emf. For half-cycle of a sinusoidal wave we have:

$$\int E_m \cdot \sin(\omega \cdot t)\, dt \tag{2.4}$$

Symbolic evaluation of the integral (2.4)

$$-\frac{E_m \cdot \cos(\omega \cdot t)}{\omega} \qquad -\left(\frac{E_m \cdot \cos(\omega \cdot t)}{\omega}\right) = \frac{-E_m}{\omega} \cdot \cos(\omega \cdot t)$$

Evaluating for ω t from 0 to π radians we obtain: Em and V_{rms} are in volts

$$\frac{-E_m}{\omega} \cdot (-1 - 1) = \frac{2 \cdot E_m}{\omega} = \frac{2 \cdot \sqrt{2} \cdot V_{rms}}{\omega} = \frac{2 \cdot \sqrt{2} \cdot Vrms}{2 \cdot \pi \cdot f} \tag{2.5}$$

$$\frac{2\sqrt{2} \cdot V_{rms}}{\omega} = N \cdot A_{Fe} \cdot \Delta B \; \cdot 10^{-5} \tag{2.6}$$

$$N \cdot A_{Fe} = \frac{2 \cdot \sqrt{2} \cdot V_{rms} \cdot 10^5}{\omega \cdot \Delta B} = \frac{2 \cdot \sqrt{2} \cdot 10^5}{\omega} \cdot \frac{V_{rms}}{\Delta B} \cdot turn - cm^2 \qquad (2.7)$$

$$\omega = 2 \cdot \pi \cdot f$$

$$f = 50 \qquad \omega = 314$$

$$N \cdot A_{Fe} = 900 \cdot \frac{V_{rms}}{\Delta B} \cdot turn - cm^2 \qquad (2.8)$$

$$f = 60 \qquad \omega = 377$$

$$N \cdot A_{Fe} = 750 \cdot \frac{V_{rms}}{\Delta B} \cdot turn - cm^2 \qquad (2.9)$$

Example 2.1

If $V_{rms} := 13800$ volts $\qquad \Delta B := 32$ kilogauss then $\qquad \dfrac{V_{rms}}{\Delta B} = 431.25$

For 50 cycles: $\qquad N \cdot A_{Fe} = 900 \cdot 431.25 = 388 \cdot 10^3 \cdot turn - cm^2 \qquad (2.10)$

For 60 cycles: $\qquad N \cdot A_{Fe} = 750 \cdot 431.25 = 323 \cdot 10^3 \cdot turn - cm^2 \qquad (2.11)$

For 400 cycles: $\qquad N \cdot A_{Fe} = 112 \cdot 431.25 = 48 \cdot 10^3 \cdot turn - cm^2 \qquad (2.12)$

The cross sectional area of the iron core times the number of turns for a 50 cycles application would be 20% larger than the 60 cycles ones, however the 400 cycles would be 85% smaller. In the electromagnetic induction equation, ϕ represents the instantaneous magnetic flux, see Eq. (1.1). The reader must realize than nothing flows around in a magnetic circuit and that the lines of flux are only a simple and convenient way to describe the magnetic field set up in the magnetic circuit. *Furthermore the flux lines are always continuous closed lines, they don't have a beginning or an end. They are closed loops. The magnetic core can change direction or cross section abruptly, but there cannot be sudden change in direction of the flux lines. So some flux lines would go outside the magnetic circuit for a short distance before it comes back. This is the so called leakage flux.* The unit of magnetic flux is the *Weber* which is the amount of flux that when set up in a magnetic circuit induces one volt-second in a one-turn-coil wound around the magnetic circuit.

1 Weber = 108 Maxwell   1 Maxwell = 1 Line

Let us consider the single-phase transformer shown in Fig. (2.1) in which the secondary coil (not shown) remains open.

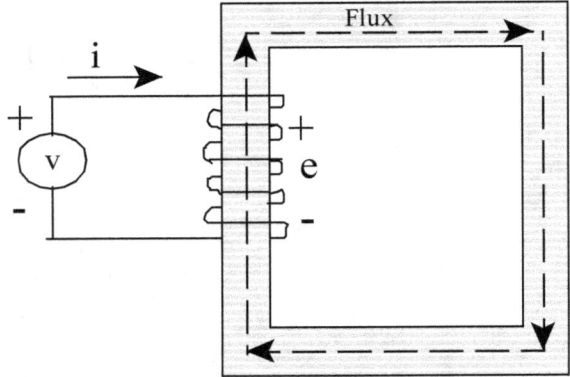

Figure 2-1 Single-phase transformer with open secundary which is not shown.

Equation (2.13) shows the instantaneous magnetic flux:

$$\phi = \phi_{max} \cdot \sin(\omega \cdot t) \tag{2.13}$$

Applying equation Eq. (2.1) we get the induced emf in the coil.

$$e = N \cdot \left(\frac{d}{dt}\phi\right) \cdot 10^{-8} = N \cdot \omega \cdot \phi_{max} \cdot \cos(\omega \cdot t) \cdot 10^{-8} \tag{2.14}$$

Let us plot the emf induced in the coil that is practically equal to voltage applied to the coil, neglecting the voltage drop in the connecting wires. See Fig. (2-2).

f = 60 cycle/sec   t = sec $\omega := 2\pi \cdot f$ $\omega = 377$ rad / sec

T = 1/f = 0.016667

$t := 0, 0.0005 .. 0.017$ $V_{rms} := 13800 \cdot volt$

$$e = E_m \cdot \sin(\omega \cdot t) \quad E_m := \sqrt{2} \cdot 13800 \quad E_m = 1.952 \times 10^4 \quad \text{volt}$$

$$e = 19520 \cdot \sin(\omega \cdot t) \tag{2.15}$$

Working with Eq. (2.14) we get:

$$e = N \cdot \omega \cdot \phi_{max} \cdot \cos(\omega \cdot t) \cdot 10^{-8} \qquad \text{Instantaneous induced volts}$$

$$e_{max} = N \cdot \omega \cdot \phi_{max} \cdot 10^{-8} \qquad \text{Maximum induced instantaneous voltage}$$

$$E_{rms} = \frac{e_{max}}{\sqrt{2}} = \frac{2\pi \, f}{\sqrt{2}} \cdot N \cdot A_{Fe} \cdot B_{max} \cdot 10^{-8}$$

$$E_{rms} = 4.44 \cdot f \cdot N \cdot A_{Fe} \cdot B_{max} \cdot 10^{-8} \cdot \text{volt} \tag{2.16}$$

N = number of turns of wire linking with the flux

B_{max} = maximum magnetic flux density in Maxwell per square cm

A_{Fe} = net cross sectional area of magnetic circuit in square cm

The volts per turn induced in the winding is:

$$\frac{E_{rms}}{N} = 4.44 \cdot f \cdot A_{Fe} \cdot B_{max} \cdot 10^{-8} \cdot \frac{\text{volt}}{\text{turn}} \tag{2.17}$$

There are many factors to consider during transformers design and manufacturing like: kind of material, thickness of core laminations, type of lamination joints to be used, acceptable noise level, hysteresis and eddy current losses, cross-section of the coil conductors circular or rectangular, insulation materials, type of cooling, corona losses and corona impact on insulation materials, and clearance from conductors to tank. Beside these and many others manufacturing factors, equations (2.7), (2.9), (2.16), and (2.17) are frequently used.

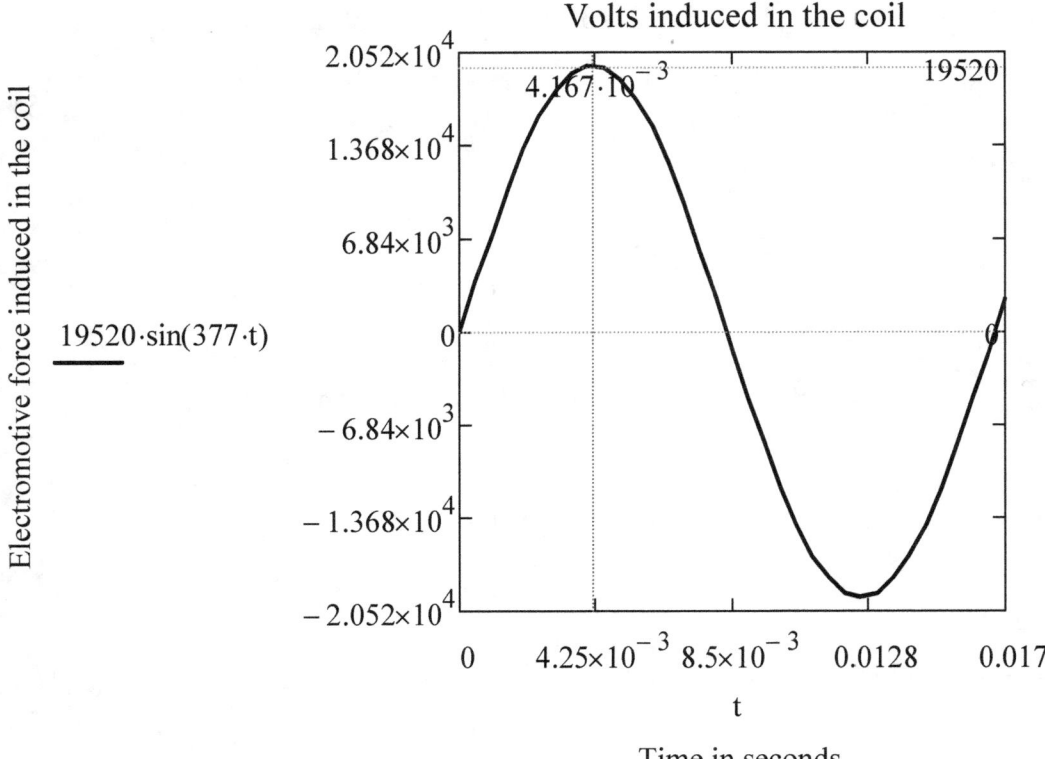

Figure 2-2 Voltage induced in the coil wound around the magnetic circuit

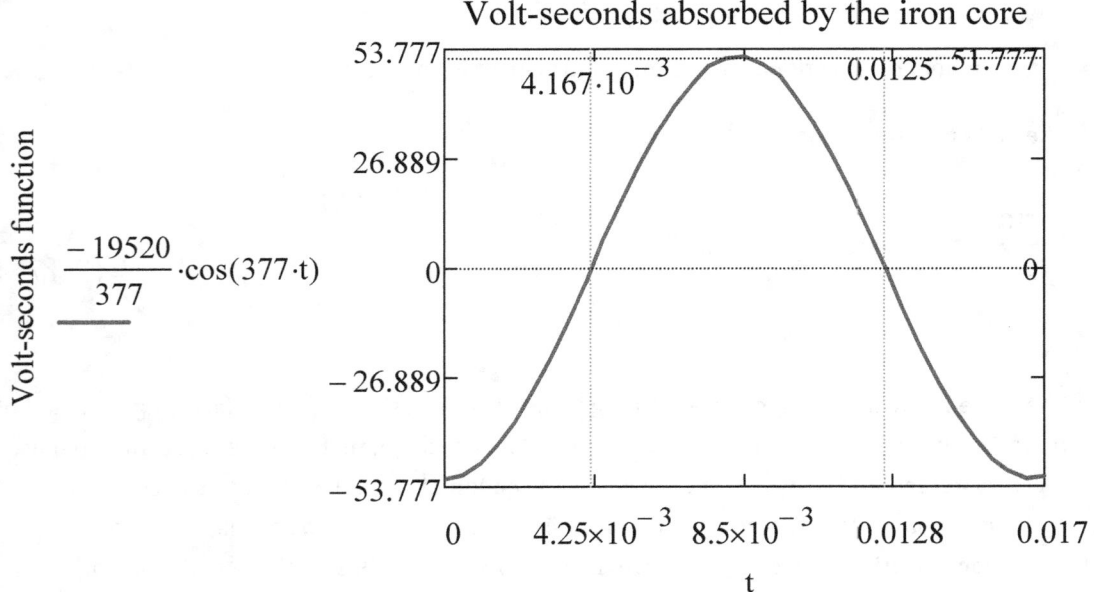

Figure 2-3 Volt-seconds applied to the magnetic circuit

2.2 Hysteresis Loss

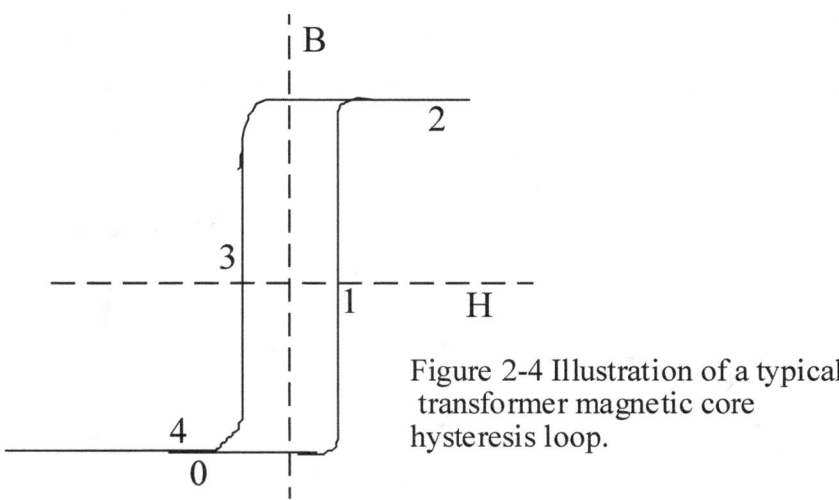

Figure 2-4 Illustration of a typical transformer magnetic core hysteresis loop.

The hysteresis loop illustrated in Fig. (2-4) is a plot of the magnetic flux density, B, vs magnetizing force, H. The hysteresis loop is produced by applying an alternating voltage of practically constant magnitude to the magnetizing coil (Transformer primary). The area inside the hysteresis loop represents the energy loss in joules per cycle per cubic meter and is expressed by Eq. (2.22). The energy delivered to the magnetizing coil per cycle of the alternating voltage is:

$$w = \int_0^T v(t) \cdot i(t)\, dt$$

(2.18)

v = instantaneous alternating voltage applied to primary winding

i = primary winding instantaneous alternating current input

t = time

T = period of the sinusoidal voltage applied to primary winding

e = instantaneous alternating voltage induced in primary winding

$e(t) = v(t)$ Neglecting the voltage drop in the primary winding and the excitation current taken by the iron core shunt reactor. See Fig. 1-8.

$$v(t) = N \cdot \left(\frac{d}{dt} \phi\ (t) \right) \qquad \phi\ (t) = A \cdot B(t) \qquad v(t) = N \cdot A \cdot \left(\frac{d}{dt} B(t) \right)$$

Here, A means area

The magnetizing force, H, is usually expressed in ampere-turn per inch, See Eq. (1.3)

$$H(t) = \frac{N \cdot i(t)}{1} \qquad\qquad i(t) = \frac{H(t) \cdot 1}{N}$$

where I means length

Substituting into (2.18)

$$v(t) \cdot i(t) = N \cdot A \cdot \left(\frac{d}{dt} B(t) \right) \cdot \frac{H(t) \cdot 1}{N} = A \cdot 1 \cdot \left(\frac{d}{dt} B(t) \right) \cdot H(t)$$

$$w = A \cdot 1 \cdot \int_0^T \left(\frac{d}{dt} B(t) \right) \cdot H(t)\ dt = A \cdot 1 \cdot \int_0^T H\ dB$$

$$(2.19)$$

$A \cdot 1$ is the volume of the core and the integral provides the area inside the hysteresis loop. So, the energy loss per cycle of the alternating voltage is equal to the core volume times the area of the B-H loop. To work with Eq. (2.19) we could use the rationalized system of unit RMKS as follows:

$$B = \frac{webers}{m^2} \qquad H = \frac{ampere - turns}{m} \qquad \int_0^T H\ dB = \frac{joules}{m^3 \cdot cycle}$$

The power loss due to hysteresis is

$$P_h = f \cdot A \cdot 1 \cdot \int_0^T H\ dB \cdot \frac{cycle}{sec} \cdot m^3 \cdot \frac{joules}{m^3 \cdot cycle} = f \cdot A \cdot 1 \cdot \int_0^T H\ dB \cdot \frac{joules}{sec}$$

$$f \cdot \frac{cycle}{sec} \qquad A \cdot 1 \cdot m^3 \qquad joule = watt - second \qquad \text{watt = joules per second}$$

$$P_h = f \cdot A \cdot l \cdot \int_0^T H\,dB \cdot watt$$

(2.20)

From Eq. (2.20) we can conclude that the hysteresis power loss is proportional to the frequency of the applied voltage, to the volume of the magnetic core, and to the area inside the hysteresis loop. To apply Eq. (2.20) is necessary to express B in weber's per square meters and H in ampere-turns per meter and the core volume in cubic meters. Sometimes, is convenient to express the power loss in the core in per-unit core volume basis. In this case Eq. (2.20) is transformed into Eq. (2.21).

$$P_h = f \cdot \int_0^T H\,dB \cdot \frac{watt}{m^3}$$

(2.21)

Writing Equation (2.21) without the frequency provides the **energy** loss in joules per cubic meter per cycle as shown on Eq. (2.22).

$$w = \int_0^T H\,dB \cdot \frac{joule}{m^3 \cdot cycle}$$

(2.22)

It is practically impossible to express analytically H as a function of B. Because core laminations magnetic properties change from one specimen to the other accordingly with heat treatment and with the fabrication process itself. So the area of the hysteresis loop is calculated graphically; and according with the instrumentation available, some engineers prefer to express B in gauss and H in Oersted.

Below we present the analysis using un rationalized cgs units; Eq. (2.19) is valid for any system of units. However, to use Eq. (2.19) with the cgs system of units we most multiply the equation by the rationalizing factor $1/4\pi$. For per unit volume computation the factor $A \cdot l$ is reduce to one cubic centimeter. So Eq. (2.19) becomes:

$$w = \frac{1}{4 \cdot \pi} \cdot \int_0^T H\,dB \cdot \frac{erg}{cm^3 \cdot cycle}$$

(2.23)

H is in oersted

B is in gauss

T is the period of the alternating voltage

Multiplying Eq. (2.23) by the core volume expressed in cubic centimeter, we obtain Eq. (2.24)

$$w = \frac{A \cdot l}{4 \cdot \pi} \cdot \int_0^T \left(\frac{d}{dt} B(t) \right) \cdot H(t) \, dt = \frac{A \cdot l}{4 \cdot \pi} \cdot \int_0^T H \, dB \cdot \frac{erg}{cycle}$$

(2.24)

Multiplying Eq. (2.24) by the frequency of the voltage applied to the magnetizing coil we obtain the energy loss per second due to hysteresis.

$$P_h = \frac{f \cdot A \cdot l}{4 \cdot \pi} \cdot \int_0^T H \, dB \cdot \frac{erg}{cycle} \cdot \frac{cycle}{sec} = \frac{f \cdot A \cdot l}{4 \cdot \pi} \cdot \int_0^T H \, dB \cdot \frac{erg}{sec}$$

(2.25)

Knowing that:

$$1 \cdot watt = \frac{Joule}{sec} = \frac{10^7 erg}{sec} \qquad 1 joule = 1.watt - second$$

Eq. (2.25) is transformed into Eq. (2.26) which provides the power loss due to hysteresis expressed in watts.

$$P_h = \frac{f \cdot A \cdot l \cdot 10^{-7}}{4 \cdot \pi} \cdot \int_0^T H \, dB \cdot watt$$

(2.26)

Equation (2.26) could be transformed as follows:

$$P_h = \frac{f \cdot A \cdot l \cdot 10^{-7}}{4 \cdot \pi} \cdot A_{loop} \cdot watt$$

(2.27)

Aloop = area of the hysteresis loop, gauss x oersted

f   = cycles per second

$A \cdot l$ = core volume in cubic centimeter

For 60 Hz application we obtain:

$$P_h = 4.775 \cdot A \cdot 1 \cdot 10^{-7} \cdot A_{loop} \cdot \text{watt} \tag{2.28}$$

2.3 Steinmetz Empirical Equation

Silicon steel laminations are in general, at the present time, imported from China or India. And often the manufactures do not provide enough information to calculate the area of the hysteresis loop. Furthermore, they don't provide the slope, coercive force, and maximum flux density of the hysteresis loop either. In general, they provide outdated, minimum possible design information. In this environment the Steinmetz empirical equation (2.29) is a good substitute. Because, it provides a simple way to calculate the hysteresis loses.

$$P_h = 10^{-7}{}_\eta \cdot f \cdot B_{max}{}^{1.6} \qquad \frac{\text{watt}}{\text{sec} \cdot \text{cm}^3} \tag{2.29}$$

η = coefficient of hysteresis loss

f = frequency of voltage applied to the coil, Hz

B_{max} = maximum magnetic flux density, gauss

The value of η depends upon the type steel used to fabricate laminations for power transformer cores. For cold rolled, grain oriented, silicon (3%) steel laminations, annealed to relieve storage, handling and manufacturing stresses, η = 0.001, for soft iron the value of η ranges from 0.002 to 0.004 and for hard cast steel is around 0.025. The experimental fact is that for a given core material tested at constant frequency and volume, the hysteresis loss depends only on the maximum flux density and its exponent.

Example 2.2

If the area inside the hysteresis loop is 30000 gauss-oersted and assuming that the applied voltage remains constant with a frequency of 60Hz, we find the energy loss due to hysteresis per cycle, per cubic centimeter using Eq. (2.23). And the power loss using Eq. (2.27).

$$P_h = \frac{1}{4\pi} \cdot A_{loop} \cdot \frac{\text{erg}}{\text{cycle} \cdot \text{cm}^3} = \frac{1}{4\pi} \cdot 3 \cdot 10^4 \cdot \frac{\text{erg}}{\text{cycle} \cdot \text{cm}^3} = 2387 \cdot \frac{\text{erg}}{\text{cycle} \cdot \text{cm}^3}$$

Using Eq. (2.27) we find the hysteresis power loss per cubic cm.

$$P_h = \frac{f \cdot 10^{-7}}{4\pi} \cdot A_{loop} \cdot \frac{watt}{cm^3} = \frac{60 \cdot 10^{-7}}{4\pi} \cdot 3 \cdot 10^4 \cdot \frac{watt}{cm^3} = 0.014 \cdot \frac{watt}{cm^3}$$

The power loss in watts per cubic centimeter is: $\quad 0.014$

Knowing that: $\qquad 1m^3 = 10^6 cm^3$

We obtain the power loss in kilowatts per cubic meter

$$cm^3 = \frac{m^3}{10^6} = 10^{-6} \cdot m^3$$

$$P_h = 0.014 \cdot \frac{watt}{cm^3} = 0.014 \cdot \frac{watt}{10^{-6} \cdot m^3}$$

$$P_h = 0.014 \cdot 10^6 \cdot \frac{watt}{m^3} = 14 \cdot \frac{kW}{m^3}$$

2.4 Magnetic Materials

The larger component of the total core loss in a power transformer is the hysteresis loss which is directly proportional to the area of the hysteresis loop. To decrease the hysteresis loss of a power transformer without decreasing the available maximum magnetic flux density, assuming that the frequency cannot be changed, is necessary to decrease the area of the hysteresis loop. This is done by precise alloying and careful heat treatment of the core material; plus rigorous control of the production process. The target is to obtain a product with low loss and high permeability.

Laminated silicon steel.

The magnetic materials used to fabricate transformer's core depends on the frequency of the application. Cores for 60 and 50 Hz large power transformers are routinely fabricated with cold rolled, grain oriented silicon steel (CRGO) laminations, shipped in the form of sheets or coils. In general, the silicon content of electrical steel laminations is around 3%. Grain oriented steels are also "decarbonized" with and allow carbon content limit of 0.06%. Another characteristic is that the grains are oriented in the rolling direction. Generally available sheet laminations grade and thickness are M-4, 0.27 mm and M-6, 0.35 mm. For coils of CRGO steels the available thickness are 0.18, 0.20, 0.27, 0.35 mm.

The reader should keep in mind that production of CRGO laminations is a high precision job that requires great care during all the production steps. In fact, laminations are affected by temperature and mechanical stresses in a complex way and their handling requires expertise and focus in the operation at hand. Finally, the CRGO laminations could suffer excessive stresses not only during fabrication but also during storage and shipment.

The laminations fabrication procedure given below is only a guide to common major steps in the fabrication procedure of different manufactures.

Laminations are cut and punched. This is the step in which burrs are developed and the grain structure damaged. They badly affect the stacking of laminations, the hysteresis loop, and noise level. Furthermore, most of the stressed introduced during this step are plastic stresses, like bending or stretching, and they could introduce a permanent deformation in the magnetic material that can be removed only by stress relief annealing.

Each lamination must be carefully "deburred" and annealed. The annealing process not only relief the mechanical stresses (elastic and plastic) but also creates an oxide film on the burrs that increase their surface resistivity which reduces the flow of inter-laminar currents. A general rule is that laminations surface resistance should be greater than 10 ohms-square centimeter. The annealing step is by itself a complicated operation that requires the right furnace. To avoid the contamination of the magnetic material alloy with harmful impurities (like carbon and oxygen) the furnace atmosphere must consists of a vacuum or inert gas, like pure dry nitrogen. Sometimes the surface finish required, resembles a glass film which is developed at very high temperature inside a   mixture of hydrogen-nitrogen atmosphere. For safety reasons the hydrogen content is maintained below ten percent .Laminations stress relief requires heating them in the range of 760 to 845 degree centigrade. In this case the furnace operator must be careful not to over-cook the laminations. Besides he must avoid large thermal gradients inside the magnetic material that could distort its shape and produce uneven alloying. To avoid contamination, some procedures require fast cooling of the furnace charge and quick removal from the furnace.

A coat of insulating film (for example Carlite 3) is applied to both sides of each sheet. The insulation coating is supposed to reduce the flow of inter-laminar eddy currents. Surface insulating films should retain their surface insulation resistant after stress relief annealing and their impact on the lamination or stacking factor should be negligible.

The cores of high voltage, large power transformers are fabricated by staking laminations of CRGO silicon steels and compressing them under even pressure. The bolts and hardware used to apply pressure must be electrically non-conducting, because otherwise they would provide an easy path for the eddy currents and the laminations would be effectively shorted out of the current path. In some cases it would be enough to use steel bolts covered with insulating film. Strong plastic bolts could be used provided they don't contaminate the transformer's oil. The volume of the laminated core changes with the compressing pressure and the lamination factor provides a way to judge how much pressure should be applied to an specific core design. The lamination or stacking factor is defined as the ratio of the theoretically calculated volume of a magnetically equivalent solid core of the same material as the laminations to the actual volume of the compressed stack For instance: Let us say that for a stack built up with M6 (35mm), CRGO steel laminations compressed at 100 psi the lamination factor is about 97.6 percent. However, the inverse of the lamination factor would be a better indicator of the compactness of the laminated core design. In this specific case it would yield that the laminated core is 2.4 percent larger. So the actual lamination factor would be: LF = Laminated Stack Volume / Solid Core Volume, where the base is the solid core volume

Non-laminated soft iron

Non-laminated soft iron is used in the fabrication of electromagnets and in others apparatus in which the magnetizing current is DC. For these types of applications soft iron is a good choice because it can reach magnetic flux density of 21.6 kilogauss at $40\,°C$ without saturating. Furthermore, soft iron does not remain magnetized when the magnetizing force is removed. However, if the magnetizing current were AC, the constant change in polarity would produce a constant reversal of the magnetic   field and induce large eddy currents in the solid soft iron core which being a metal has low resistivity. As a consequence of heat generated by the eddy currents (RI2), the temperature of the soft iron core will constantly increase and could eventually become red-hot if it is not force cooled by air or oil. Besides there would be an unnecessary waste of energy.

2.5 Dielectric materials

In most insulating materials the ratio of the electric flux density to the electrical field intensity is constant. Symbolically:

$$\varepsilon = \frac{D}{\xi} \qquad\qquad (2.30)$$

D is in Coulomb per square meter

ξ is in volt per meter

The dielectric constant, ε, is expressed in Coulomb per volt-meter or in Farad per meter as show below.

$$\frac{\dfrac{coulomb}{m^2}}{\dfrac{volt}{m}} = \frac{coulomb}{volt \cdot meter} = \frac{farad}{meter}$$

The dielectric constant of a vacuum is ε 0 = 8.85 x 10-12 farad per meter. In general, insulating materials have dielectric constants values in the range of two to six time the value for vacuum. Frequently, the dielectric constant is given in term of the *specific* dielectric constant which is the dielectric constant of the material divided by the dielectric constant of vacuum. Symbolically:

$$\varepsilon_s = \frac{\varepsilon_m}{\varepsilon_0} \tag{2.31}$$

ε_m is the dielectric constant of the material

ε_0 is the dielectric constant of vacuum

ε_s is the specific dielectric constant of the material

In static electric fields the electric flux starts on positive charges and end on negative ones. So one coulomb of electric flux lines stars from each positive coulomb of electric charge, and ends on a negative coulomb of electric charge. Similarly, *in time varying electric fields* each coulomb of positive charge has one coulomb of electric flux starting on it, and each coulomb of negative charge has one coulomb of electric flux ending on it. Besides, the flux lines in a time varying electric field could also form closed loops with no beginning or ending and therefore these flux lines are in addition to the ones created by the fields of the charges.

The *dielectric strength* of a material is the field intensity at which the material is no longer an insulator. And will quickly break down. In practical applications insulators should not be stressed beyond one half their dielectric strength. Meaning a safety factor of at least 2 most be used in high-voltage applications. Air has a low dielectric strength, 3000 kV/meter, so insulating material should not contain air bubbles because they could become ionized by the electric field and attack and damage the insulating material. The insulating material used in electrical windings of transformers, motors and many other equipment are often vacuum impregnated to avoid imbedded air bubbles. Actually, the air is removed by evacuating the entire winding. Next, while still under vacuum, the entire winding is submerged in a hot liquid of insulating material. After turning off the vacuum machine normal atmospheric pressure is admitted into the chamber, forcing the insulating material into the empty spaces between coil's wires. Finally, after cooling and drying, the insulating material hardens and becomes an effective insulator and seal against moisture.

In power transformers the windings are usually operated immersed in transil oil, which provide additional insulation of the windings. Furthermore, often the oil is circulated and cooled by a combination of pump and radiators, which cools the innermost part of the transformer. In large HV transformers, the oil is usually covered with a layer of inert gas to minimize dielectric materials damage due to corona discharge inside the transformer. The corona discharge is produced when the line-to-line or line-to-ground voltages exceeds the corona threshold which is highly dependent on temperature and humidity. The ionized air molecules, near HV transmission line conductors, produces the corona effect which is manifested by typical hissing noises, the smell of ozone, and a red or blue glow, all of these is accompanied by corona induces currents through the ionized air along the line conductors which are the cause of significant losses.

2.6 Standard Markings of Single-Phase Transformer Terminals.

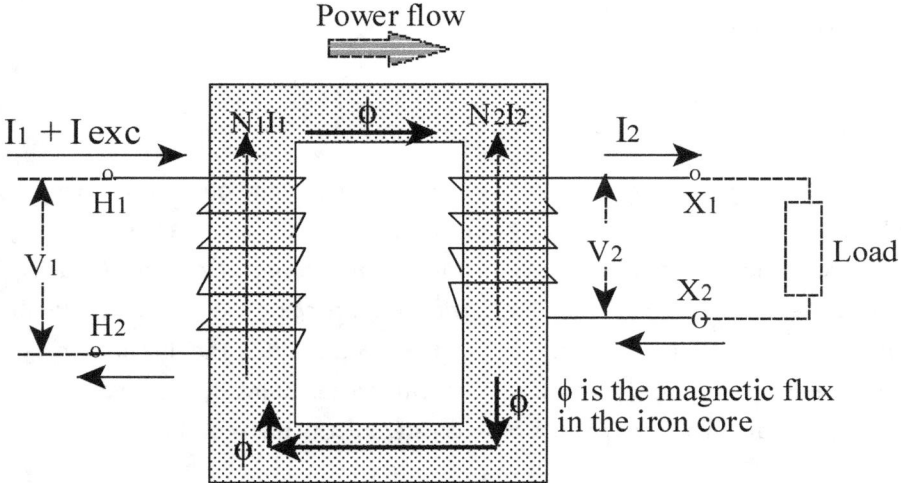

Figure 2-5 Illustration of the standard winding markings of a single-phase transformer

In Fig. 2-5 the terminals of the high voltage winding are marked H1 and H2, and the terminals of the low voltage winding are marked X1 and X2. The load component of the input current coming into H1 produces a magnetomotive force, N1I1 which nullify the magnetomotive force (mmf) produced by the load current flowing through the secondary winding. Both mmf are equal and of opposite direction, and both are proportional to the load current. This nullification is due to the Lenz law. So the secondary current must flows out of terminal X1 when the primary current is flowing into terminal H1. *The resultant flux, linking both windings is produced by the magnetizing component, Imag, of the primary current. Symbolically.*

$$\overrightarrow{I_P} = \overrightarrow{I_1} + \overrightarrow{I_{exc}}$$ Total primary current

$$\overrightarrow{I_{exc}} = \overrightarrow{\left(I_{mag} + I_{cL}\right)}$$ Magnetizing plus core losses currents

The primary winding mmf balances the secondary winding mmf.

$$\overrightarrow{\left(N_1 \cdot I_1\right)} + \overrightarrow{\left(N_2 \cdot I_2\right)} = 0$$

$$N_1 \cdot I_1 = N_2 \cdot I_2 \qquad \textbf{It is less confusing to work only with the magnitudes}$$

I_1 is the magnitude of the load component of the primary current

I_2 is the magnitude of the secondary or load current

$$I_1 = \frac{N_2}{N_1} \cdot I_2 \quad a = \frac{N_1}{N_2} \quad I_1 = \frac{I_2}{a} \qquad I_2 = a \cdot I_1 \tag{2.32}$$

For the primary and secondary voltages the following applies neglecting impedances

$$\frac{|V_1|}{N_1} = \frac{|V_2|}{N_2}$$

$$\frac{|V_1|}{|V_2|} = \frac{N_1}{N_2} = a \qquad V_1 = a \cdot V_2 \qquad V_2 = \frac{V_1}{a} \tag{2.33}$$

For some applications the transformation ratio, **a**, is given as a number greater than one; In fact, the transformation ratio is greater than one for a step-down transformer and less than one for a step-up transformer. *As shown in Fig. 2-5 the load component of the primary current must be in phase with the secondary current because the mmf (sine waves) they produce must cancel each other at all times. This means that H1 and X1 are positive (or negative) at the same* *time with respect to H2 and X2 respectively. So both currents are in phase.* However, reversing the secondary current direction while the primary current remains the same, makes the primary and secondary current phasors 180 degrees out of phase with respect to each other. In conclusion, the primary and secondary winding current phasors could be either in phase or 180 degrees out of phase, depending on the assumed positive current directions in the primary and secondary windings. Same thing happens with the primary and secondary voltages where the voltage per turn in each winding must be the same, because the same flux links with both windings. Furthermore, the induced voltage phasor in each winding are in phase or 180 degrees out of phase depending on what terminal is selected as positive when calculating the voltage drop in each winding. Although this appears confusing, the true is that you could not make a fatal mistake, because the Lenz law would make it works for you, regardless of how you connect the terminals to the driving voltage and to the load. However, your circuit calculations could be wrong. Figure 2-6 provides another method of marking the primary and secondary windings. A dot is placed on one end of each coil, so that current flowing from the dotted end to the unmarked end of each coil induces mmf in the same direction. In Fig. 2-6 the dots indicate that the dotted ends are positive at the same time with respect to the no-dotted ends. Meaning that the primary and secondary current are in phase. In conclusion Fig. 2-6 shows the positive direction for the primary and secondary current.

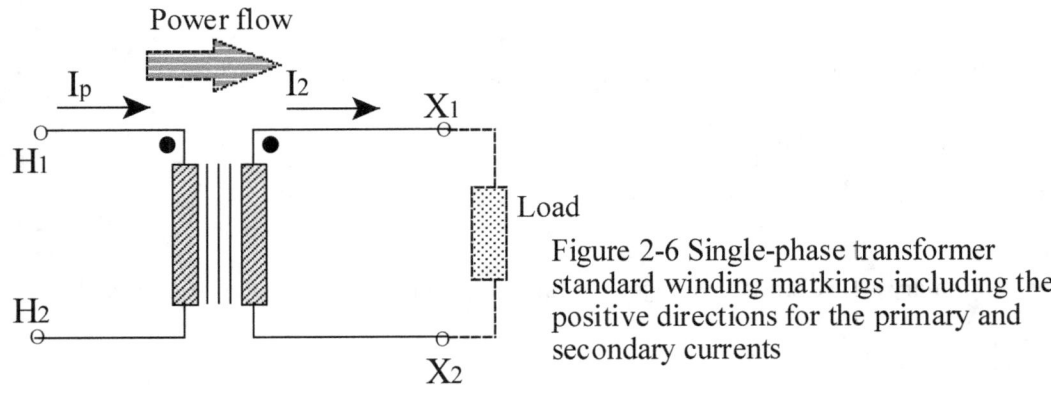

Figure 2-6 Single-phase transformer standard winding markings including the positive directions for the primary and secondary currents

2.7 Standard Markings of Three-Phase Transformer Terminals.

The high voltage terminals of three-phase transformers are marked H1, H2, H3 and the low voltage terminals are marked X1, X2, X3. In Y-Y or Δ-Δ connected transformers the markings are placed in such a way that voltages to the neutral from terminals H1, H2, H3 are in phase with voltages to neutral from X1, X2, X3 respectively. Fig. 2-7 illustrates the connection diagram for a Y-Y connected three-phase transformer as well as its phasor diagram.

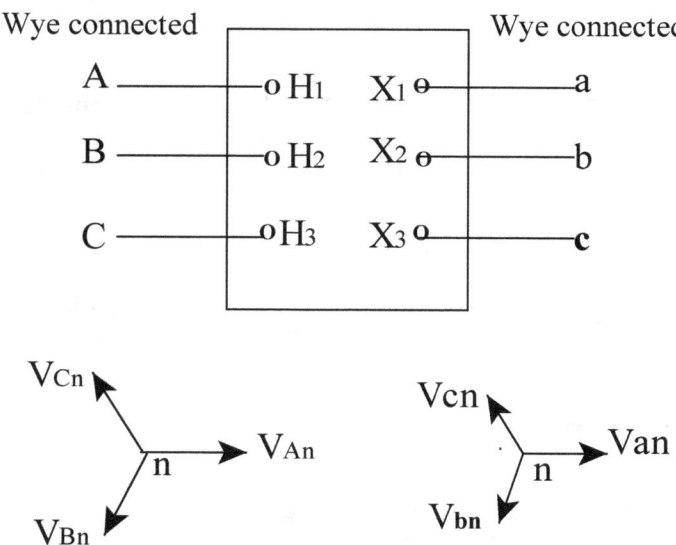

Figure 2-7 Connection and phasor diagrams for Y-Y connected three-phase transformer.

The connection diagram for Y-Δ connected three-phase transformer is given in Fig. 2-8

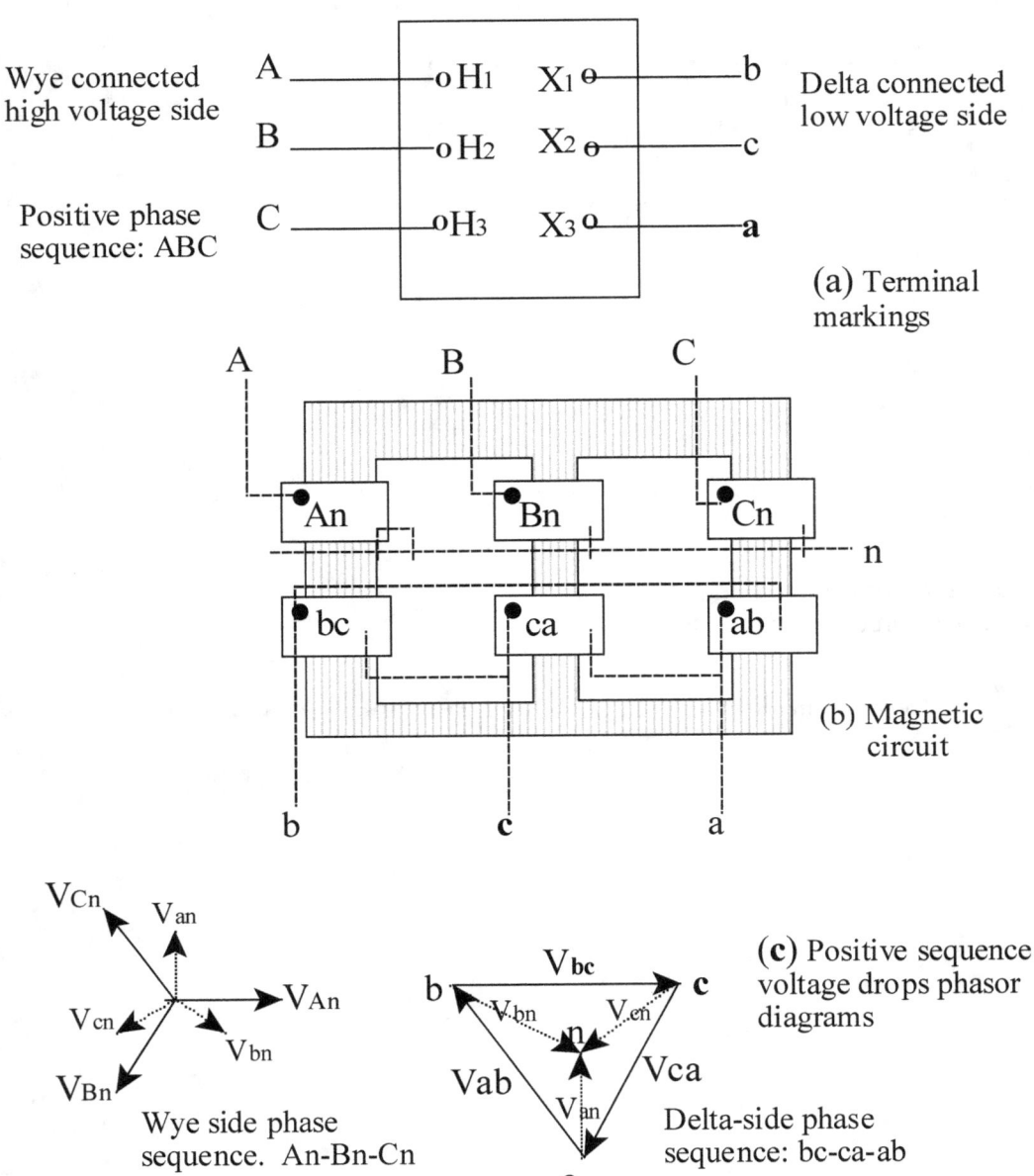

Wye connected high voltage side

Positive phase sequence: ABC

Delta connected low voltage side

(a) Terminal markings

(b) Magnetic circuit

(c) Positive sequence voltage drops phasor diagrams

Wye side phase sequence. An-Bn-Cn

Delta-side phase sequence: bc-ca-ab

Figure 2-8 Three-phase transformer, connections and voltage drop phasor diagrams.

As depicted in Fig. 2-8b the phasors representing the line to line voltages in the delta-side must be in phase with the corresponding line to neutral voltage phasors in the wye-side, independently of the sequence. *The reader must realize that coils that are wound on the same core-leg have their voltages in phase, like Bn and ca. And that the legs sinusoidal magnetizing currents follow the sequence of the three-phase input driving voltage.* Fig. 2.8c shows the line to line voltages in the delta side in function of the line to neutral voltages. See Eqns. (2.34) - (2.36).

$$\overrightarrow{V_{bc}} = \overrightarrow{V_{bn}} + \overrightarrow{V_{nc}} = \overrightarrow{V_{bn}} - \overrightarrow{V_{cn}} \qquad (2.34)$$

$$\overrightarrow{V_{ca}} = \overrightarrow{V_{cn}} + \overrightarrow{V_{na}} = \overrightarrow{V_{cn}} - \overrightarrow{V_{an}} \qquad (2.35)$$

$$\overrightarrow{V_{ab}} = \overrightarrow{V_{an}} + \overrightarrow{V_{nb}} = \overrightarrow{V_{an}} - \overrightarrow{V_{bn}} \qquad (2.36)$$

In Fig. 2-8a line **b** is connected to terminal **X1,** line **c** to **X2** and line **a** to **X3.** In this way, as required by ANSI, the line to neutral voltage at H1 leads the line to neutral voltage at X1 by 30 degrees, same for H2 and X2 and H3 and X3. See Fig. 2-8c. Up to now we have analyzed only the positive sequence case, working now with the negative sequence we obtain the phasor diagrams illustrated in Fig. 2.9. The positive voltage phase sequence in the wye-side of Fig. 2.8 is ABC and negative sequence is ACB.

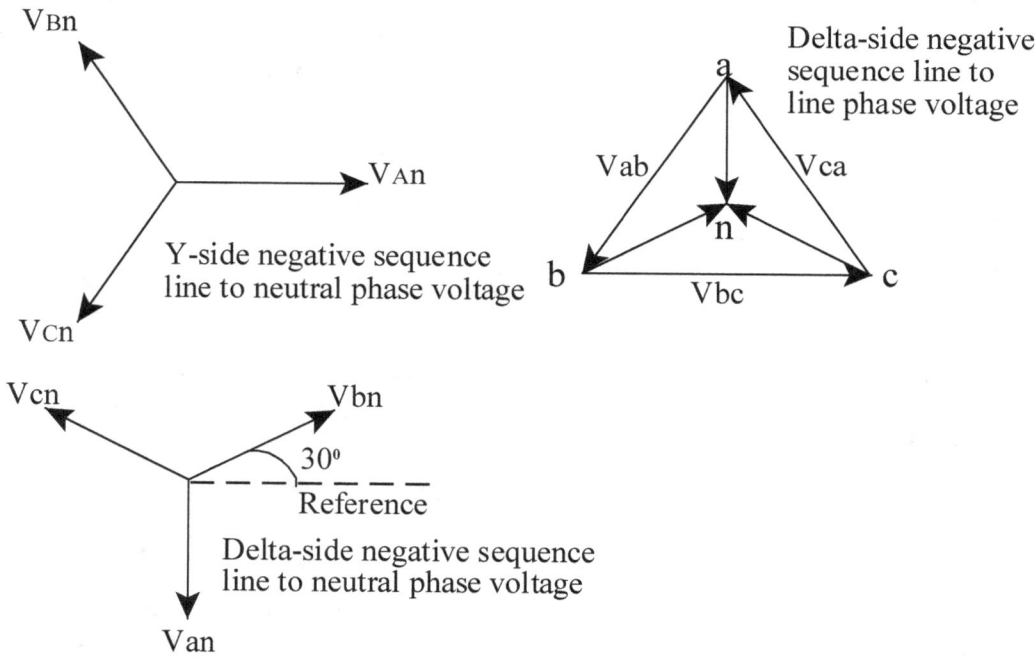

Figure 2-9 Illustraton of the negative sequence phasor diagrams of the Y-Δ transformer shown in Fig. 2-8.

The negative sequence line to line voltages in the delta side are provided below in function of the line to neutral voltages.

$$\overrightarrow{V_{bc}} = \overrightarrow{V_{bn}} + \overrightarrow{V_{nc}} = \overrightarrow{V_{bn}} - \overrightarrow{V_{cn}} \tag{2.37}$$

$$\overrightarrow{V_{ca}} = \overrightarrow{V_{cn}} + \overrightarrow{V_{na}} = \overrightarrow{V_{cn}} - \overrightarrow{V_{an}} \tag{2.38}$$

$$\overrightarrow{V_{ab}} = \overrightarrow{V_{an}} + \overrightarrow{V_{nb}} = \overrightarrow{V_{an}} - \overrightarrow{V_{bn}} \tag{2.39}$$

The phasor diagrams of the positive sequence, see of Fig 2-8c, shows that Van leads VAn by 90 degrees and Fig. 2-9 shows that negative sequence Van lags VAn by 90 degrees

Figure 2-10 shows the three-phase wiring diagram for the same transformer depicted in Fig. 2-8.

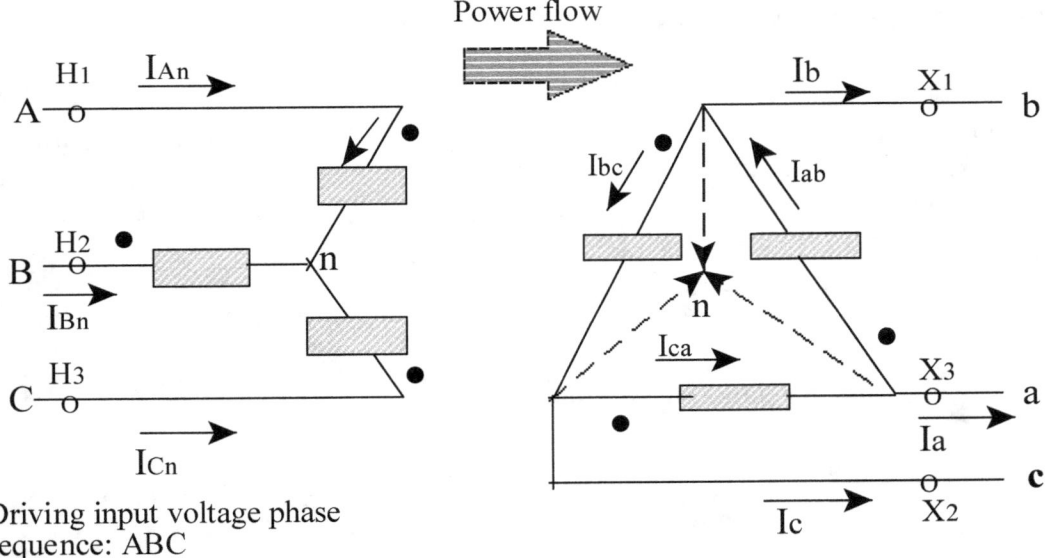

Driving input voltage phase sequence: ABC

Figure 2-10 shows the three-phase wiring diagram for the wye-delta transformer depicted on Fig. 2-8

The current pairs of phasors IAn - Ibc, IBn - Ica, and ICn - Iab can only be180 degrees out of phase because their respective windings are mounted in the same transformer's core leg as illustrated in Fig. 2-8 and the primary and secondary windings are wound in the same direction **(all currents directions are away from the dots when going through the winding).** Fig. 2-11 shows the currents phase relations for the Y-Δ three-phase transformer. For both the positive and negative sequences current components.

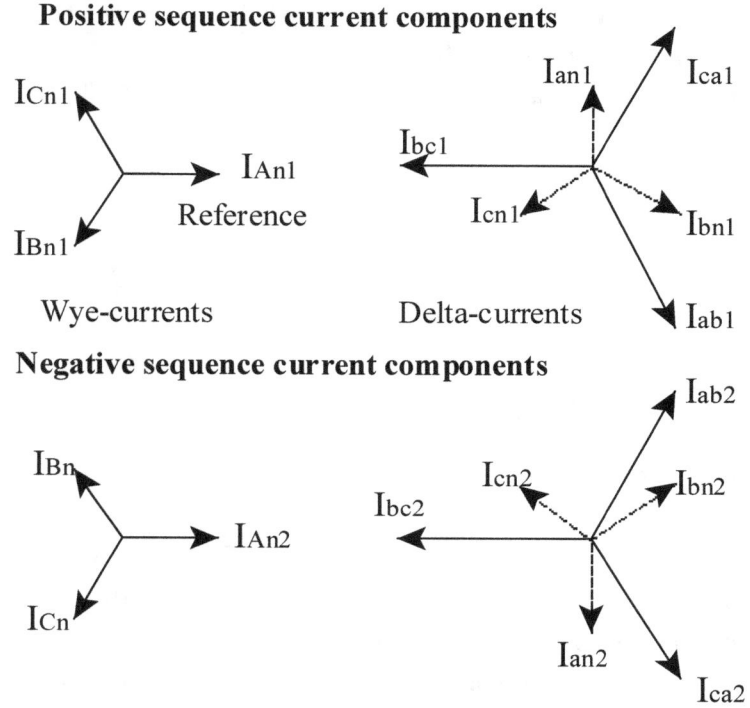

Positive sequence current components

Wye-currents   Delta-currents

Negative sequence current components

Figure 2-11 Currents phase relation for a Y-Δ transformer

In Fig. 2-11 it is shown that Ian1 leads IAn1 by 90 degrees and that Ian2 lags IAn2 by 90 degrees. From here on, in symmetrical components, digit 1 means positive sequence and digit 2 means negative sequence. The results reached so far could be expressed symbolically in per unit and neglecting the magnetizing current and the transformer impedance as follows:

$$\overrightarrow{I_{an1}} = j \cdot \overrightarrow{I_{An1}} \tag{2.40}$$

$$\overrightarrow{I_{an2}} = -j \cdot \overrightarrow{I_{An2}} \tag{2.41}$$

$$\overrightarrow{V_{an1}} = j \cdot \overrightarrow{V_{An1}} \tag{2.42}$$

$$\overrightarrow{V_{an2}} = -j \cdot \overrightarrow{V_{An2}} \tag{2.43}$$

Equations (2.40) to (2.43) were deducted assuming that in the Y-Δ three-phase transformer the high voltage side was the Y-side, and that the power was flowing from the high to the low voltage sides. Never the less Eqns. (2.40) to (2.43) are valid for both directions of the power flow and both types of primary winding connections Y or Δ. They are indeed very valuable equations when solving symmetrical components problems involving three-phase transformers Y-Δ or Δ -Y connected. The operator j rotates the phasor to which is applied by 90 degrees.

CHAPTER 3

Symmetrical Components

3.1 Introduction

The method of symmetrical components provides a powerful tool for the analysis of unbalanced electrical networks, especially those containing polyphase electrical machinery. This method was published in the 1918 Transactions of the A.I.E.E., Vol. 37 by Charles Legeyt Fortescue. Since then it has become an important tool in the analysis of unbalanced electrical circuits.

Although, the method of symmetrical components is applicable to any unbalanced polyphase electrical network we limit our discussion to three-phase electrical networks. The Fortescue theorem said that any asymmetrical three-phase system of phasors can be resolved into to three symmetrical system of phasors. These phasors are coplanar vectors representing sinusoidal electrical quantities, like voltages and currents, and they rotate with an angular velocity of 377 radians per second for 60 Hz power systems. The phasors representing voltages or currents in a three-phase unbalanced electrical power systems can be resolved into three balanced set of phasors, each set consists of three phasors of equal magnitude, with a time-phase sequence or phase separation in time, or order of rotation as described below:

1. The positive-sequence set: is a symmetric set of phasors composed of three phasors of equal magnitude, displaced on time by 120 degrees from each other and having the same phase sequence or order of rotation as the original asymmetric system of phasors.

2. The negative-sequence set: is a symmetric set of phasors composed of three phasors of equal magnitude, displaced on time by 120 degrees from each other and having the opposite phase sequence or order of rotation as the original unbalanced system of phasors.

3. The zero-sequence set: is a symmetric set of phasors composed of three phasors of equal magnitude and in phase (no separated on time) with each other. Therefore they have the same time-phase position with respect to any reference axis. This set of uniphased phasors is called the zero-sequence set. The reader should understand that the zero-sequence components also rotates at 377 radians per second.

We will not give a rigorous proof of the Fortesque theorem, instead we will show that any asymmetric three-phase system of phasors could be represented by three symmetric systems or sets, as listed above, and that the phasor addition of the corresponding components, one from each set, results in the original system of phasors. Furthermore, the application of the method to electrical power systems requires to indicate in a circuit diagram of the original three-phase electrical network, the positive circuit directions for voltages and currents.

3.2 Asymmetric Three-Phase System of Phasors and their Symmetrical Components.

In general, the phasors composing an asymmetric three-phase system could have any magnitude and any phase angle position with respect to each other. Figures 3-1d shows a set of asymmetric phasors. And figures 3-1a through 3-1c show the three sets of symmetric phasors which are the symmetrical components of the original set of asymmetric phasors.

To discuss the system in general terms, without voltage or current connotations, we designated the asymmetric phasors shown in Fig. 3-1d as Ua, Ub, Uc rather than V or I. And arbitrarily assigned to them the abc phase sequence.

Figure 3-1a shows the positive sequence symmetrical components also with sequence abc because they must have the same sequence as the original asymmetric set. The positive sequence system consists of three phasors which are completely determined once we know the magnitude and phase position of any one of them. The positive sequence phasors are designated as Ua1, Ub1, Uc1 respectively, meaning that Ua1 is the positive sequence component of Ua and similarly for Ub1 and Uc1.Figure 3-1b shows the negative-sequence components with sequence acb and Fig. 3-1c shows the zero-sequence components .This system consists of three phasors in sink which each other.

64

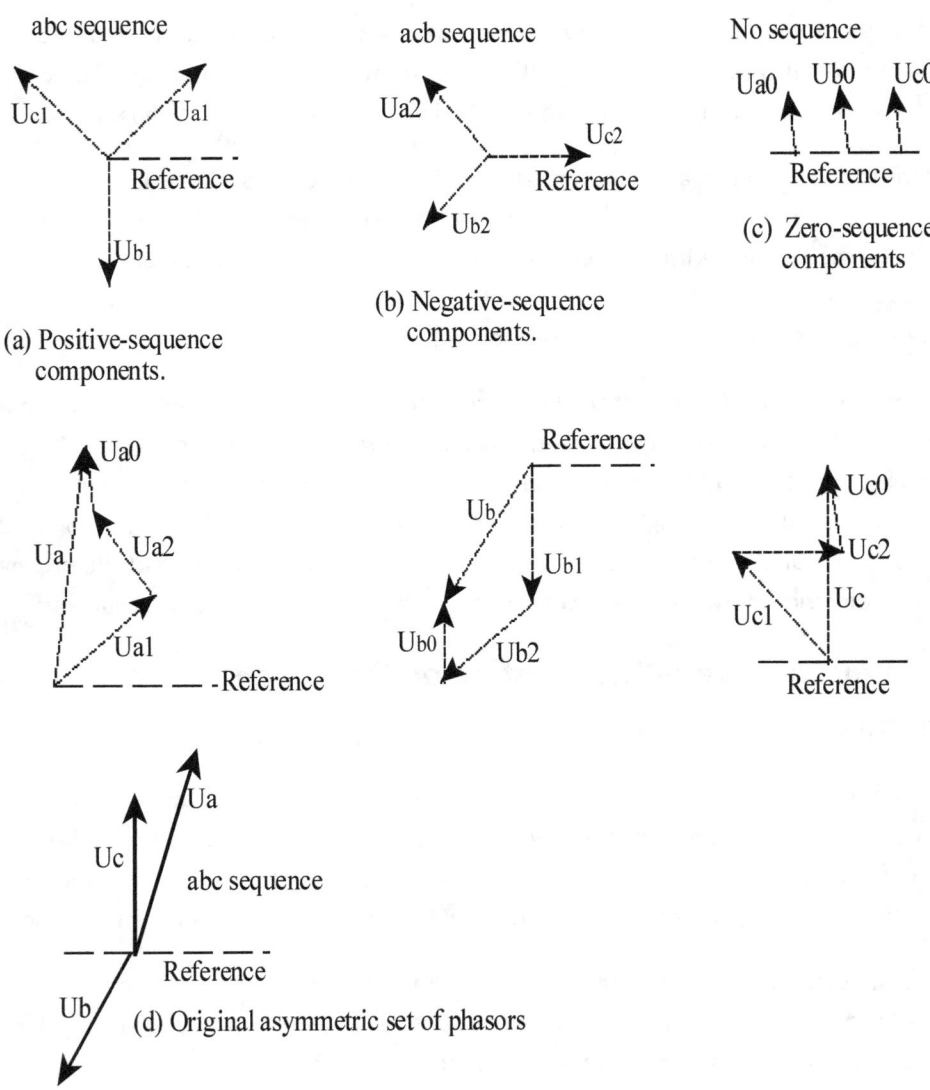

Figure 3-1 Diagrams Illustrating an asymmetrical set of phasors and its symmetrical components .

The original unbalanced phasors expressed in terms of their symmetrical components are:

$$U_a = U_{a0} + U_{a1} + U_{a2} \tag{3.1}$$

$$U_b = U_{b0} + U_{b1} + U_{b2} \tag{3.2}$$

$$U_c = U_{c0} + U_{c1} + U_{c2} \tag{3.3}$$

The reader must realize that any single phasor of any symmetrical set, can't exists by itself. To be equivalent to the original phasor, it must be accompanied by the others symmetrical phasors of same phase but different sequences that make up the original phasor. However, some symmetrical phasors could be eliminated in accordance with the circuit connections. In general symmetrical components could be related to each other by the exponential operators given below:

$$\varepsilon^{j\cdot\theta} = \cos(\theta) + j\cdot\sin(\theta)$$

This operator produce rotation in the positive direction, counter clock wise.

$$\varepsilon^{-j\cdot\theta} = \cos(\theta) - j\cdot\sin(\theta)$$

This operator produce rotation in the negative direction, clock wise.

These complex numbers have unit magnitude and are the conjugates of each other. Any phasor multiplied by any of these two complex operators is rotated by plus or minus θ radians. For the specific case of the symmetrical components of three-phase power systems it is convenient to relate each other by means of the **a** operator which is a phasor (complex number) of magnitude one and 120 degrees angle. Symbolically:

$$a = \varepsilon^{j\cdot120°}$$

When **a** multiplies another phasor, rotates it by 120 degrees or $2\pi/3$ radians in the positive or counterclockwise direction. Table 3-1 provides the value of several functions of **a**.

$$a = 1\cdot\varepsilon^{j\cdot120°} = \cos(120°) + j\cdot\sin(120°) = -0.5 + j\cdot0.866 = \frac{-1}{2} + j\cdot\frac{\sqrt{3}}{2}$$

$$a^2 = 1\cdot\varepsilon^{j\cdot240°} = \cos(240°) + j\cdot\sin(240°) = -0.5 - j\cdot0.866 = \frac{-1}{2} - j\cdot\frac{\sqrt{3}}{2} \quad \text{Conjugate of a}$$

$$a^3 = 1\cdot\varepsilon^{j\cdot360°} = \cos(360°) + j\cdot\sin(360°) = 1 + 0$$

$$a^4 = 1\cdot\varepsilon^{j\cdot120°} = \cos(120°) + j\cdot\sin(120°) = -0.5 + j\cdot0.866 = \frac{-1}{2} + j\cdot\frac{\sqrt{3}}{2} \quad \text{Equal to a}$$

$$1 + a = 0.5 + j\cdot0.866 = -a^2$$

$$1 - a = 1.5 - j \cdot 0.866 = \sqrt{3} \cdot \varepsilon^{-j \cdot 30°}$$

$$1 + a^2 = 0.5 - j \cdot 0.866 = -a = 1 \cdot \varepsilon^{-j \cdot 60°}$$

$$1 - a^2 = 1.5 + j \cdot 0.866 = \sqrt{3} \cdot \varepsilon^{j \cdot 30°}$$

$$a + a^2 = -1$$

$$a - a^2 = j \cdot \sqrt{3} = \sqrt{3} \cdot \varepsilon^{j \cdot 90°}$$

$$a^2 - a = -j \cdot \sqrt{3} = \sqrt{3} \cdot \varepsilon^{j \cdot 270°} = \sqrt{3} \cdot \varepsilon^{-j \cdot 90°}$$

$$1 + a + a^2 = 0 + j \cdot 0$$

$$1 + a^3 + a^3 = 3$$

$$1 + a^2 + a^4 = 0$$

Table 3-1 Values of several functions of a

The operator j has magnitude one and produces 90 degrees rotation in the positive direction. Symbolically:

$$j = \varepsilon^{j \cdot 90°} = \cos(90°) + j \cdot \sin(90°) = 0 + j$$

$$j^2 = \varepsilon^{j \cdot 180} = \cos(180°) + j \cdot \sin(180°) = -1 + j \cdot 0 = -1$$

Any phasor multiplied by -1 is rotated by 180 degrees.

Any phasor multiplied by the operator j squared is rotated by 180 degrees. Both operators -1 and j squared produce the same effect and therefore they are equal. Symbolically:

$$j^2 = -1 \qquad j = \sqrt{-1}$$

Similarly, there are three possible values for the operator **a**, the cubic roots of 1.

$$a^3 = 1 \quad a = \sqrt[3]{1} \qquad a = \cos\left(\frac{2 \cdot k \cdot \pi}{3}\right) + j \cdot \sin\left(\frac{2 \cdot k \cdot \pi}{3}\right) \qquad \text{Roots of the } \mathbf{a} \text{ operator}$$

$k = 0, 1, 2$ All other values of k, positive or negative, produce repetitive values of the roots.

Any phasor multiplied by the operator **a** should be rotated by 120 degrees, this correspond to $k = 1$

$$k := 1 \qquad a = \cos\left(\frac{2 \cdot k \cdot \pi}{3}\right) + j \cdot \sin\left(\frac{2 \cdot k \cdot \pi}{3}\right) = \frac{-1}{2} + j \cdot \frac{\sqrt{3}}{2} \qquad j := \sqrt{-1}$$

$$\left| \left(\frac{-1}{2} + j \cdot \frac{\sqrt{3}}{2} \right)^2 \right| = 1 \qquad \operatorname{atan}\left(\frac{\frac{\sqrt{3}}{2}}{\frac{-1}{2}} \right) = -60 \cdot^{\circ}$$

$$\operatorname{atan} = -60^{\circ} + n \cdot 180^{\circ}$$

Where n is any integer, for $n = 1$ $\operatorname{atan} = 120^{\circ}$

Therefore:

$$a = \varepsilon^{j \cdot \frac{2\pi}{3}} = \varepsilon^{j \cdot 120^{\circ}} \qquad \text{<<<<}$$

In equations (3.1 - 3.3), repeated here for clarity, the number of unknown voltages can be reduced by substituting all the symmetrical components of Ub and Uc as indicated in Eqns. (3.4 - 3.5) and Fig. 3-1

$$U_a = U_{a0} + U_{a1} + U_{a2} \tag{3.1}$$

$$U_b = U_{b0} + U_{b1} + U_{b2} \tag{3.2}$$

$$U_c = U_{c0} + U_{c1} + U_{c2} \tag{3.3}$$

Where:

$$U_{b0} = U_{a0} \qquad U_{b1} = a^2 \cdot U_{a1} \qquad U_{b2} = a \cdot U_{a2} \tag{3.4}$$

$$U_{c0} = U_{a0} \qquad U_{c1} = a \cdot U_{a1} \qquad U_{c2} = a^2 \cdot U_{a2} \tag{3.5}$$

$$U_a = U_{a0} + U_{a1} + U_{a2} \tag{3.6}$$

$$U_b = U_{a0} + a^2 \cdot U_{a1} + a U_{a2} \tag{3.7}$$

$$U_c = U_{a0} + a \cdot U_{a1} + a^2 \cdot U_{a2}$$

(3.8)

Equations (3.6 - 3.8) are expressed as function of the **a** operator and the symmetrical components of phase a. It could as well been written for the symmetrical components of phase b or c. Writing Eqns. (3.6 - 3.8) in matrix format we get the matrix equation (3.9) which provides the original phasors as a function of their symmetrical components. And is valid for any set of three phasors. It doesn't need to be the phasors of a power system.

$$\begin{pmatrix} U_a \\ U_b \\ U_c \end{pmatrix} = \begin{pmatrix} 1 & 1 & 1 \\ 1 & a^2 & a \\ 1 & a & a^2 \end{pmatrix} \cdot \begin{pmatrix} U_{a0} \\ U_{a1} \\ U_{a2} \end{pmatrix}$$

(3.9)

Equation (3.10) provides the transformation matrix for a three phasors system. Pre multiplying the symmetrical component matrix by the A-matrix we obtain the original phasor matrix. It converts the symmetrical component phasors into the original phasors.

$$A := \begin{pmatrix} 1 & 1 & 1 \\ 1 & a^2 & a \\ 1 & a & a^2 \end{pmatrix}$$

(3.10)

The conjugate of the transformation matrix is:

$$\overline{A} = \begin{pmatrix} 1 & 1 & 1 \\ 1 & a & a^2 \\ 1 & a^2 & a \end{pmatrix}$$ Remember, a and a2 are conjugate

(3.11)

The inverse of the A matrix is:

$$A^{-1} = \begin{pmatrix} 0.3333 + 4.6669j \times 10^{-6} & 0.3333 - 2.3335j \times 10^{-6} & 0.3333 - 2.3335j \times 10^{-6} \\ 0.3333 - 2.3335j \times 10^{-6} & -0.1667 + 0.2887j & -0.1667 - 0.2887j \\ 0.3333 - 2.3335j \times 10^{-6} & -0.1667 - 0.2887j & -0.1667 + 0.2887j \end{pmatrix}$$

In the A^{-1} the imaginary part of the elements on the first row and first column are practically zero and therefore could be neglected.

$$A^{-1} = \begin{pmatrix} 0.3333 & 0.3333 & 0.3333 \\ 0.3333 & -0.1667 + 0.2887j & -0.1667 - 0.2887j \\ 0.3333 & -0.1667 - 0.2887j & -0.1667 + 0.2887j \end{pmatrix}$$

$$3 \cdot A^{-1} = \begin{pmatrix} 1 & 1 & 1 \\ 1 & -0.5 + 0.866j & -0.5 - 0.866j \\ 1 & -0.5 - 0.866j & -0.5 + 0.866j \end{pmatrix}$$

Using Table 3-1 we express the inverse of matrix A as a function of the **a** operator. Symbolically:

$$3 \cdot A^{-1} = \begin{pmatrix} 1 & 1 & 1 \\ 1 & a & a^2 \\ 1 & a^2 & a \end{pmatrix} \qquad A^{-1} = \frac{1}{3} \cdot \begin{pmatrix} 1 & 1 & 1 \\ 1 & a & a^2 \\ 1 & a^2 & a \end{pmatrix} \qquad \text{Inverse of the A matrix} \qquad (3.12)$$

From Eqns (3.11, 3.12) we conclude that the inverse of the A matrix is equal to one third of the conjugate of the transformation matrix. Symbolically:

$$A^{-1} = \frac{1}{3} \cdot \overline{A} \tag{3.13}$$

Pre multiplying both sides of Eqn. (3.9) by A^{-1} and having in account that the

product of a matrix and its inverse is equal to one:

$$\frac{1}{3} \cdot \overline{A} \cdot \begin{pmatrix} U_a \\ U_b \\ U_c \end{pmatrix} = \frac{1}{3} \cdot \overline{A} \cdot A \cdot \begin{pmatrix} U_{a0} \\ U_{a1} \\ U_{a2} \end{pmatrix} \qquad\qquad \frac{1}{3} \cdot \overline{A} \cdot \begin{pmatrix} U_a \\ U_b \\ U_c \end{pmatrix} = \begin{pmatrix} U_{a0} \\ U_{a1} \\ U_{a2} \end{pmatrix}$$

$$\begin{pmatrix} U_{a0} \\ U_{a1} \\ U_{a2} \end{pmatrix} = \frac{1}{3} \cdot \begin{pmatrix} 1 & 1 & 1 \\ 1 & a & a^2 \\ 1 & a^2 & a \end{pmatrix} \cdot \begin{pmatrix} U_a \\ U_b \\ U_c \end{pmatrix} \tag{3.14}$$

From Eqn. (3.14) we obtain the symmetrical components in function of the original unbalanced set of phasors.

$$U_{a0} = \frac{1}{3} \left(U_a + U_b + U_c \right) \tag{3.15}$$

$$U_{a1} = \frac{1}{3}\left(U_a + a \cdot U_b + a^2 \cdot U_c\right)$$

(3.16)

$$U_{a2} = \frac{1}{3}\left(U_a + a^2 \cdot U_b + a \cdot U_c\right)$$

(3.17)

Once you know any symmetrical component of a set, then you can easily find the remaining symmetrical components using Eqns. (3.4 - 3.5). *Equation (3.15) shows that if the phasor sum of the unbalanced phasors were zero, then the zero- sequence components could not exit. Furthermore if the U phasors represent the line-to-line voltages of a three-phase power system, and because the sum of the line-to-line voltages of a three-phase power system, balanced or unbalanced, is always zero, then there could not be zero sequence components present in the line-to-line voltages of any three-phase power system.* However, the sum of the line-to-neutral voltages could be different from zero and therefore they could contain zero sequence components. The reader must keep in mind that although the above analytical discussion of the Fortescue method was developed in function of the U phasors, it is valid for any set of phasors, voltage or current. Below we provide, as functions of their symmetrical components, the most important voltage and current equations applicable to a three-phase power systems. From Eqns. (3.1 - 3.17) we get:

$$V_a = V_{a0} + V_{a1} + V_{a2}$$

(3.18)

$$V_b = V_{b0} + V_{b1} + V_{b2}$$

(3.19)

$$V_c = V_{c0} + V_{c1} + V_{c2}$$

(3.20)

$$V_a = V_{a0} + V_{a1} + V_{a2}$$

(3.21)

$$V_b = V_{a0} + a^2 \cdot V_{a1} + a V_{a2}$$

(3.22)

$$V_c = V_{a0} + a \cdot V_{a1} + a^2 \cdot V_{a2}$$

(3.23)

$$\begin{pmatrix} V_a \\ V_b \\ V_c \end{pmatrix} = A \cdot \begin{pmatrix} V_{a0} \\ V_{a1} \\ V_{a2} \end{pmatrix}$$

$$V_{ao} = \frac{1}{3}\left(V_a + V_b + V_c\right)$$

(3.24)

$$V_{a1} = \frac{1}{3}\left(V_a + a \cdot V_b + a^2 \cdot V_c\right)$$

(3.25)

$$V_{a2} = \frac{1}{3}\left(V_a + a^2 \cdot V_b + a \cdot V_c\right)$$

(3.26)

$$\begin{pmatrix} V_{a0} \\ V_{a1} \\ V_{a2} \end{pmatrix} = \frac{1}{3} \cdot \begin{pmatrix} 1 & 1 & 1 \\ 1 & a & a^2 \\ 1 & a^2 & a \end{pmatrix} \cdot \begin{pmatrix} V_a \\ V_b \\ V_c \end{pmatrix}$$

$$I_a = I_{a0} + I_{a1} + I_{a2}$$

(3.27)

$$I_b = I_{b0} + I_{b1} + I_{b2}$$

(3.28)

$$I_c = I_{c0} + I_{c1} + I_{c2}$$

(3.29)

$$I_a = I_{a0} + I_{a1} + I_{a2}$$

(3.30)

$$I_b = I_{a0} + a^2 \cdot I_{a1} + a \cdot I_{a2}$$

(3.31)

$$I_c = I_{a0} + a \cdot I_{a1} + a^2 \cdot I_{a2}$$

(3.32)

$$\begin{pmatrix} I_a \\ I_b \\ I_c \end{pmatrix} = A \cdot \begin{pmatrix} I_{a0} \\ I_{a1} \\ I_{a2} \end{pmatrix}$$

$$I_{a0} = \frac{1}{3}\left(I_a + I_b + I_c\right)$$

(3.33)

$$I_{a1} = \frac{1}{3}\left(I_a + a \cdot I_b + a^2 \cdot I_c\right)$$

(3.34)

$$I_{a2} + \frac{1}{3}\left(I_a + a^2 \cdot I_b + a \cdot I_c\right)$$

(3.35)

$$\begin{pmatrix} I_{a0} \\ I_{a1} \\ I_{a2} \end{pmatrix} = \frac{1}{3} \cdot \begin{pmatrix} 1 & 1 & 1 \\ 1 & a & a^2 \\ 1 & a^2 & a \end{pmatrix} \cdot \begin{pmatrix} I_a \\ I_b \\ I_c \end{pmatrix}$$

3.3 Three-phase Electrical Power Systems

This section presentation is tailored to electrical power systems that are normally balanced and become unbalanced due to the unexpected occurrence of an open circuit type fault or an unsymmetrical short-circuit, or just the connection of unbalanced three-phase loads or a single phase loads. These types of events produce voltages and currents unbalances in the power system network whether the network itself is balanced or not. Figure 3-2 is an illustration of a balanced three-phase power system in which the lines self impedances are equal and the mutual impedances between lines are also equal. This implies that the three lines conductors are identical and equidistant (equilateral triangular configuration) from each other or that the transmission lines have been totally transposed to compensate for the different line separations and therefore different mutual impedances. However, in practice, the way that inductive reactances are now computed removes the need to consider the coupling between transmission lines conductors. In network analysis, transmission line reactances are calculated as follows:

Each node self admittance is equal to the sum of all the admittances terminating on that node.

The mutual admittance between a pair of nodes is equal to the negative of the sum of all admittances connected directly between the pair of nodes considered.

To simplify the computations and make advantageous the use of symmetrical components it is necessary to eliminate the off-diagonal elements from the sequence impedance matrix of the three-phase power system being considered. Below we discuss the same problem using the traditional concepts of mutual self and mutual reactance.

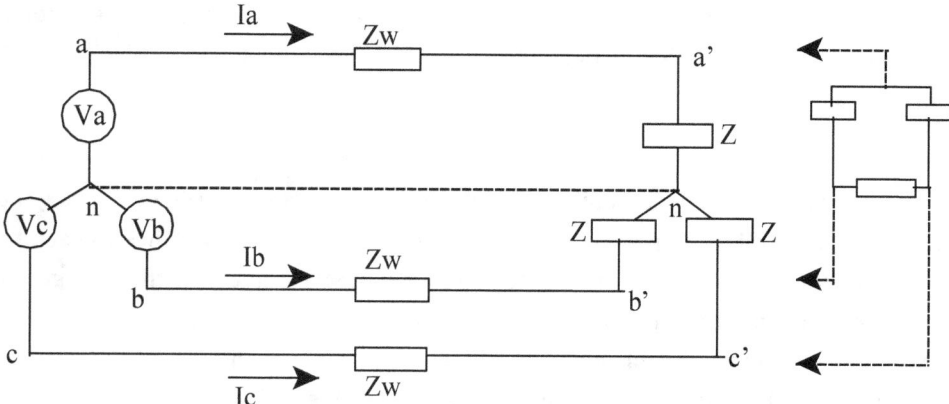

Figure 3-2 Ilustration of a balanced three-phase circuit

As shown in Fig. 3-2 the line wires of all the phases have the same Zw self impedance and the mutual impedances Zm between pair of wires are also assumed to be equal because the transmission line is totally transposed. Let me remind the reader that the mutual impedance between phases a and b is equal to the voltage induced in phase **a** per ampere of current circulating in phase **b** when phase c is open circuited. Besides, all the generator terminal voltages must have the same magnitude.

$$V_a = Z_w \cdot I_a + Z_m \cdot (I_b + I_c) \qquad\qquad I_a + I_b + I_c = 0 \qquad\qquad I_b + I_c = -I_a$$

$$V_a = Z_w \cdot I_a - Z_m \cdot I_a = (Z_w - Z_m) I_a \qquad\qquad Z_w - Z_m = \frac{V_a}{I_a} = Z_{a1} \qquad\qquad (3.36)$$

Equation (3.36) provides the phase **a** impedance to the positive sequence current indicated by the subscript 1. The positive sequence impedance is defined as the impedance of the wire minus the mutual impedance between wires. The positive sequence impedance for each phase are:

$$Z_{a1} = \frac{V_a}{I_a}$$

Phase **a** impedance to positive sequence current

$$Z_{b1} = \frac{V_b}{I_b}$$

Phase **b** impedance to positive sequence current

$$Z_{c1} = \frac{V_c}{I_c}$$

Phase **c** impedance to positive sequence current

Because in the system depicted in Fig. 3-2 all the self and mutual impedances are equal, then the positive sequence impedance of all the phases are the same. Symbolically:

$$Z_{a1} = Z_{b1} = Z_{c1}$$

The reader must realize than in real power systems many circuit components, like generators, transformers, transmission lines, breakers, etc. present different impedance to each sequence current. However, if the load connected to the circuit is balanced as it is the voltages applied to it then there cannot be *negative or zero* sequence components in the line currents. However, because unpredictable external events can unbalance the power system we must consider the negative and zero sequence impedances as well. Therefore, we can write the following matrix equations for the lines voltage drops.

$$\begin{pmatrix} V_{a,a'} \\ V_{b,b'} \\ V_{c,c'} \end{pmatrix} = \begin{pmatrix} Z_{a,a} & Z_{a,b} & Z_{a,c} \\ Z_{b,a} & Z_{b,b} & Z_{b,c} \\ Z_{c,a} & Z_{c,b} & Z_{c,c} \end{pmatrix} \cdot \begin{pmatrix} I_a \\ I_b \\ I_c \end{pmatrix} \tag{3.37}$$

$$\begin{pmatrix} Z_{a,a} & Z_{a,b} & Z_{a,c} \\ Z_{b,a} & Z_{b,b} & Z_{b,c} \\ Z_{c,a} & Z_{c,b} & Z_{c,c} \end{pmatrix}$$ Is the phase impedance matrix of the original network.

Replacing the elements of the impedance matrix as follows:

$$Z_{a,b} = Z_{b,a} = Z_{a,c} = Z_{c,a} = Z_{b,c} = Z_{c,b} = Z_m \qquad\qquad Z_{a,a} = Z_{b,b} = Z_{c,c} = Z_w$$

Equation (3.37) becomes:

$$\begin{pmatrix} V_{a,a'} \\ V_{b,b'} \\ V_{c,c'} \end{pmatrix} = \begin{pmatrix} Z_w & Z_m & Z_m \\ Z_m & Z_w & Z_m \\ Z_m & Z_m & Z_w \end{pmatrix} \cdot \begin{pmatrix} I_a \\ I_b \\ I_c \end{pmatrix} \tag{3.38}$$

Remember any set of phasors could be expressed in terms of their symmetrical components.

From equations (3.27 - 3.29) we know that:

$$\begin{pmatrix} V_{a,a'} \\ V_{b,b'} \\ V_{c,c'} \end{pmatrix} = A \cdot \begin{pmatrix} V_{a,0} \\ V_{a,1} \\ V_{a,2} \end{pmatrix} \qquad\qquad A = \begin{pmatrix} 1 & 1 & 1 \\ 1 & a^2 & a \\ 1 & a & a^2 \end{pmatrix}$$

Similarly:

$$\begin{pmatrix} I_a \\ I_b \\ I_c \end{pmatrix} = A \cdot \begin{pmatrix} I_{a,0} \\ I_{a,1} \\ I_{a,2} \end{pmatrix}$$

Substituting in Eqn. (3.38) we obtain:

$$A \cdot \begin{pmatrix} V_{a,0} \\ V_{a,1} \\ V_{a,2} \end{pmatrix} = \begin{pmatrix} Z_w & Z_m & Z_m \\ Z_m & Z_w & Z_m \\ Z_m & Z_m & Z_w \end{pmatrix} \cdot \left[A \cdot \begin{pmatrix} I_{a,0} \\ I_{a,1} \\ I_{a,2} \end{pmatrix} \right]$$

(3.39)

From equation (3.11 and 3.13) we know that:

$$A^{-1} = \frac{1}{3} \cdot \overline{A} \qquad\qquad \overline{A} = \begin{pmatrix} 1 & 1 & 1 \\ 1 & a & a^2 \\ 1 & a^2 & a \end{pmatrix}$$

Pre multiplying both sides of Eqn. (3.39) by A-1 we get:

$$\begin{pmatrix} 1 & 0 & 0 \\ 0 & 1 & 0 \\ 0 & 0 & 1 \end{pmatrix} \cdot \begin{pmatrix} V_{a,0} \\ V_{a,1} \\ V_{a,2} \end{pmatrix} = \frac{1}{3} \cdot \overline{A} \cdot \left[\begin{pmatrix} Z_w & Z_m & Z_m \\ Z_m & Z_w & Z_m \\ Z_m & Z_m & Z_w \end{pmatrix} \cdot A \cdot \begin{pmatrix} I_{a,0} \\ I_{a,1} \\ I_{a,2} \end{pmatrix} \right] \qquad A^{-1} \cdot A = \begin{pmatrix} 1 & 0 & 0 \\ 0 & 1 & 0 \\ 0 & 0 & 1 \end{pmatrix}$$

$$\begin{pmatrix} V_{a,0} \\ V_{a,1} \\ V_{a,2} \end{pmatrix} = \frac{1}{3} \cdot \begin{pmatrix} 1 & 1 & 1 \\ 1 & a & a^2 \\ 1 & a^2 & a \end{pmatrix} \cdot \begin{pmatrix} Z_w & Z_m & Z_m \\ Z_m & Z_w & Z_m \\ Z_m & Z_m & Z_w \end{pmatrix} \cdot \begin{pmatrix} 1 & 1 & 1 \\ 1 & a^2 & a \\ 1 & a & a^2 \end{pmatrix} \cdot \begin{pmatrix} I_{a,0} \\ I_{a,1} \\ I_{a,2} \end{pmatrix}$$

(3.40)

Remember you cannot change the order of the factors. Matrix algebra is not commutative. Working on the right side of the equal sign of Eqn. (3.40), we get:

$$\begin{pmatrix} 1 & 1 & 1 \\ 1 & a & a^2 \\ 1 & a^2 & a \end{pmatrix} \cdot \begin{pmatrix} Z_w & Z_m & Z_m \\ Z_m & Z_w & Z_m \\ Z_m & Z_m & Z_w \end{pmatrix}$$

$$\begin{pmatrix} 2 \cdot Z_m + Z_w & 2 \cdot Z_m + Z_w & 2 \cdot Z_m + Z_w \\ Z_m \cdot a^2 + Z_m \cdot a + Z_w & Z_m \cdot a^2 + Z_w \cdot a + Z_m & Z_w \cdot a^2 + Z_m \cdot a + Z_m \\ Z_m \cdot a^2 + Z_m \cdot a + Z_w & Z_w \cdot a^2 + Z_m \cdot a + Z_m & Z_m \cdot a^2 + Z_w \cdot a + Z_m \end{pmatrix}$$

$$\begin{pmatrix} 2 \cdot Z_m + Z_w & 2 \cdot Z_m + Z_w & 2 \cdot Z_m + Z_w \\ Z_m \cdot a^2 + Z_m \cdot a + Z_w & Z_m \cdot a^2 + Z_w \cdot a + Z_m & Z_w \cdot a^2 + Z_m \cdot a + Z_m \\ Z_m \cdot a^2 + Z_m \cdot a + Z_w & Z_w \cdot a^2 + Z_m \cdot a + Z_m & Z_m \cdot a^2 + Z_w \cdot a + Z_m \end{pmatrix} \cdot \begin{pmatrix} 1 & 1 & 1 \\ 1 & a^2 & a \\ 1 & a & a^2 \end{pmatrix}$$

$$\begin{bmatrix} 6 \cdot Z_m + 3 \cdot Z_w & (2 \cdot Z_m + Z_w) \cdot (a^2 + a + 1) & (2 \cdot Z_m + Z_w) \cdot (a^2 + a + 1) \\ (2 \cdot Z_m + Z_w) \cdot (a^2 + a + 1) & Z_m \cdot a^4 + 2 \cdot Z_w \cdot a^3 + 3 \cdot Z_m \cdot a^2 + 2 \cdot Z_m \cdot a + Z_w & (a^2 + a + 1) \cdot (a^2 \cdot Z_w + 2 \cdot a \cdot Z_m - a \cdot Z_w + Z_w) \\ (2 \cdot Z_m + Z_w) \cdot (a^2 + a + 1) & (a^2 + a + 1) \cdot (a^2 \cdot Z_w + 2 \cdot a \cdot Z_m - a \cdot Z_w + Z_w) & Z_m \cdot a^4 + 2 \cdot Z_w \cdot a^3 + 3 \cdot Z_m \cdot a^2 + 2 \cdot Z_m \cdot a + Z_w \end{bmatrix}$$

Using Table 3-1 and simplifying, the above matrix is reduced to:

$$1 + a + a^2 = 0 \qquad a^4 = a \qquad a^3 = 1$$

$$\begin{pmatrix} 6 \cdot Z_m + 3 \cdot Z_w & 0 & 0 \\ 0 & Z_m \cdot a^4 + 2 \cdot Z_w \cdot a^3 + 3 \cdot Z_m \cdot a^2 + 2 \cdot Z_m \cdot a + Z_w & 0 \\ 0 & 0 & Z_m \cdot a^4 + 2 \cdot Z_w \cdot a^3 + 3 \cdot Z_m \cdot a^2 + 2 \cdot Z_m \cdot a + Z_w \end{pmatrix}$$

$$\begin{pmatrix} 6 \cdot Z_m + 3 \cdot Z_w & 0 & 0 \\ 0 & 3 \cdot Z_m \cdot a^2 + 3 \cdot Z_m \cdot a + 3 \cdot Z_w & 0 \\ 0 & 0 & 3 \cdot Z_m \cdot a^2 + 3 \cdot Z_m \cdot a + 3 \cdot Z_w \end{pmatrix}$$

$$\frac{1}{3} \cdot \begin{pmatrix} 6 \cdot Z_m + 3 \cdot Z_w & 0 & 0 \\ 0 & 3 \cdot Z_m \cdot a^2 + 3 \cdot Z_m \cdot a + 3 \cdot Z_w & 0 \\ 0 & 0 & 3 \cdot Z_m \cdot a^2 + 3 \cdot Z_m \cdot a + 3 \cdot Z_w \end{pmatrix}$$

$$\begin{pmatrix} 2 \cdot Z_m + Z_w & 0 & 0 \\ 0 & Z_m \cdot a^2 + Z_m \cdot a + Z_w & 0 \\ 0 & 0 & Z_m \cdot a^2 + Z_m \cdot a + Z_w \end{pmatrix} \qquad a^2 + a = -1$$

Eqn. (3.41) is a diagonal matrix and is the symmetrical component impedance matrix of the circuit depicted in Fig. 3-2.

$$\begin{pmatrix} Z_w + 2 \cdot Z_m & 0 & 0 \\ 0 & Z_w - Z_m & 0 \\ 0 & 0 & Z_w - Z_m \end{pmatrix}$$

(3.41)

Substituting in Eqn. (3.40) we finally obtain:

$$\begin{pmatrix} V_{a,0} \\ V_{a,1} \\ V_{a,2} \end{pmatrix} = \begin{pmatrix} Z_w + 2 \cdot Z_m & 0 & 0 \\ 0 & Z_w - Z_m & 0 \\ 0 & 0 & Z_w - Z_m \end{pmatrix} \cdot \begin{pmatrix} I_{a,0} \\ I_{a,1} \\ I_{a,2} \end{pmatrix}$$

(3.42)

$$\begin{pmatrix} Z_w + 2 \cdot Z_m & 0 & 0 \\ 0 & Z_w - Z_m & 0 \\ 0 & 0 & Z_w - Z_m \end{pmatrix} \cdot \begin{pmatrix} I_{a,0} \\ I_{a,1} \\ I_{a,2} \end{pmatrix} = \begin{bmatrix} \left(2 \cdot Z_m + Z_w\right) I_{a,0} \\ -\left(Z_m - Z_w\right) I_{a,1} \\ -\left(Z_m - Z_w\right) I_{a,2} \end{bmatrix}$$

(3.43)

$$V_{a,0} = \left(Z_w + 2 \cdot Z_m\right) I_{a,0}$$

(3.44)

$$V_{a,1} = \left(Z_w - Z_m\right) I_{a,1}$$

(3.45)

$$V_{a,2} = \left(Z_w - Z_m\right)I_{a,2}$$ (3.46)

The transformation of the impedance matrix, see Eqn. (3.37), from the original network format to the symmetrical components format yields equations (3.44 - 3.46)) which simplify the computations of the line voltage drops symmetrical sequence components, from which we can obtain the line voltage drops in the original circuit.

3.4 Three-Phase Power Systems with Asymmetric Line Impedances.

The unbalance in an electrical power system could be produced by circuit impedance unbalances, by applied voltages unbalances (very rare), by external sources unbalance like: connected load unbalances, or unbalance produced by open wires or non-symmetrical short circuits. All these external sources of unbalance are capable of producing voltage and current unbalances in the three-phase circuit even when the circuit is normally perfectly balanced.

First, let us consider the unbalance created when the line impedances are not equal. Figure 3-3 illustrates this case.

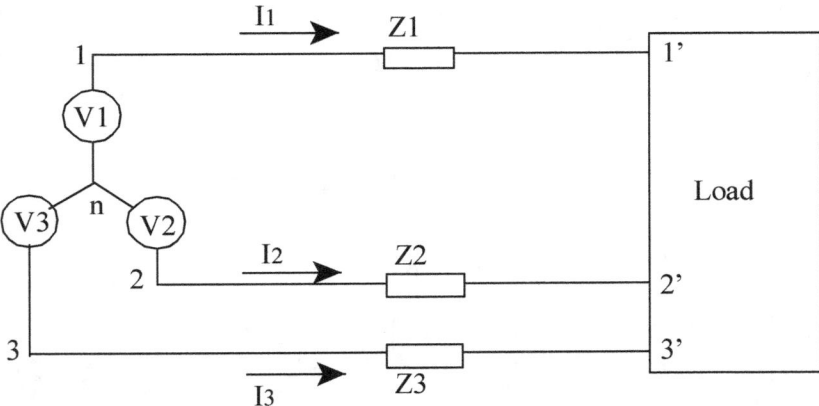

Figure 3-3 Ilustration of a three-phase circuit with unbalanced line impedances

In Fig. 3-3 the circuit asymmetry is produced by the unequal line impedances where the self line impedances are unequal. *Assuming that there is no coupling between wires and therefore no mutual impedances we can write the following matrix equations for the lines voltage drops.*

$$\begin{pmatrix} V_{1,1'} \\ V_{2,2'} \\ V_{3,3'} \end{pmatrix} = \begin{pmatrix} Z_1 & 0 & 0 \\ 0 & Z_2 & 0 \\ 0 & 0 & Z_3 \end{pmatrix} \cdot \begin{pmatrix} I_1 \\ I_2 \\ I_3 \end{pmatrix}$$ (3.47)

Where the impedance matrix of the actual electrical circuit depicted in Fig. (3.3) which has unequal self line impedances and no mutual impedances is:

$$\begin{pmatrix} Z_1 & 0 & 0 \\ 0 & Z_2 & 0 \\ 0 & 0 & Z_3 \end{pmatrix}$$

From equations (3.30 - 3.32) we know that:

$$\begin{pmatrix} I_1 \\ I_2 \\ I_3 \end{pmatrix} = A \cdot \begin{pmatrix} I_{1,0} \\ I_{1,1} \\ I_{1,2} \end{pmatrix} \qquad A = \begin{pmatrix} 1 & 1 & 1 \\ 1 & a^2 & a \\ 1 & a & a^2 \end{pmatrix}$$

From equation (3.21 - 3.23) we know that:

$$\begin{pmatrix} V_{1,1'} \\ V_{2,2'} \\ V_{3,3'} \end{pmatrix} = A \cdot \begin{pmatrix} V_{1,0} \\ V_{1,1} \\ V_{1,2} \end{pmatrix}$$

So equation (3.47) becomes:

$$A \cdot \begin{pmatrix} V_{1,0} \\ V_{1,1} \\ V_{1,2} \end{pmatrix} = \begin{pmatrix} Z_1 & 0 & 0 \\ 0 & Z_2 & 0 \\ 0 & 0 & Z_3 \end{pmatrix} \cdot \left[A \cdot \begin{pmatrix} I_{1,0} \\ I_{1,1} \\ I_{1,2} \end{pmatrix} \right] \qquad (3.48)$$

From equation (3.13) we know that:

$$A^{-1} = \frac{1}{3} \cdot \overline{A} \qquad \overline{A} = \begin{pmatrix} 1 & 1 & 1 \\ 1 & a & a^2 \\ 1 & a^2 & a \end{pmatrix}$$

Pre multiplying both sides of Eqn. (3.48) by A-1 we get:

$$A^{-1}\cdot\left[A\cdot\begin{pmatrix}V_{1,0}\\V_{1,1}\\V_{1,2}\end{pmatrix}\right]=\frac{1}{3}\cdot\overline{A}\cdot\left[\begin{pmatrix}Z_1&0&0\\0&Z_2&0\\0&0&Z_3\end{pmatrix}\cdot\left[A\cdot\begin{pmatrix}I_{1,0}\\I_{1,1}\\I_{1,2}\end{pmatrix}\right]\right] \qquad A^{-1}\cdot A=\begin{pmatrix}1&0&0\\0&1&0\\0&0&1\end{pmatrix}$$

$$\begin{pmatrix}1&0&0\\0&1&0\\0&0&1\end{pmatrix}\cdot\begin{pmatrix}V_{1,0}\\V_{1,1}\\V_{1,2}\end{pmatrix}=\frac{1}{3}\cdot\begin{pmatrix}1&1&1\\1&a&a^2\\1&a^2&a\end{pmatrix}\cdot\begin{pmatrix}Z_1&0&0\\0&Z_2&0\\0&0&Z_3\end{pmatrix}\cdot\begin{pmatrix}1&1&1\\1&a^2&a\\1&a&a^2\end{pmatrix}\cdot\begin{pmatrix}I_{1,0}\\I_{1,1}\\I_{1,2}\end{pmatrix}$$

(3.49)

Working on the right side of the equal sign of Eqn. (3.49), we get:

$$\begin{pmatrix}1&1&1\\1&a&a^2\\1&a^2&a\end{pmatrix}\cdot\begin{pmatrix}Z_1&0&0\\0&Z_2&0\\0&0&Z_3\end{pmatrix}=\begin{pmatrix}Z_1&Z_2&Z_3\\Z_1&a\cdot Z_2&a^2\cdot Z_3\\Z_1&a^2\cdot Z_2&a\cdot Z_3\end{pmatrix}$$

$$\left[\begin{pmatrix}Z_1&Z_2&Z_3\\Z_1&a\cdot Z_2&a^2\cdot Z_3\\Z_1&a^2\cdot Z_2&a\cdot Z_3\end{pmatrix}\cdot\begin{pmatrix}1&1&1\\1&a^2&a\\1&a&a^2\end{pmatrix}\right]$$

$$\begin{pmatrix}Z_1+Z_2+Z_3&Z_2\cdot a^2+Z_3\cdot a+Z_1&Z_3\cdot a^2+Z_2\cdot a+Z_1\\Z_3\cdot a^2+Z_2\cdot a+Z_1&a^3\cdot Z_2+a^3\cdot Z_3+Z_1&Z_3\cdot a^4+Z_2\cdot a^2+Z_1\\Z_2\cdot a^2+Z_3\cdot a+Z_1&Z_2\cdot a^4+Z_3\cdot a^2+Z_1&a^3\cdot Z_2+a^3\cdot Z_3+Z_1\end{pmatrix}\cdot\begin{pmatrix}I_{1,0}\\I_{1,1}\\I_{1,2}\end{pmatrix}$$

(3.50)

$$\frac{1}{3}\cdot\begin{pmatrix}Z_1+Z_2+Z_3&Z_2\cdot a^2+Z_3\cdot a+Z_1&Z_3\cdot a^2+Z_2\cdot a+Z_1\\Z_3\cdot a^2+Z_2\cdot a+Z_1&a^3\cdot Z_2+a^3\cdot Z_3+Z_1&Z_3\cdot a^4+Z_2\cdot a^2+Z_1\\Z_2\cdot a^2+Z_3\cdot a+Z_1&Z_2\cdot a^4+Z_3\cdot a^2+Z_1&a^3\cdot Z_2+a^3\cdot Z_3+Z_1\end{pmatrix}$$

(3.51)

Matrix (3.51) is the sequence impedance matrix of the circuit depicted on Fig. (3.3) and it is a non-diagonal and non-symmetric matrix. When a matrix is not diagonal it means that there are mutual coupling between the phase sequence networks. And non-symmetric matrix are characterized by different off-diagonal terms, meaning different mutual coupling between the phase sequence networks.

On a perfectly balanced system, which is with equal line impedances and no mutual impedances, positive sequence currents produce only positive sequence voltage drops, negative sequence currents produce only negative sequence voltage drops, and zero sequence currents produce only zero sequence voltage drops. In this ideal system there is not interaction between the phase sequence networks, and therefore they can be analyzed separately. For this to occur the sequence impedance matrix must be diagonal. In this way a PPS voltage drop is due to the flow of PPS current only, and similarly for the NPS and ZPS circuits. Where: PPS = Positive Phase Sequence, NPS = Negative Phase Sequence, ZPS = Zero Phase Sequence.

Symbolically, From Eqn. (3.50) we get:

$$
\begin{bmatrix}
\left(Z_1 + Z_2 + Z_3\right)I_{1,0} + \left(Z_2{\cdot}a^2 + Z_3{\cdot}a + Z_1\right){\cdot}I_{1,1} + \left(Z_3{\cdot}a^2 + Z_2{\cdot}a + Z_1\right){\cdot}I_{1,2} \\
\left(a^3{\cdot}Z_2 + a^3{\cdot}Z_3 + Z_1\right){\cdot}I_{1,1} + \left(Z_3{\cdot}a^4 + Z_2{\cdot}a^2 + Z_1\right){\cdot}I_{1,2} + \left(Z_3{\cdot}a^2 + Z_2{\cdot}a + Z_1\right){\cdot}I_{1,0} \\
\left(Z_2{\cdot}a^4 + Z_3{\cdot}a^2 + Z_1\right){\cdot}I_{1,1} + \left(a^3{\cdot}Z_2 + a^3{\cdot}Z_3 + Z_1\right){\cdot}I_{1,2} + \left(Z_2{\cdot}a^2 + Z_3{\cdot}a + Z_1\right){\cdot}I_{1,0}
\end{bmatrix}
$$

Substituting in Eqn. (3.49) we get:

$$
V_{1,0} = \frac{\left(Z_2{\cdot}a^2 + Z_3{\cdot}a + Z_1\right){\cdot}I_{1,1}}{3} + \frac{\left(Z_3{\cdot}a^2 + Z_2{\cdot}a + Z_1\right){\cdot}I_{1,2}}{3} + \frac{\left(Z_1 + Z_2 + Z_3\right)I_{1,0}}{3}
$$

(3.52)

$$
V_{1,1} = \frac{\left(a^3{\cdot}Z_2 + a^3{\cdot}Z_3 + Z_1\right){\cdot}I_{1,1}}{3} + \frac{\left(Z_3{\cdot}a^4 + Z_2{\cdot}a^2 + Z_1\right){\cdot}I_{1,2}}{3} + \frac{\left(Z_3{\cdot}a^2 + Z_2{\cdot}a + Z_1\right){\cdot}I_{1,0}}{3}
$$

(3.53)

$$
V_{1,2} = \frac{\left(Z_2{\cdot}a^4 + Z_3{\cdot}a^2 + Z_1\right){\cdot}I_{1,1}}{3} + \frac{\left(a^3{\cdot}Z_2 + a^3{\cdot}Z_3 + Z_1\right){\cdot}I_{1,2}}{3} + \frac{\left(Z_2{\cdot}a^2 + Z_3{\cdot}a + Z_1\right){\cdot}I_{1,0}}{3}
$$

(3.54)

Equations (3.52 - 3.54) show that in the case illustrated in Fig. 3-3 the voltages drop of any one sequence is a function of all the three sequence currents. *So when the lines impedances are unbalanced there is no practical advantage in introducing symmetrical components to calculate the line voltage drops.* However, if the impedances were equal, meaning:

$$Z_1 = Z_2 = Z_3 = Z$$

Then equations (3.52 - 3.54) reduce to:

$$V_{1,0} = \frac{\left(a^2 + a + 1\right)Z \cdot I_{1,1} + \left(a^2 + a + 1\right)Z \cdot I_{1,2} + 3 \cdot Z \cdot I_{1,0}}{3} = Z \cdot I_{1,0}$$

$$V_{1,1} = \frac{\left(a^3 + a^3 + 1\right)Z \cdot I_{1,1} + \left(a^4 + a^2 + 1\right)Z \cdot I_{1,2} + \left(a^2 + a + 1\right)Z \cdot I_{1,0}}{3} = Z \cdot I_{1,1}$$

$$V_{1,2} = \frac{\left(a^4 + a^2 + 1\right)Z \cdot I_{1,1} + \left(a^3 + a^3 + 1\right)Z \cdot I_{1,2} + \left(a^2 + a + 1\right)Z \cdot I_{10}}{3} = Z \cdot I_{1,2}$$

$$V_{1,0} = Z \cdot I_{1,0} \quad V_{1,1} = Z \cdot I_{1,1} \quad V_{1,2} = Z \cdot I_{1,2} \tag{3.55}$$

Equations (3.55) show that if the line impedances are balanced, each sequence component of the unbalanced line current flowing through each line impedance produces a voltage drop of the same sequence than the current component sequence. For instance: the negative sequence component of the current only produces the negative sequence component of the voltage drop. In contrast, if the line impedances are unequal, see Eqns. (3.52 - 3.54), then: the voltage drop at any sequence is a function of the three current sequences. In real engineering computation, totally transposed transmission lines are usually considered as having balanced series impedances which includes the effect of the mutual reactance between lines. The reader must be aware, that transposing a line would increase the installation cost and produce construction delays, so in practice most transmission lines are not transposed. However, some ones are transposed only at convenient spans, like before reaching the transmission line terminal at a substation. And of course, there are no perfectly transposed lines. Equation (3.49) is valid provided that the mutual impedances between lines are neglected. Otherwise, the impedance matrix would have non-zero off-diagonal elements. And therefore Eqns. (3.52 - 3.54) would have additional terms. However, in electrical networks design the lines reactances are calculated in such a way that eliminates the need for considering the inductive coupling between lines. Network lines reactances are calculated as follows:

Each node (bus) self admittance equals to the sum of all admittances terminating on that node. Each mutual admittances between pair of nodes equals the negative of the sum of all admittances connected directly between the pair of nodes considered.

The reader must realize that unbalance loads introduce network unbalances and that is one of the reasons that all loads are neglected in short circuit calculation. Furthermore, an open phase (broken phase-wire) produces a very large system unbalance, which could be represented by introducing an infinity impedance in the affected line. Furthermore, the symmetric components theory requires that all three phase sequence circuits (PPS, NPS, and ZPS) be separated from each other with no mutual coupling between them.

3.5 Complex Power Computations from Symmetrical Components.

In any polyphase electrical power system, balanced or not, the total power consumed is equal to the sum of the powers consumed in each phase. In general, the complex power flowing into a three-phase power system, balanced or unbalanced, could be symbolically expressed as follows:

$S = P + j \cdot Q$ Total power flowing into the system.

Where P is the real power and Q is the reactive power.

$$S = S_a + S_b + S_c = V_a \cdot \overline{I_a} + V_b \cdot \overline{I_b} + V_c \cdot \overline{I_c} \tag{3.56}$$

Where Va, Vb, and Vc are line to neutral voltages and Ia, Ib, and Ic are phase currents, which in a wye connected generator are equal to the line currents. The analysis that follows is applicable to the circuit shown in Fig. 3-2 but it is also valid for an unbalanced three-phase power system, wye or delta connected, provided the proper voltages and currents are used to calculate the power.

$$S = \begin{pmatrix} V_a & V_b & V_c \end{pmatrix} \begin{pmatrix} \overline{I_a} \\ \overline{I_b} \\ \overline{I_c} \end{pmatrix} = \begin{pmatrix} V_a \\ V_b \\ V_c \end{pmatrix}^T \cdot \begin{pmatrix} \overline{I_a} \\ \overline{I_b} \\ \overline{I_c} \end{pmatrix} \tag{3.57}$$

In a conjugate matrix every element is the conjugate of the same element in the original matrix. Applying Equation (3.9) we obtain the original line to neutral voltages as a function of their symmetrical components.

$$\begin{pmatrix} Va \\ V_b \\ V_c \end{pmatrix} = \begin{pmatrix} 1 & 1 & 1 \\ 1 & a^2 & a \\ 1 & a & a^2 \end{pmatrix} \cdot \begin{pmatrix} V_{a0} \\ V_{a1} \\ V_{a2} \end{pmatrix} \qquad \begin{pmatrix} 1 & 1 & 1 \\ 1 & a^2 & a \\ 1 & a & a^2 \end{pmatrix} = A = A^T \qquad a^2 = \overline{a} \tag{3.58}$$

The transpose of the product of two matrices is equal to the product of the transposes of the matrices but in reverse order.

$$\begin{pmatrix} V_a \\ V_b \\ V_c \end{pmatrix}^T = \begin{pmatrix} V_{a0} \\ V_{a1} \\ V_{a2} \end{pmatrix}^T \cdot \begin{pmatrix} 1 & 1 & 1 \\ 1 & a^2 & a \\ 1 & a & a^2 \end{pmatrix}^T \tag{3.59}$$

$$\begin{pmatrix} V_a \\ V_b \\ V_c \end{pmatrix}^T = V^T \cdot A^T \qquad V = \begin{pmatrix} V_{a0} \\ V_{a1} \\ V_{a2} \end{pmatrix} \text{ Voltage as a function of its sym. components} \tag{3.60}$$

Similarly the line currents as a function of their symmetrical components are:

$$\begin{pmatrix} I_a \\ I_b \\ I_c \end{pmatrix} = \begin{pmatrix} 1 & 1 & 1 \\ 1 & a^2 & a \\ 1 & a & a^2 \end{pmatrix} \cdot \begin{pmatrix} I_{a0} \\ I_{a1} \\ I_{a2} \end{pmatrix} \tag{3.61}$$

$$\overline{\begin{pmatrix} I_a \\ I_b \\ I_c \end{pmatrix}} = (\overline{A} \cdot I) \qquad I = \begin{pmatrix} I_{a0} \\ I_{a1} \\ I_{a2} \end{pmatrix} \tag{3.62}$$

From Eqns. (3.57 - 3.62) we obtain:

$$S = \left(V^T \cdot A^T \right) \overline{(A \cdot I)} = V^T \cdot A^T \cdot \overline{A} \cdot \overline{i} \tag{3.63}$$

It has been establish that: $\quad A^T = A \quad$ And that $\quad a^2 = \overline{a} \quad$ or that: $\quad a = \overline{a^2}$

Substituting in Eq. (3.63) we get:

$$S = \left(V_{a0} + V_{a1} + V_{a2} \right) \begin{pmatrix} 1 & 1 & 1 \\ 1 & a^2 & a \\ 1 & a & a^2 \end{pmatrix} \cdot \overline{\begin{pmatrix} 1 & 1 & 1 \\ 1 & a & a^2 \\ 1 & a^2 & a \end{pmatrix}} \cdot \begin{pmatrix} I_{a0} \\ I_{a1} \\ I_{a2} \end{pmatrix} \tag{3.64}$$

$$\begin{pmatrix} 1 & 1 & 1 \\ 1 & a^2 & a \\ 1 & a & a^2 \end{pmatrix} \cdot \begin{pmatrix} 1 & 1 & 1 \\ 1 & a & a^2 \\ 1 & a^2 & a \end{pmatrix} = \begin{pmatrix} 1+1+1 & 1+a+a^2 & 1+a^2+a \\ 1+a^2+a & 1+a^3+a^3 & 1+a^4+a^2 \\ 1+a+a^2 & 1+a^2+a^4 & 1+a^3+a^3 \end{pmatrix} \tag{3.65}$$

Using Table 3-1 the above product is simplified to:

$$\begin{pmatrix} 3 & 0 & 0 \\ 0 & 3 & 0 \\ 0 & 0 & 3 \end{pmatrix} = 3 \cdot \begin{pmatrix} 1 & 0 & 0 \\ 0 & 1 & 0 \\ 0 & 0 & 1 \end{pmatrix} \quad \text{Three times the unit matrix} \quad 3 \cdot U \tag{3.66}$$

Substituting in Eqn. (3.64) we get:

$$S = 3 \cdot \left(V_{a0} + V_{a1} + V_{a2} \right) \begin{pmatrix} \overline{I}_{a0} \\ \overline{I}_{a1} \\ \overline{I}_{a2} \end{pmatrix} \tag{3.67}$$

$$S = 3 \cdot V_{a0} \cdot \overline{I}_{a0} + 3 \cdot V_{a1} \cdot \overline{I}_{a1} + 3 \cdot V_{a2} \cdot \overline{I}_{a2} \tag{3.68}$$

Equation 3.68 is written for phase a, but it could have been written for phase b or c. For convenience and to simplify the equation the subscript a is usually dropped. Symbolically:

$$S = 3 \cdot V_0 \cdot \overline{I}_0 + 3 \cdot V_1 \cdot \overline{I}_1 + 3 \cdot V_2 \cdot \overline{I}_2 \quad <<< \tag{3.69}$$

In this chapter, when the phase subscript doesn't appear it must be assumed that the equation refers to phase a. Equations (3.64 and 3.69) show how to compute the complex power of an unbalance three-phase power circuit using the symmetrical components of the voltages and currents.

3.6 Sequence Impedances of Circuit Elements.

Electrical circuits or networks are formed by many devices or components, the ones that have a significant impact on the value of the circuit sequence impedances are: synchronous generators and motors, transmission lines and transformers.

The impedance of non-rotating, symmetrical linear elements are independent of the phase sequence or order, assuming that the voltages applied to them are balanced. *This is the reason why transmission lines, have identical positive and negative sequence impedances. However, the impedance of a transmission line to zero sequence currents is much larger than the impedance it present to positive and negative sequence currents, because the zero sequence current path is different and the magnetic field created by the three zero sequence currents are in phase. The impedances that rotating machinery present to any of the three sequence currents is in generally different for each sequence.* For instance, the impedance to the negative sequence current is different to the impedance to the positive sequence current, because the magnetomotive force (mmf) produced by the negative sequence armature current rotates in opposite direction to the rotor rotation, while the mmf produced by the positive sequence armature current rotates in the same direction than the rotor. Furthermore, the magnetic flux produced by the armature positive-sequence current is stationary with respect to the rotor and both rotate at synchronous speed. When on the other hand the flux produced by the armature negative-sequence current sweep the surface of the rotor at double synchronous speed inducing currents in the field and damper windings wound on the rotor. According to the Lenz law, these induced currents are of such direction that the mmf's they produced must oppose the change in magnetic flux. The two mmf's, one produced by negative-sequence current and the other produced by the induced currents in the field and damper windings are in opposition to each other and in balance. In short, the induced mmf's oppose the cause that produced them. Actually, the currents induced in these windings prevent the sweeping flux from penetrating the rotor's core. *As a consequence of the sweeping action the mmf produced by the negative-sequence armature current constantly changes its position with respect to direct and quadrature axes of the rotor. And that is why the negative-sequence reactance is approximately equal to the average of direct and quadrature subtransient reactances.*

Let us assume that only zero sequence currents flow in the armature windings of a three-phase machine. In this machine the magnitude of the currents and mmf's, in all the phases, are at a maximum at the same time. However, the windings are placed on the surface of the armature in such a way that the point of maximum mmf for any specific winding is displaced 120 electrical degrees in space from the maximum mmf point of other two. Assuming that the magnitude of the mmf produced by the zero sequence current of each phase has a perfectly sinusoidal distribution in space then we would have threes sinusoidal mmf's evenly distributed around the armature and such that their sum at any point would be zero. As a consequence no air gap flux would be produced by the zero sequence currents, except for the leakage flux and the flux produced by the end turns. *So theoretically, the only zero sequence reactance is due to leakage and end turns.* However, in a real case the windings are not perfectly distributed and they do no create perfect sinusoidal mmf's, therefore the sum of the mmf's is not exactly zero at all the points. The air gap flux due to the non-zero-sum of the zero mmf's produces a small increase of the zero sequence reactance of synchronous machines.

In transmission lines the return path to the generator's neutral of the zero sequence component of the load current is long, unpredictable and complex, and could include earth, overhead ground wires and buried counterpoise wires. Because the three zero sequence currents are equal in magnitude and phase, the magnetic field they create is very different from the magnetic fields created by the positive or negative sequence current. The faster the flux concatenations with the zero current path changes, the larger is the inductance of the path and its reactance. In general, the zero-sequence inductive reactance of overhead transmission lines is around 2 to 3.5 times larger than its positive sequence reactance. The actual value changes with the number of ground wires and number of circuit installed in each transmission line tower. Transformers installed in three-phase power circuits, could be single-phase or three-phase units. Single-phase units are used for single-phase transformation or in bank of three for three-phase transformation. However, three-phase transformer units are by far the preferred type for high voltage three-phase transformation applications. Because they are cheaper, have smaller foot-print, occupied less space, and are more efficient. *Transformers are symmetrical non rotating static devices and as such their impedances are independent of the phase sequence of the voltages applied to them, assuming that the applied voltages are balanced. Therefore, for any type of power transformers, it is generally assumed that the series impedance to all symmetrical components sequences are equal. Furthermore, because--the resistive part of the impedance is small compared to the reactance, especially for HV one-megawatt and larger units--the value of the impedance is considered to be equal to only the value of the reactance.* The reader should be aware that in some type of computations it is necessary to include in the transformer model shunt admittances, to take into account the losses due to the exciting or magnetizing current and the hysteresis losses. For balanced wye and delta connected loads the positive, the negative and the zero-sequence transformer's impedances are all equal.

CHAPTER 4

Symmetrical Sequence Networks

4.1 Introduction

Electrical circuits or networks are formed by many devices or components, the ones that have a significant impact on the overall value of the circuit sequence impedances are: synchronous generators and motors, transmission lines, transformers and coupling capacitance to ground.

The impedance of nonrotating, symmetrical linear components are independent of the phase sequence or order of phase rotation, assuming that the voltages applied to them are balanced. This is the reason why transmission lines, have identical positive and negative sequence impedances. However, the impedance of a transmission line to zero sequence currents is much larger than the impedance it presents to positive and negative sequence currents, because the zero sequence current path is different and the magnetic field created by the three zero sequence currents are in phase.

The impedances that rotating machinery present to any of the three sequence currents is in generally different for each sequence. For instance, the impedance of synchronous generators to the negative sequence current is different to the impedance it presents to positive sequence currents, because the magnetomotive force (mmf) produced by the negative sequence armature current rotates in opposite direction to the rotor rotation, while the mmf produced by the positive sequence armature current rotates in the same direction than the rotor. Furthermore, the magnetic flux produced by the armature positive-sequence current is stationary with respect to the rotor and both rotate at synchronous speed. When on the other hand the flux produced by the armature negative-sequence current sweep the surface of the rotor at double synchronous speed inducing currents in the field and damper windings wound on the rotor. According to the Lenz law, these induced currents are of such direction that the mmf's they produced oppose the change in magnetic flux. The two mmf's, one produced by negative-sequence current and the other produced by the induced currents in the field and damper windings are in opposition to each other and in balance.

In short, the induce mmf's oppose the cause that produced them. Actually, the currents induced in these windings prevent the sweeping flux from penetrating the rotor's core.

As a consequence of the sweeping action the mmf produced by the negative-sequence armature current constantly changes its position with respect to direct and quadrature axes of the rotor. And that is why the negative-sequence reactance is approximately equal to the average of direct and quadrature subtransient reactances.

Let us assume that only zero sequence currents flow in the armature windings of a three-phase machine. In this machine the magnitude of the currents and mmf's, in all the phases, are at a maximum at the same time. However, the windings are placed on the surface of the armature in such a way that the point of maximum mmf for any specific winding is displaced 120 electrical degrees in space from the maximum mmf point of other two. Assuming that the magnitude of the mmf produced by the zero sequence current of each phase has a perfectly sinusoidal distribution in space then we would have threes sinusoidal mmf's evenly distributed around the armature and such that their sum at any point would be zero. As a consequence no air gap flux would be produced by the zero sequence currents, except for the leakage flux and the flux produced by the end turns. So theoretically, the only zero sequence reactance is due to leakage and the end turns. However, in a real case the windings are not perfectly distributed and they do no create perfect sinusoidal mmf, therefore the sum of the mmf's is not exactly zero at all the points. The air gap flux due to the non-zero-sum of the zero mmf's produces a small increase of the zero sequence reactance of synchronous machines.

In transmission lines the return path to the generator's neutral of the zero sequence component of the load current is long, unpredictable, and complex and could include earth, overhead ground wires and buried counterpoise wires. Because the three zero sequence currents are equal in magnitude and phase, the magnetic field they create is very different from the magnetic fields created by the positive or negative sequence current. The faster the flux concatenations with the zero current path changes, the larger is the inductance of the path and its reactance. In general, the zero-sequence inductive reactance of overhead transmission lines is around 2 to 3.5 times larger than its positive sequence reactance. The actual value changes with the number of ground wires and number of circuit in the same tower.

Transformers installed in three-phase power circuits, could be single-phase or three-phase units. Single-phase units are used for single-phase transformation or in bank of three for three-phase transformation. However, three-phase transformer units are by far the preferred type for high voltage three-phase transformation applications. Because they are cheaper, have smaller foot-print, occupied less space, and are more efficient. Transformers are symmetrical non rotating static devices and as such their impedances are independent of the phase sequence of the voltages applied to them, assuming that the applied voltages are balanced. Therefore, for any type of power transformers, it is generally assumed that the series impedance to all symmetrical components sequences are equal. Furthermore, because--the resistive part of the impedance is small compared to the reactance, especially for HV one-megawatt and larger units--the value of the impedance is considered to be equal to only the value of the reactance. The reader should be aware that in some type of computations it is necessary to include in the transformer model shunt admittances, to take into account the losses due to the exciting or magnetizing current and the hysteresis losses. For balanced wye and delta connected loads the positive, the negative and the zero-sequence impedances are all equal.

4.2 Sequence Networks Where the Symmetrical Components Live.

To simplify calculations, balanced three-phase power systems are analyzed in a per phase basis. As if the calculations were made for a single-phase network. Furthermore, using the symmetrical components method, see section 3.1 and 3.2, we could replace any asymmetric (unbalanced) three-phase system of voltage or current phasors with three symmetric separated sets of phasors. These three sets of phasors are called: Positive Phase Sequence (PPS), Negative Phase Sequence (NPS), and Zero Phase Sequence (ZPS). *To each of these sets of phasors correspond a phase sequence network that only contains currents, voltages, and impedances of the same sequence as the namesake.* **The reader must realize that if the original three-phase network is assumed symmetric or perfectly balanced, as is done in short circuit studies, then the three sequence networks are separated with no mutual coupling between them. Actually in this perfectly balanced system the NPS and ZPS networks cannot carry any current or voltage. If a fault or any other unbalance producing event occurs in a balanced three phase network, then all three phase sequence networks would potentially** carry current and voltage and all of them will share the common point of unbalance. As a result the three phase sequence networks must be connected at the point of unbalance or where the short circuit occurs. The reader must realize that the currents and voltages suddenly appearing in NPS and ZPS networks are due to the unbalanced condition suddenly imposed in the original three-phase network and it has nothing to do with the voltage sources existing in the PPS network which is designated as the only active phase sequence network. Finally, each phase sequence network consists of impedances or susceptance's of the same namesake as the network being considered.

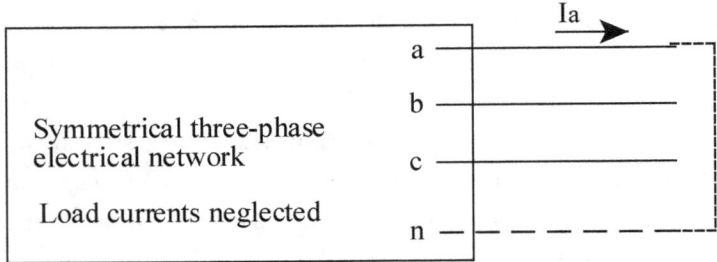

Figure 4-1 Balanced three phase electrical network with phase **a** short circuited to grounded neutral

The line currents and phase voltages in Fig.4-1 are:

$$I_a = I_a \qquad V_a = 0 \tag{4.1}$$

$$I_b = 0 \qquad V_b = V_b \tag{4.2}$$

$$I_c = 0 \qquad V_c = V_c \tag{4.3}$$

To express the line currents and phase voltages of Fig. 4-1 in terms of symmetrical components we use equations (3.15 - 3.17).

$$I_{a0} = \frac{1}{3}\left(I_a + 0 + 0\right) \tag{4.4}$$

$$I_{a1} = \frac{1}{3}\left(I_a + 0 + 0\right) \tag{4.5}$$

$$I_{a2} = \frac{1}{3}\left(I_a + 0 + 0\right) \tag{4.6}$$

$$I_{a0} = I_{a1} = I_{a2} = \frac{1}{3} \cdot I_a \qquad \textit{Current symmetrical components} \qquad (4.7)$$

$$V_{a0} = \frac{1}{3}\left(0 + V_b + V_c\right) \qquad (4.8)$$

$$V_{a1} = \frac{1}{3}\left(0 + a \cdot V_b + a^2 \cdot V_c\right) \qquad (4.9)$$

$$V_{a2} = \frac{1}{3}\left(0 + a^2 \cdot V_b + a \cdot V_c\right) \qquad (4.10)$$

From Table 3-1 we get:

$$a = \varepsilon^{j \cdot 120°} \qquad\qquad a^2 = \varepsilon^{j \cdot 240°}$$

Substituting in Equations (4.8 - 4.10) we obtain the voltage symmetrical components

$$V_{a0} = \frac{1}{3}\left(V_b + V_c\right) \qquad (4.11)$$

$$V_{a1} = \frac{1}{3}\left(\varepsilon^{j \cdot 120°} \cdot V_b + \varepsilon^{j \cdot 240°} \cdot V_c\right) = \frac{1}{3}\left[\varepsilon^{j \cdot 120°} \cdot V_b + \varepsilon^{j \cdot (-120°)} \cdot V_c\right] \qquad (4.12)$$

$$V_{a2} = \frac{1}{3}\left[\varepsilon^{j \cdot (-120°)} \cdot V_b + \varepsilon^{j \cdot 120°} \cdot V_c\right] \qquad (4.13)$$

4.3 Sequence Impedances of Network Components

The impedance of the sequence networks is determined by the impedances of the devices and components that make-up the sequence network under consideration. In linear and symmetrical *static* circuits the positive and negative sequence impedances are identical because the impedances of these circuits are independent of the time sequence of the voltage phases, if the applied voltage are balanced.

Transmission lines are considered static devices, they don't move, however the impedance they present to zero sequence currents is different from their impedance to positive and negative sequence currents. Zero sequence currents flowing in a transmission line, consists of identical (same magnitude and phase angle) currents in each phase and their return path includes overhead ground wires and the earth. Actually, the magnetic field produced by the zero sequence currents is very different from the one produced by the positive or negative sequence current. *This is the reason why the zero sequence inductive reactance of overhead transmission lines is from to two to three and a half times larger than their positive or negative reactances.*

Power transformers are static devices but their circuitry includes nonlinear elements, namely their magnetic cores. *However, it is an accepted practice to consider that the series impedance at all sequences are equal regardless of the type of transformer.* In large power, high voltage applications the preferred transformers are three-phase transformers with core type magnetic circuits. They offer lower initial cost, smaller foot print and smaller space requirements, and higher efficiency. Other possibilities for three-phase transformation are: three-phase transformers with shell type magnetic or a bank of single phase transformers. Table 4-1 provides typical range of values of the inductive reactance of large three-phase power transformers. *The transformers winding resistances are so small that could be neglected.* In figures 4-2 and 4-3 we provide the most relevant, for short circuit studies, equivalent circuit of a three-phase iron-core transformer; assuming that the transformer is connected to a symmetrical network and that it carries a balanced load or at no-load condition.

The sequence impedances (reactances) of rotating machines, like synchronous generators and motors, are different for each current sequence. *In the classical model where $X_d \approx X_q$ the positive sequence inductive reactance is by definition the same as the generator direct axis reactance. See equations (4.4 - 4.6) and Table 4-2.*

The impedance to negative sequence currents is harder to interpret because the mmf produced by a negative sequence current rotates in opposite direction to the rotor and therefore the flux it creates also rotates in opposite direction to the rotor. By contrast the flux generated by the positive sequence current is in sync with the rotor. So, the negative sequence flux induce currents in the rotor field and damper windings, which in turn, by the Lenz law, creates mmf's that oppose the mmf produced by the armature's negative sequence currents. The result is that the armature's flux produced by the negative sequence current do not penetrate into the rotor core. However, as stated before, the flux produced by the negative sequence current rotates around the rotor and in doing so, it constantly changes its position with respect to the direct and quadrature rotor axes. This flux path is similar to the one occurring in the subtransient period. *So, the negative sequence reactance should have a value approximately equal to the average of the direct and quadrature subtransient reactances.*

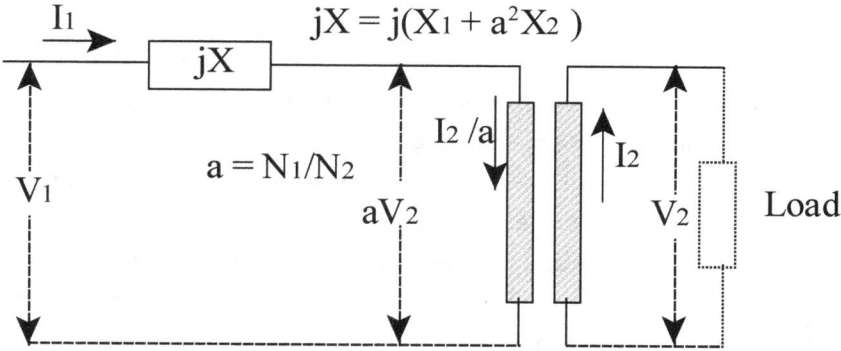

Figure 4-2 Iron core transformer with resistances and excitation shunt circuit neglected and secondary reactance referred to the primary side.

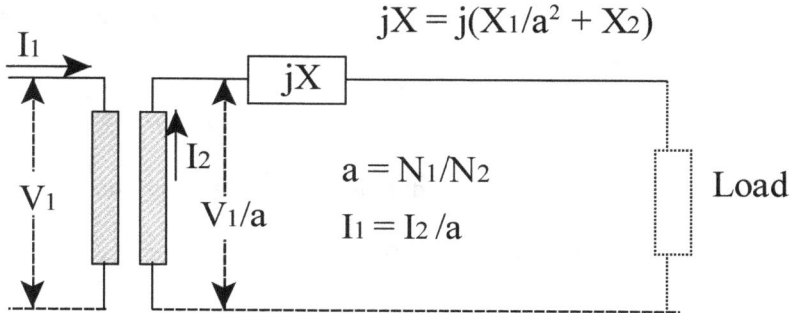

Figure 4-3 Iron core transformer with resistances and excitation shunt circuit neglected and transformer primary reactance referred to the secondary side.

The magnetomotive forces produced by the zero sequence currents circulating in the three armature windings of a three-phase generator reach simultaneously their maximum value, however the windings are wounds in such way around the circumference of the armature that the point in space at which the maximum magnetomotive force of any of the phases is reached is separated by 120 electrical degrees from the points in space at which the other two phases reach the maximum value of their magnetomotive forces. Assuming that the mmf produced by the zero sequence current circulating in each phase winding has a perfect sinusoidal distribution in space, then at any point of the armature inside-perimeter, the sum of the three mmf's would be zero. Therefore no magnetic flux due to the zero sequence currents would cross the air gap at any point. So the zero sequence inductive reactance of any phase winding would only be due to the flux leaking from the iron path of the armature into the surrounding air and to the flux produced by the armature windings end-turns. Besides, the sum of the mmf produced by the zero sequence currents is not actually zero and the resulting small mmf produces a small addition to the zero sequence reactance. The sequence impedances of some common devices and components are given below.

Transformers. Large power transformers are all three-phase units. Because their lower initial cost, higher efficiency, smaller foot print, simpler bussing, and oil spill containment design. It is an established practice to neglect the transformer winding resistances and therefore the impedance is considered equal to the reactance. However, the reader most realize that the resistance cannot be neglected when calculating losses. *In general, for all types of transformers, it is assumed that for a given transformer the value of all sequence impedances are equal.* Although, the zero sequence impedance value could be different than the positive and negative sequence values. Table 4-1 provides the reactance range of large three-phase transformers.

Synchronous Generators. Table 4-2 provides the range of typical line to neutral reactance values for some types of synchronous generators. Four different values are provided for the positive sequence reactance, as follows:

Xd   direct-axis, synchronous reactance effective during steady state conditions.

Xq   quadrature axis, synchronous reactance effective during steady state conditions.

Xd" sub-transient reactance, effective during the first 100 milliseconds of the transient.

X_d' transient reactance, effective between 100 and 5000 milliseconds after transient initiation

In general, the negative sequence reactance of synchronous generators is equal to the sub-transient reactance, except for the case of salient pole generators without damper windings. And the values of the zero sequence reactance are smaller than the values of the positive and negative sequence reactance.

X_2 negative sequence reactance X_0 zero sequence reactance

The value listed in Table 4-2 are applicable to stability and short circuit computations. For normal operating computations, during steady state conditions the generator's synchronous reactance is designated as X_s and it has the same value as X_d before the magnetic circuit gets saturated due to high armature winding currents.

Transmission lines. For perfectly transposed transmission lines, the positive and negative sequence reactances are equal. However, the zero sequence reactance must be different because the zero sequence component of the line currents returns to the generator neutral via earth and overhead ground wires. Furthermore, the magnetic field produced by the zero sequence currents (all are in-phase) is different from the magnetic field created by the positive or negative sequence currents. As a result the zero sequence inductive reactance of overhead transmission lines is much larger, from 2 to 3.5 larger than the positive sequence reactance. The reader should be aware that ground wires (overhead and underground) form many parallel returns for the zero sequence current and therefore they decrease the zero sequence impedance. The resistance of the zero sequence circuit is also larger because the return circuit includes earth and wires (not necessarily copper wires). Never the less it is usually neglected.

Range of three-phase transformer reactances
in per unit of rated transformer MVA

Line to line system voltage in kV	Forced air cooled. In per-unit	Forced oil cooled. In per-unit
34.5	0.05-0.08	0.09-0.14
69	0.06-0.10	0.10-0.16
115	0.06-0.11	0.10-0.20
138	0.06-0.13	0.10-0.22
161	0.06-0.14	0.11-0.25
230	0.07-0.16	0.12-0.27
345	0.08-0.17	0.13-0.28
500	0.10-0.20	0.16-0.34
700	0.11-0.21	0.19-0.35

Table 4-1 Typical inductive reactance range of 25 MVA and larger, 60Hz three-phase power transformers.

Per unit, 60Hz Reactances of Three-Phase Synchronous Generators and Synchronous Condensers

| Machine | Positive Sequence | | | | Negative Sequence | Zero Sequence |
	X_d unsaturated iron core	X_q at rated current	X_d' at rated voltage	X_d'' at rated voltage	X_2 at rated current	X_0 at rated current
Two pole generators	0.95-1.45	0.92-1.42	0.12-0.21	0.07-0.14	0.07-0.14	0.01-0.08
Four pole generators	1.00-1.45	0.92-1.42	0.20-0.28	0.12-0.17	0.12-0.17	0.015-0.14
Salient-pole generators with dampers	0.60-1.50	0.40-0.80	0.20-0.50	0.13-0.32	0.13-0.32	0.03-0.23
Salient-pole generators without dampers	0.60-1.50	0.40-0.80	0.20-0.50	0.20-0.50	0.35-0.65	0.03-0.24
Air cooled Synchronous. Condensers	1.25-2.20	0.95-1.30	0.30-0.50	0.19-0.30	0.18-0.40	0.025-0.150
Hydrogen cooled Syn. Condensers	1.50-2.65	1.10-1.55	0.36-0.60	0.23-0.36	0.22-0.48	0.030-0.18

Table 4-2 typical reactances range of 60Hz three-phase synchronous machines

If the reader wants to express the transformer reactances given on Table 4-1 in a different base, then, assuming that the system is balanced, he can use the following formula:

$$Z_2 = Z_1 \cdot \frac{MVA_{b2}}{MVA_{b1}} \cdot \left(\frac{kV_{b1}}{kV_{b2}} \right)^2 \qquad \text{impedance in base 2 given impedance in base 1.} \quad (4.14)$$

MVA_b Is base three-phase megavolt-amperes

kV_b Is line-to-line kilovolts

Z_1 Ohms line to neutral impedance in base 1

Z_2 Ohms line to neutral impedance in base 2

Transformer reactances usually are toward the lower end of the range provided. Decreasing the transformer reactance usually means higher efficiency, because it means less copper in the windings and consequently smaller copper losses, but also means larger short circuit currents. However, reducing the reactance increases the maximum power that could be safely transmitted, and also increases the steady state stability limit. All these factors should be carefully considered when selecting the reactance of large transformers. The reader must be aware that the zero sequence reactance values given in Table 4-2 changes critically with the armature windings pitch. The lowest value of the range is for 2/3 pitch windings.

The transient and subtransient reactances values given in Table 4-2 for salient pole generators are such that the *lowest* value of the range usually is for the *higher rotor speed* and the *highest* value is usually associated with *low speed* rotors.

4.4 Sequence Networks

The phasors representing voltages or currents in a three-phase unbalanced electrical power systems can be resolved into three independent balanced set of phasors. These three set of phasors are designated as: the positive-sequence, the negative-sequence and zero sequence sets. We are considering a special case of unbalance in which the network hardware components (generators, wires, transformers, etc.) are all balanced, but the occurrence of an unexpected event introduce unbalanced currents or voltages in the network. Before the unbalance trigger event the network is balanced and is represented by three separated networks where each of the symmetrical sequence components live its individual life. In this condition only the positive sequence network contains a circulating current. The current in the other two networks is zero. During the event the networks lose their pristine individuality and become connected at the trigger point.

The common method of solution represents each hardware device by three single-phase networks, the positive, the negative and the zero sequence networks. Then all device networks are sorted and assembled according with their sequence into three containing networks which are tied together in accordance with the specific type of unbalancing event. Rotating machinery such as AC generators and motors all have positive sequence generated or induced internal voltages which only are present in the positive sequence network. In symmetrical power systems the positive and negative sequence impedances of static components are identical. And to convert a positive sequence network into a negative sequence network the only thing required is to omit the emf's (assuming balanced generated emf's), and to change the impedance (no often necessary) belonging to any rotational equipment included in the positive sequence network. Caution: *the impedance connected between a generator neutral and ground cannot be included in the positive or negative sequence networks. Only the zero sequence current flows through the grounding impedance.*

Three-phase balanced or symmetrical networks can be analyzed on a per phase basis using single-phase circuit analysis concepts and techniques. However, in section 3.4 we demonstrated that if the line impedances are unequal, see equations (3.52 - 3.55), then, the line voltage drop at any one sequence is a function of the three current sequences. So, to simplify the analysis and to be able to use single-phase analysis concepts, it is necessary to consider transmission lines as balanced (with equal line impedances). To solve the problem of unequal coupling between phases produced by unequal conductors spacing, we consider that all transmission lines are perfectly transposed. However, perfectly transposed transmission lines do not exists; besides most transmission lines in the United State of America are no transposed. Furthermore, in electrical network design and in engineering studies the procedure used to calculate the lines reactances eliminates the need for considering the inductive coupling between lines. The procedure used is as follows: *Each bus self admittance is equal to the sum of all admittances terminating on that bus. And each mutual admittance between pair of busses equals the negative of the sum of all admittances connected directly between the pair of busses considered.*

4.5 Generators Sequence Networks

Figure 4-4 shows a reactance grounded generator with lines current produced by a short circuit in its output terminals. Depending on the type of fault occurring in the output terminals, one or two of the line currents could be zero. If ground is part of the short circuit path then there would be a current flowing into the generator neutral.

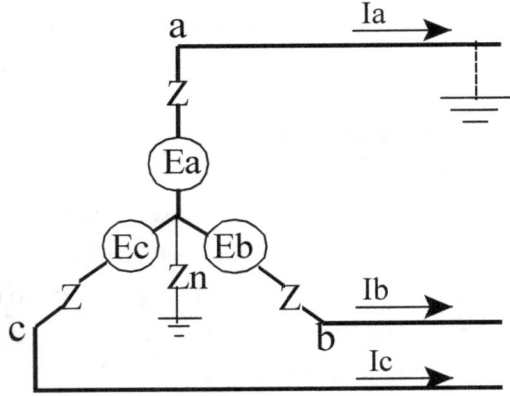

Figure 4-4 Reactance of grounded generator with some or all the output terminals short circuited.

The sequence components of the short circuit currents illustrated in Fig. 4-4 are shown in Fig. 4-5. Each current sequence component circulates inside their own sequence network which is drafted as single-phase only. The reader should notice the following:

1. The single-phase sequence networks shown in Fig. 4-5 represent the original three-phase network before the unbalance trigger event. Figure 4-5 was drafted for phase a, but we could have used anyone of the three phases. In symmetrical components work, if you know one

phase current of a set, then you could very easily determined the other two phase currents in the same set.

2. The positive sequence network is the only one that has an **emf.** This generated voltage is equal to the positive-sequence no-load terminal voltage to neutral.

3. The generator neutral is the phasor's reference (reference bus) for the positive and negative sequence networks

4. The generator ground is the phasor's reference (reference bus) for the zero sequence network.

5. Since only zero sequence component currents flows through the ground impedance, the neutral potential in the positive and negative sequence networks is the same than the ground potential. But the neutral potential in the zero sequence network is $3I_{a0}Z_n$. Which implies that in normal operating conditions the potential of the generator neutral is equal to the ground potential or the reference bus potential.

6. The positive sequence generator network only contains the generator's positive sequence impedance. Similarly, the negative sequence network only contains the generator's negative sequence impedance. However, the zero sequence network contains the generator's zero sequence impedance plus three times the grounding impedance. The purpose of the grounding impedance is to limit the zero-sequence current during a fault to ground in any one of the phases.

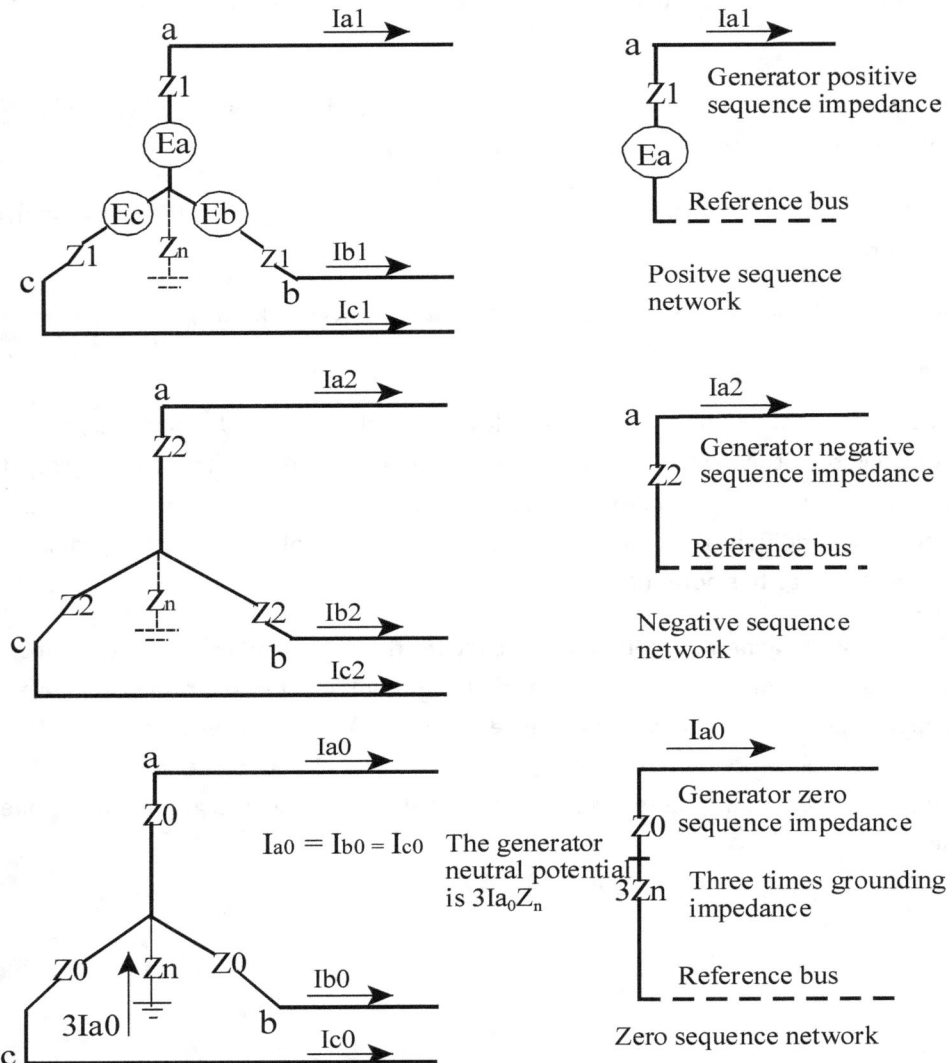

Figure 4-5 Sequence networks and sequence components of the short circuit current for the generator illustrated in Fig. 4-4.

The symmetrical components of the voltage drop from terminal a of phase **a** to the reference bus are:

$$V_{a1} = E_a - I_{a1} \cdot Z_1 \qquad\qquad (4.15)$$

$$V_{a2} = -I_{a2} \cdot Z_2 \qquad\qquad (4.16)$$

$$V_{a0} = -I_{a0} \cdot \left(3 \cdot Z_n + Z_0 \right) \qquad\qquad (4.17)$$

Equations (4.15) - (4.17) are applicable to any three-phase generator delivering unbalance currents. If the generator is unloaded before the fault, Ea is the positive-sequence no-load terminal voltage. But if the generator is delivering current before the fault, then Ea must be calculated as the generated emf minus the voltage drop, in the subtransient, transient or synchronous reactance, produced by the current existing before the fault. *From Eqns. (4.15 to 4.17) we can compute the sequence current components for different types of faults.* In matrix format Eqns. (4.15 to 4.17) become:

$$
\begin{pmatrix} V_{a0} \\ V_{a1} \\ V_{a2} \end{pmatrix} = \begin{pmatrix} 0 \\ E_a \\ 0 \end{pmatrix} - \begin{pmatrix} Z_0 & 0 & 0 \\ 0 & Z_1 & 0 \\ 0 & 0 & Z_2 \end{pmatrix} \cdot \begin{pmatrix} I_{a0} \\ I_{a1} \\ I_{a2} \end{pmatrix}
$$

(4.18)

Equation 4.18 together with the conditions imposed by the short circuit will be used to calculate I_{a1} as a function of Ea, Z1, Z2, and Z0.

4.6 Generators Inductive Reactances

The inductive reactance of rotating machines is a function of several variables, and time is one of them. In contrast the inductive reactance of static devices, like wires and transformers, is a single value that practically remains constant. In circuit analysis, we first select, according with the application, the appropriate type of inductive reactance among three possible types: subtransient, transient, and synchronous. For stability, short circuit, and protection analysis is best to use the subtransient value of the generator's reactance. Although some times, where delayed tripping of circuit breakers is used, it is better to use the transient value. The computation of the generator's reactance during transient conditions is complicated and far from precise. So it is best to use the value provided by the manufacturer which is based on factory's test, or to select a value from an application table. Table 4-2 shows typical reactances values of three-phase synchronous machines. The reader needs to realize that most of the time manufactures conduct the reactance tests at reduce power because he does not have available at the plant the power required for a full power test. The test is conducted at reduced terminal voltage and no-load condition and a three-phase symmetrical fault is applied to the generators terminals. The test results are interpreted based on the generator's classical model and neglecting the small resistance of the armature windings. And because the small duration of the test, it is fair to assume that the generator's terminal voltages, the DC field current and the angular speed of the drive remain constant during the test. Equation (4.19 - 4.21) are applied to compute the generator's reactances.

Direct axis subtransient reactance:

$$ X_d{}^{''} = \frac{E_g}{I^{''}} $$

(4.19)

Direct axis transient reactance:
$$X_d' = \frac{E_g}{I'}$$
(4.20)

Direct axis synchronous reactance:
$$X_d = \frac{E_g}{I}$$
(4.21)

I = Rms value of the steady state component of the short circuit current in amperes

I' = Rms value of the transient component of the short circuit current in amperes.

I'' = Rms value of the subtransient component of the short circuit current in amperes.

Eg = Rms value of the line to neutral voltage of one phase at no load condition. Equal to the emf generated on that phase in volts (neglecting the drop produced by the magnetizing current).

If the generator classic model is use to determine the reactances, then it means that the quadrature reactances are considered equal to the direct axis reactances.

4.7 Zero Sequence Networks

To draft and understand the zero sequence networks the reader must be familiar with the following points.

- Zero-sequence currents operates in single-phase mode in a three-phase power systems because anywhere in the system, the three zero sequence currents have the same magnitude and phase angle.

- To flow, a zero sequence current needs a return path as part of the power system closed circuit.

- The reference for a zero-sequence voltage is the ground potential at the point in the power system where the voltage is specified.

- The ground may be part of the zero-sequence currents return path, which implies that the ground potential is not uniform. Furthermore, the potential of the reference bus of the zero-sequence network is not uniform.

- The zero-sequence impedance of a transmission line includes the impedance of the ground and ground wires. However, the return circuit-segment of the zero-sequence network is considered to have has zero impedance, and it is the reference bus of the system. And therefore, voltages measured with respect to the reference bus of the zero-sequence network gives the voltage to ground.

- Again, the system reference bus which is the return circuit of the zero-sequence network has zero impedance.

In a wye connected three-phase circuit, see Fig. 4-5a, with no connection between the neutral of the wye and ground or to another neutral point located elsewhere in the system, the phasor sum of the three currents is zero. Which implies that there is not zero sequence component in any of the three line currents, because they are in phase, and therefore their sum can never be zero. This fact is interpreted by assuming that the power system impedance to zero sequence current is infinity beyond the neutral point, which is represented as an open circuit in the zero sequence network between the neutral of the wye and the reference bus. However, if the neutral of wye is directly connected to ground, as illustrated in Fig. 4-5b, then a zero sequence impedance must be inserted in the zero sequence network between the neutral point and the reference bus. In Fig. 4-5c the impedance Zn was inserted between the neutral and ground and therefore the impedance 3Zn must be inserted in the zero sequence network between the neutral and the reference bus.

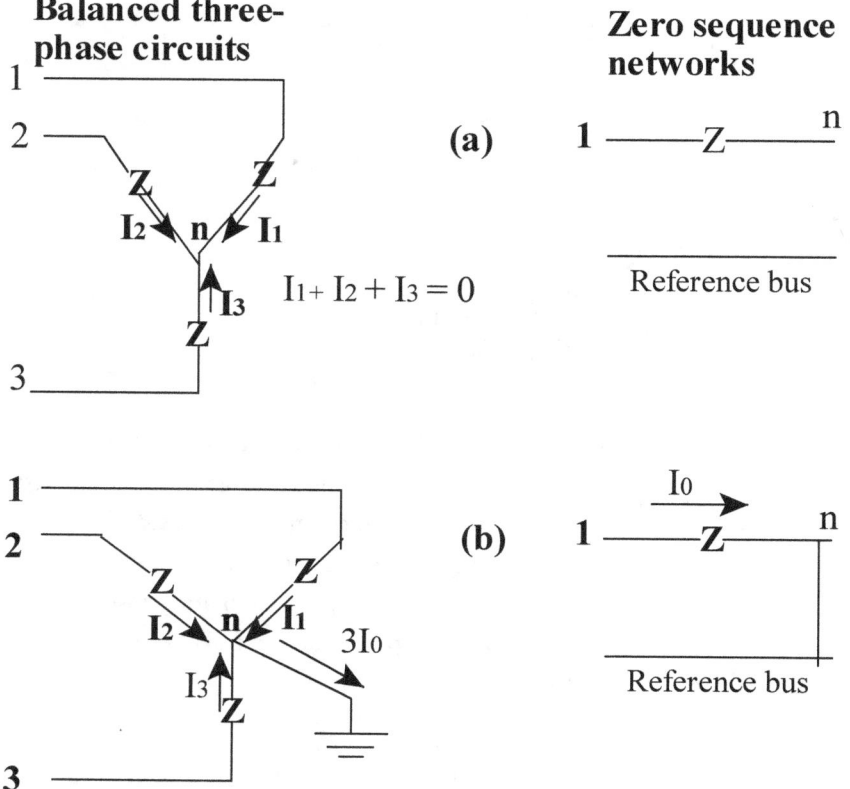

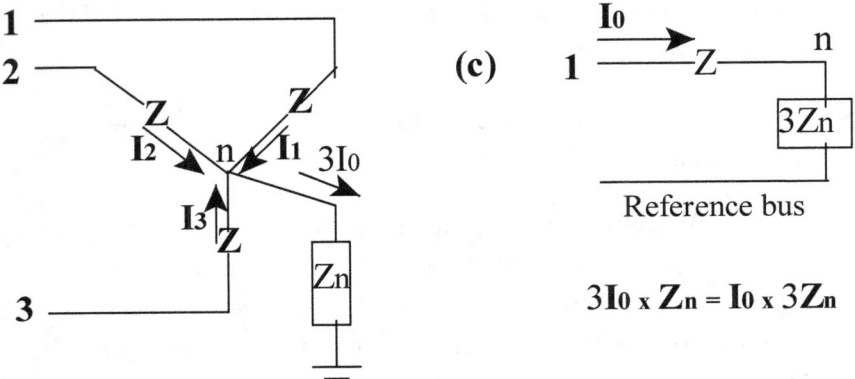

$$3I_0 \times Z_n = I_0 \times 3Z_n$$

Figure 4-5 Zero-sequence networks of a wye connected load with and without grounding impedance. Let us now consider delta connected load circuits. Fig. 4-6 illustrates the three-phase circuit of a Δ -connected load and its zero-sequence network.

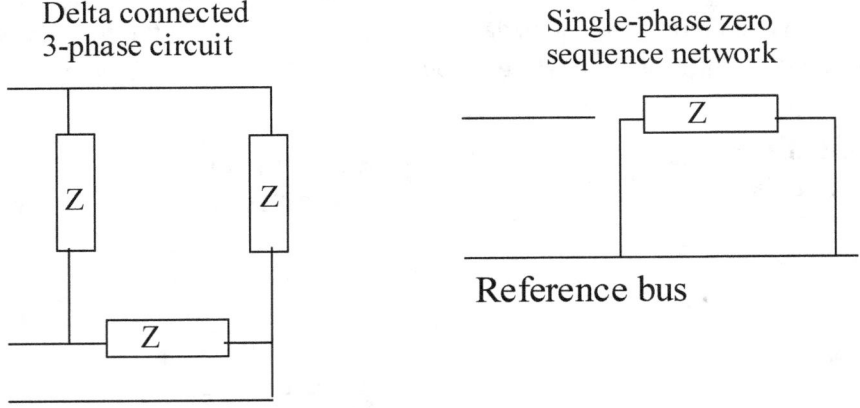

Figure 4-6 Three-phase delta connected load and the zero sequence network of any of its phases

A delta connected circuit does not provide a return path for the zero sequence line currents, this is represented in the zero sequence network as an open circuit. However, zero sequence currents could circulate inside the delta, provided they are induced from an outside source or produced by zero sequence generated voltages. Zero sequence voltages generated in the phases of the delta are canceled by the voltage drop produced by the circulating zero sequence current in the zero sequence impedance of each phase. So, no zero sequence voltage difference exists between the delta terminals.

CHAPTER 5

Solution of Asymmetrical Faults Applying Symmetrical Components.

Asymmetrical faults produce current and voltage unbalance in the power system. They are produced by asymmetrical short circuits, asymmetrical faults through impedances (or arc) or just open conductors. And they are classified as: single line-to-ground faults, line-to-line faults, and double line-to-ground fault. Three-phase faults occur very seldom and they are always considered to be symmetrical faults. Open conductor faults could be due to actual conductor breaking or to the action of fuses that may not open all three-conductors simultaneously.

The symmetrical components method is the preferred choice to analyze asymmetrical faults. The first type of problem is to solve short circuit faults occurring at the output terminals of an unloaded AC generator. See Fig. 5-1

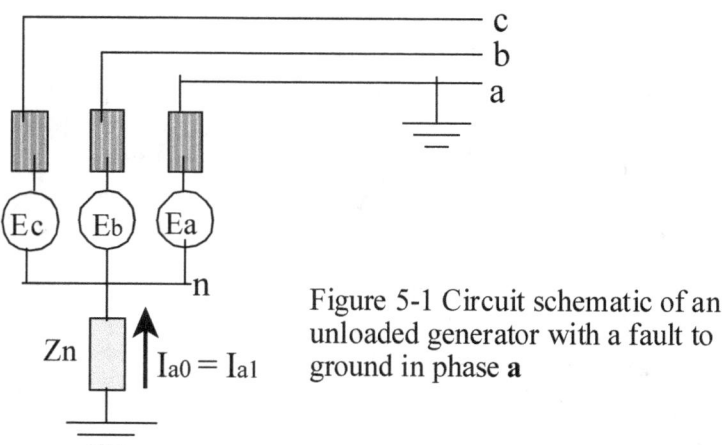

Figure 5-1 Circuit schematic of an unloaded generator with a fault to ground in phase **a**

5.1 Single Line to Ground Short Circuit on an Unloaded Generator

The symmetrical components of the fault current are calculated using Eqns. (3.34 and 4.18) plus the equations expressing the conditions imposed by the fault. Symbolically:

$$\overrightarrow{I_b} = 0 \qquad \overrightarrow{I_c} = 0 \qquad \overrightarrow{V_a} = 0$$

Equation (3.34) provides the symmetrical components of the short circuit current in function of the original unbalanced line currents. Symbolically

$$\begin{pmatrix} \overrightarrow{I_{a0}} \\ \overrightarrow{I_{a1}} \\ \overrightarrow{I_{a2}} \end{pmatrix} = \frac{1}{3} \cdot \begin{pmatrix} 1 & 1 & 1 \\ 1 & a & a^2 \\ 1 & a^2 & a \end{pmatrix} \cdot \begin{pmatrix} \overrightarrow{I_a} \\ 0 \\ 0 \end{pmatrix} \tag{5.1}$$

$$\overrightarrow{I_{a0}} = \frac{1}{3} \cdot \overrightarrow{I_a} \qquad \overrightarrow{I_{a1}} = \frac{1}{3} \cdot \overrightarrow{I_a} \qquad \overrightarrow{I_{a2}} = \frac{1}{3} \cdot \overrightarrow{I_a} \tag{5.2}$$

$$\overrightarrow{I_a} = 3 \cdot \overrightarrow{I_{a1}} = 3 \cdot \overrightarrow{I_{a2}} = 3 \cdot \overrightarrow{I_{a0}} \tag{5.3}$$

From Eq. (5.2) we conclude that the three symmetrical components of the short circuit current in phase **a** are equal to one third of the short circuit current. Symbolically:

$$\overrightarrow{I_{a1}} = \overrightarrow{I_{a2}} = \overrightarrow{I_{a0}} = \frac{\overrightarrow{I_a}}{3} \tag{5.4}$$

Substituting in Eq. (4.18) we obtain:

$$\begin{pmatrix} \overrightarrow{V_{a0}} \\ \overrightarrow{V_{a1}} \\ \overrightarrow{V_{a2}} \end{pmatrix} = \begin{pmatrix} 0 \\ \overrightarrow{E_a} \\ 0 \end{pmatrix} - \begin{pmatrix} \overrightarrow{Z_0} & 0 & 0 \\ 0 & \overrightarrow{Z_1} & 0 \\ 0 & 0 & \overrightarrow{Z_2} \end{pmatrix} \cdot \begin{pmatrix} \overrightarrow{I_{a1}} \\ \overrightarrow{I_{a1}} \\ \overrightarrow{I_{a1}} \end{pmatrix} \text{ simplify} \rightarrow \begin{pmatrix} V_{a0} \\ V_{a1} \\ V_{a2} \end{pmatrix} = \begin{pmatrix} -I_{a1} \cdot Z_0 \\ E_a - I_{a1} \cdot Z_1 \\ -I_{a1} \cdot Z_2 \end{pmatrix} \tag{5.5}$$

$$\begin{pmatrix} 1 & 1 & 1 \end{pmatrix} \cdot \begin{pmatrix} \overrightarrow{V_{a0}} \\ \overrightarrow{V_{a1}} \\ \overrightarrow{V_{a2}} \end{pmatrix} = \begin{pmatrix} 1 & 1 & 1 \end{pmatrix} \cdot \begin{bmatrix} \overrightarrow{(-Z_0 \cdot I_{a1})} \\ \overrightarrow{(E_a - Z_1 \cdot I_{a1})} \\ \overrightarrow{(-Z_2 \cdot I_{a1})} \end{bmatrix} \tag{5.6}$$

$$\overrightarrow{(V_{a0} + V_{a1} + V_{a2})} = \overrightarrow{(-Z_0 \cdot I_{a1} + E_a - Z_1 \cdot I_{a1} - Z_2 \cdot I_{a1})} \tag{5.7}$$

$$\overrightarrow{V_a} := 0 \qquad\qquad \text{Phase a is grounded}$$

The addition of the symmetric components of phase **a** voltage most be zero.

$$\overrightarrow{(V_{a0} + V_{a1} + V_{a2})} = 0$$

Going back to Eq. (5.7) we obtain:

$$\overrightarrow{\left(E_a - I_{a1}\left(Z_0 + Z_1 + Z_2\right)\right)} = 0$$

$$I_{a1} = \frac{E_a}{\left(Z_0 + Z_1 + Z_2\right)} \qquad\qquad (5.8)$$

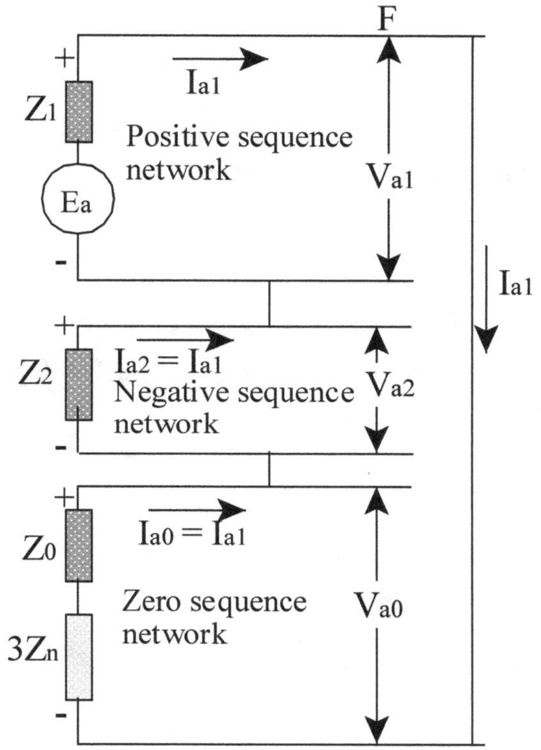

Figure 5-2 Sequence networks of a generator at no-lad condition experiencing single line to ground fault

5.2 Line to Line Short Circuit on an Unloaded Generator

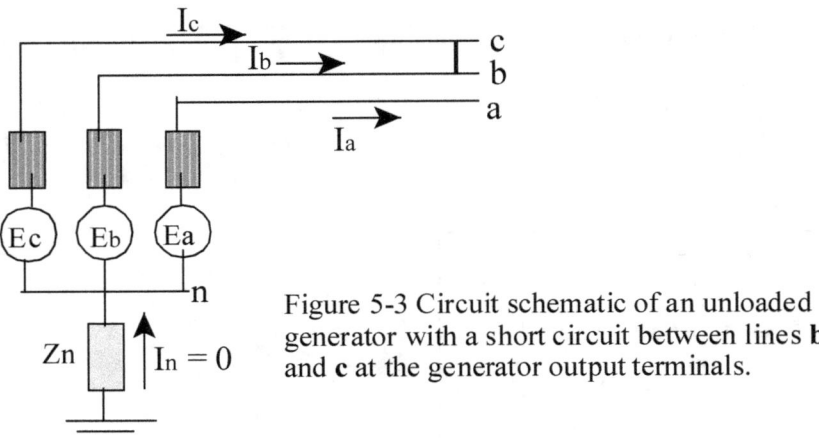

Figure 5-3 Circuit schematic of an unloaded generator with a short circuit between lines **b** and **c** at the generator output terminals.

Figure 5-3 shows an unloaded generator with a short circuit between lines b and c. Since there is not a fourth wire or ground return, there are no zero sequence current in Fig. 5-3. The conditions imposed by the short circuit are:

$$\vec{V_b} = \vec{V_c} \qquad \vec{I_a} = 0 \qquad \vec{I_b} = -\vec{I_c} \qquad \vec{I_b} + \vec{I_c} = 0 \qquad (5.9)$$

Applying Eq. (3.25) knowing that Vb = Vc we obtain the symmetrical components of Va in function of the line voltages under short circuit conditions.

$$\begin{pmatrix} \vec{V_{a0}} \\ \vec{V_{a1}} \\ \vec{V_{a2}} \end{pmatrix} = \frac{1}{3} \cdot \begin{pmatrix} 1 & 1 & 1 \\ 1 & a & a^2 \\ 1 & a^2 & a \end{pmatrix} \cdot \begin{pmatrix} \vec{V_a} \\ \vec{V_b} \\ \vec{V_b} \end{pmatrix} \qquad (5.10)$$

$$\vec{V_{a0}} = \frac{1}{3}\left(\vec{V_a} + \vec{V_b} + \vec{V_b} \right) \qquad (5.11)$$

$$\vec{V_{a1}} = \frac{1}{3}\left(\vec{V_a} + a \cdot \vec{V_b} + a^2 \cdot \vec{V_b} \right) \qquad (5.12)$$

$$\vec{V_{a2}} = \frac{1}{3}\left(\vec{V_a} + a^2 \cdot \vec{V_b} + a \cdot \vec{V_b} \right) \qquad (5.13)$$

$$\vec{V_{a0}} = \frac{1}{3}\left(\vec{V_a} + 2 \cdot \vec{V_b} \right)$$

$$\overrightarrow{V_{a1}} = \frac{1}{3}\overline{\left(V_a + V_b\overrightarrow{\left(a + a^2\right)}\right)} = \frac{1}{3}\overrightarrow{\left(V_a - V_b\right)} \qquad a + a^2 = -1 \qquad a^2 + a = -1$$

$$\overrightarrow{V_{a2}} = \frac{1}{3}\overline{\left(V_a + V_b\overrightarrow{\left(a^2 + a\right)}\right)} = \frac{1}{3}\overrightarrow{\left(V_a - V_b\right)} \qquad \overrightarrow{V_{a1}} = \overrightarrow{V_{a2}} \tag{5.14}$$

From Eq. (5.14) it is obvious that $V_{a1} = V_{a2}$ and we also know that $I_a = 0$ and

$I_b = -I_c$. So the symmetrical components of the short circuit current are obtaining by substituting in Eq. (3.34). The reader should not confuse the symmetrical components of the line currents with the real line currents.

$$\begin{pmatrix} \overrightarrow{I_{a0}} \\ \overrightarrow{I_{a1}} \\ \overrightarrow{I_{a2}} \end{pmatrix} = \frac{1}{3} \cdot \begin{pmatrix} 1 & 1 & 1 \\ 1 & a & a^2 \\ 1 & a^2 & a \end{pmatrix} \cdot \begin{pmatrix} 0 \\ \overrightarrow{-I_c} \\ \overrightarrow{I_c} \end{pmatrix} \tag{5.15}$$

Substituting in Eq. (5.15) the short circuit conditions. From equations (5.9), (3.34), and section 3-2 we obtain:

$$\overrightarrow{I_{a0}} = \frac{1}{3}\left(0 - \overrightarrow{I_c} + \overrightarrow{I_c}\right) = 0 \tag{5.16}$$

$$\overrightarrow{I_{a1}} = \frac{1}{3}\left(0 - a \cdot \overrightarrow{I_c} + a^2 \cdot \overrightarrow{I_c}\right) = \frac{1}{3} \cdot \overrightarrow{I_c}\left(a^2 - a\right) = \frac{\sqrt{3}}{3} \cdot \overrightarrow{I_c}(-j) \tag{5.17}$$

$$\overrightarrow{I_{a2}} = \frac{1}{3}\left(0 - a^2\overrightarrow{I_c} + a \cdot \overrightarrow{I_c}\right) = \frac{1}{3}\overrightarrow{I_c} \cdot \left(a - a^2\right) = \frac{\sqrt{3}}{3}\overrightarrow{I_c} \cdot j \tag{5.18}$$

$$\overrightarrow{I_{a2}} = -\overrightarrow{I_{a1}} \tag{5.19}$$

In Fig. (5-3) the grounding impedance is not zero, but Ia0 is zero, see Eq. (5.16), so Va0 must be zero. And from Eqns. (5.14) and (5.19) we know that V.a1 = V.a2 and I.a2 = -I.a1 Substituting in Eq. (4.18) we obtain:

$$\begin{pmatrix} 0 \\ \vec{V}_{a1} \\ \vec{V}_{a1} \end{pmatrix} = \begin{pmatrix} 0 \\ \vec{E}_a \\ 0 \end{pmatrix} - \begin{pmatrix} \vec{Z}_0 & 0 & 0 \\ 0 & \vec{Z}_1 & 0 \\ 0 & 0 & \vec{Z}_2 \end{pmatrix} \cdot \begin{pmatrix} 0 \\ \vec{I}_{a1} \\ -\vec{I}_{a1} \end{pmatrix}$$

(5.20)

$\vec{E}_a$ is a rise in voltage

$$\begin{pmatrix} 0 \\ \vec{E}_a \\ 0 \end{pmatrix} - \begin{pmatrix} \vec{Z}_0 & 0 & 0 \\ 0 & \vec{Z}_1 & 0 \\ 0 & 0 & \vec{Z}_2 \end{pmatrix} \cdot \begin{pmatrix} 0 \\ \vec{I}_{a1} \\ -\vec{I}_{a1} \end{pmatrix} \text{ simplify} \rightarrow \begin{pmatrix} 0 \\ E_a - I_{a1} \cdot Z_1 \\ I_{a1} \cdot Z_2 \end{pmatrix}$$

To simplify the notation some phasor's arrows were neglected

$$\begin{pmatrix} 1 & 1 & -1 \end{pmatrix} \cdot \begin{pmatrix} 0 \\ V_{a1} \\ V_{a1} \end{pmatrix} = \begin{pmatrix} 1 & 1 & -1 \end{pmatrix} \cdot \begin{pmatrix} 0 \\ E_a - Z_1 \cdot I_{a1} \\ Z_2 \cdot I_{a1} \end{pmatrix}$$

$$\begin{pmatrix} 1 & 1 & -1 \end{pmatrix} \cdot \begin{pmatrix} 0 \\ V_{a1} \\ V_{a1} \end{pmatrix} \text{ simplify} \rightarrow 0$$

$$\begin{pmatrix} 1 & 1 & -1 \end{pmatrix} \cdot \begin{pmatrix} 0 \\ E_a - Z_1 \cdot I_{a1} \\ Z_2 \cdot I_{a1} \end{pmatrix} \text{ simplify} \rightarrow E_a - I_{a1} \cdot Z_1 - I_{a1} \cdot Z_2$$

$$0 = \vec{E}_a - \vec{I}_{a1} \cdot \left(\vec{Z}_1 + \vec{Z}_2 \right) \qquad \vec{I}_{a1} = \frac{\vec{E}_a}{\vec{Z}_1 + \vec{Z}_2}$$

(5.21)

Applying Eq. (5.19) we obtain:

$$\vec{I}_{a2} = \frac{-\vec{E}_a}{\vec{Z}_1 + \vec{Z}_2} \tag{5.22}$$

To calculate the short circuit currents in lines b and c we use Eqns. (5.9), (5.16), (3.31), (3.32) and Table 3-1 as follows:

$$j := \sqrt{-1}$$

$$\vec{I}_a = \vec{I}_{a0} + \vec{I}_{a1} + \vec{I}_{a2} = 0 \tag{5.23}$$

$$\vec{I}_b = \vec{I}_{a0} + a^2 \cdot \vec{I}_{a1} + a \cdot \vec{I}_{a2} = \varepsilon^{j \cdot 240°} \cdot \vec{I}_{a1} + \varepsilon^{j \cdot 120°} \cdot \vec{I}_{a2} \tag{5.24}$$

$$\vec{I}_c = \vec{I}_{a0} + a \cdot \vec{I}_{a1} + a^2 \cdot \vec{I}_{a2} = \varepsilon^{j \cdot 120°} \cdot \vec{I}_{a1} + \varepsilon^{j \cdot 240°} \cdot \vec{I}_{a2} \tag{5.25}$$

Once the symmetrical components of the line current, see Eqns. (5.21) and (5.22), are calculated then use Eqns. (5.24) and (5.25) to calculate the line currents. The reader should not confuse the operator **a** with the subscript a.

The conditions imposed by the short circuit and expressed by Eqns. (5.9) and (5.19) are specifically applicable to line to line short circuits. These equations also determine the connections of the sequence networks. Furthermore, Eq. (5.14) requires that the positive and negative networks be in parallel.

See Fig. (5-4) where the reader surely will note that there is not zero sequence network, because Z_0 does not participate in Eqns. (5.21) and (5.22). As a consequence, and because the zero sequence network is not present $I_{a2} = -I_{a1}$. Symbolically, this condition is expressed by Eq. (5.19).

112

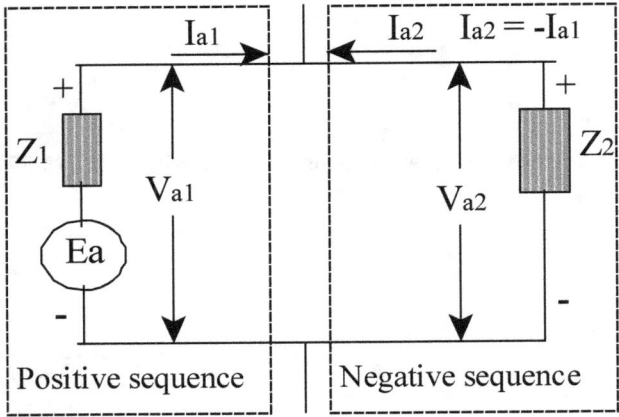

Figure 5-4 Sequence networks of an ungrounded and unloaded generator with a line-to-line short circuit between lines b and c

5-3 Double line short circuit to ground

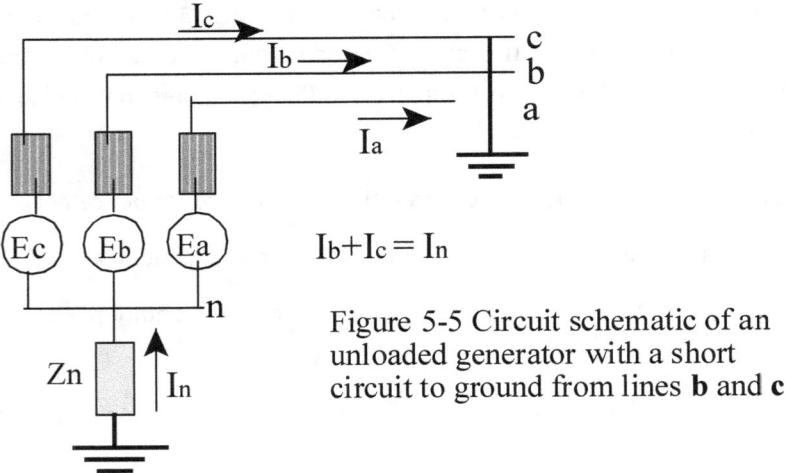

$I_b + I_c = I_n$

Figure 5-5 Circuit schematic of an unloaded generator with a short circuit to ground from lines **b** and **c**

Figure 5-5 shows line b and c shorted to ground at the generator output terminals. The conditions imposed by the short circuit are:

$$\vec{V_b} = 0 \qquad \vec{V_c} = 0 \qquad \vec{I_a} = 0 \qquad\qquad (5.26)$$

The symmetrical sequence components of phase **a** voltage are given by matrix Eq. (3.25) and the short circuit conditions.

$$
\begin{pmatrix} \vec{V}_{a0} \\ \vec{V}_{a1} \\ \vec{V}_{a2} \end{pmatrix} = \frac{1}{3} \cdot \begin{pmatrix} 1 & 1 & 1 \\ 1 & a & a^2 \\ 1 & a^2 & a \end{pmatrix} \cdot \begin{pmatrix} \vec{V}_a \\ 0 \\ 0 \end{pmatrix} \text{ simplify } \rightarrow \begin{pmatrix} V_{a0} \\ V_{a1} \\ V_{a2} \end{pmatrix} = \begin{pmatrix} \dfrac{V_a}{3} \\ \dfrac{V_a}{3} \\ \dfrac{V_a}{3} \end{pmatrix}
\tag{5.27}
$$

Two matrices are equal if their elements are one-to-one equal. To simplify the arrows will not be used with the phasor notation. So from Eq. (5.27) we conclude that:

$$
V_{a0} = V_{a1} = V_{a2} = \frac{V_a}{3}
\tag{5.28}
$$

Applying Eq. (4.18) we get

$$
\begin{pmatrix} V_{a0} \\ V_{a1} \\ V_{a2} \end{pmatrix} = \begin{pmatrix} 0 \\ E_a \\ 0 \end{pmatrix} - \begin{pmatrix} Z_0 & 0 & 0 \\ 0 & Z_1 & 0 \\ 0 & 0 & Z_2 \end{pmatrix} \cdot \begin{pmatrix} I_{a0} \\ I_{a1} \\ I_{a2} \end{pmatrix} \text{ simplify } \rightarrow \begin{pmatrix} V_{a0} \\ V_{a1} \\ V_{a2} \end{pmatrix} = \begin{pmatrix} -I_{a0} \cdot Z_0 \\ E_a - I_{a1} \cdot Z_1 \\ -I_{a2} \cdot Z_2 \end{pmatrix}
\tag{5.29}
$$

From Equation (5.29) we know that:

$$
V_{a1} = E_a - Z_1 \cdot I_{a1}
\tag{5.30}
$$

Where Ea is a raise in voltage, not a voltage drop.

Substituting in Eq. (4.18) the results of Eqs. (5.28), (5.29), and (5.30) we obtain:

$$
\begin{pmatrix} E_a - Z_1 \cdot I_{a1} \\ E_a - Z_1 \cdot I_{a1} \\ E_a - Z_1 \cdot I_{a1} \end{pmatrix} = \begin{pmatrix} -I_{a0} \cdot Z_0 \\ E_a - I_{a1} \cdot Z_1 \\ -I_{a2} \cdot Z_2 \end{pmatrix}
\tag{5.31}
$$

Multiplying both sides of (5.31) by Z^{-1} we convert the voltage matrix into a current matrix.

$$Z = \begin{pmatrix} Z_0 & 0 & 0 \\ 0 & Z_1 & 0 \\ 0 & 0 & Z_2 \end{pmatrix} \qquad Z^{-1} = \begin{pmatrix} \dfrac{1}{Z_0} & 0 & 0 \\ 0 & \dfrac{1}{Z_1} & 0 \\ 0 & 0 & \dfrac{1}{Z_2} \end{pmatrix} \qquad Z \cdot Z^{-1} = 1$$

$$\begin{pmatrix} \dfrac{1}{Z_0} & 0 & 0 \\ 0 & \dfrac{1}{Z_1} & 0 \\ 0 & 0 & \dfrac{1}{Z_2} \end{pmatrix} \cdot \begin{pmatrix} E_a - Z_1 \cdot I_{a1} \\ E_a - Z_1 \cdot I_{a1} \\ E_a - Z_1 \cdot I_{a1} \end{pmatrix} \rightarrow \begin{pmatrix} \dfrac{E_a - I_{a1} \cdot Z_1}{Z_0} \\ \dfrac{E_a - I_{a1} \cdot Z_1}{Z_1} \\ \dfrac{E_a - I_{a1} \cdot Z_1}{Z_2} \end{pmatrix}$$

$$\begin{pmatrix} \dfrac{1}{Z_0} & 0 & 0 \\ 0 & \dfrac{1}{Z_1} & 0 \\ 0 & 0 & \dfrac{1}{Z_2} \end{pmatrix} \cdot \begin{pmatrix} -I_{a0} \cdot Z_0 \\ E_a - I_{a1} \cdot Z_1 \\ -I_{a2} \cdot Z_2 \end{pmatrix} \rightarrow \begin{pmatrix} -I_{a0} \\ \dfrac{E_a - I_{a1} \cdot Z_1}{Z_1} \\ -I_{a2} \end{pmatrix}$$

$$\begin{pmatrix} \dfrac{E_a - I_{a1} \cdot Z_1}{Z_0} \\ \dfrac{E_a - I_{a1} \cdot Z_1}{Z_1} \\ \dfrac{E_a - I_{a1} \cdot Z_1}{Z_2} \end{pmatrix} = \begin{pmatrix} -I_{a0} \\ \dfrac{E_a - I_{a1} \cdot Z_1}{Z_1} \\ -I_{a2} \end{pmatrix} \qquad (5.32)$$

From (5.28) we know that the symmetrical components of the voltage are equal therefore the symmetrical components of the currents must be equal. Using Eq. (5.32) we obtain

$$I_{a0} = \frac{E_a - Z_1 \cdot I_{a1}}{Z_0} \qquad I_{a1} = \frac{E_a - Z_1 \cdot I_{a1}}{Z_1} \qquad I_{a2} = \frac{E_a - Z_1 \cdot I_{a1}}{Z_2}$$

and knowing from the fault conditions that: $\vec{I}_{a0} + \vec{I}_{a1} + \vec{I}_{a2} = \vec{I}_a = 0$

$$\vec{I}_a = \frac{\vec{E}_a}{\vec{Z}_1} = 0 \tag{5.33}$$

$$\frac{E_a - Z_1 \cdot I_{a1}}{Z_0} + \frac{E_a - Z_1 \cdot I_{a1}}{Z_1} + \frac{E_a - Z_1 \cdot I_{a1}}{Z_2} = \frac{E_a}{Z_1} \tag{5.34}$$

$$I_{a1} = \frac{E_a(Z_0 + Z_2)}{(Z_0 \cdot Z_1 + Z_0 \cdot Z_2 + Z_1 \cdot Z_2)} = \frac{E_a}{\dfrac{Z_0 \cdot Z_1}{Z_0 + Z_2} + \dfrac{Z_0 \cdot Z_2}{Z_0 + Z_2} + \dfrac{Z_1 \cdot Z_2}{Z_0 + Z_2}} \tag{5.35}$$

$$\frac{Z_0 \cdot Z_1}{Z_0 + Z_2} + \frac{Z_0 \cdot Z_2}{Z_0 + Z_2} + \frac{Z_1 \cdot Z_2}{Z_0 + Z_2} \quad \text{simplify} \;\rightarrow\; Z_1 + Z_2 - \frac{Z_2^2}{Z_0 + Z_2}$$

$$Z_2 - \frac{Z_2^2}{Z_0 + Z_2} \quad \text{simplify} \;\rightarrow\; \frac{Z_0 \cdot Z_2}{Z_0 + Z_2}$$

Applying Eq. (5.35) we obtain

$$I_{a1} = \frac{E_a}{Z_1 + \dfrac{Z_0 \cdot Z_2}{Z_0 + Z_2}}$$

$$\tag{5.36}$$

To calculate the short circuit currents in lines b and c see Eqns. (5.28) and (5.30).

$$V_{a0} = V_{a2} = V_{a1} = E_a - I_{a1} \cdot Z_1 \qquad (5.37)$$

From Eq. (5.29) we obtain

$$V_{a2} = -Z_2 \cdot I_{a2} \qquad I_{a2} = \frac{-V_{a2}}{Z_2} \qquad (5.38)$$

$$V_{a0} = -Z_0 \cdot I_{a0} \qquad I_{a0} = \frac{-V_{a0}}{Z_0} \qquad (5.39)$$

From Eqns. (3.30 to 3.32) we obtain the short circuit current $I_b + I_c$

$$\begin{pmatrix} I_a \\ I_b \\ I_c \end{pmatrix} = \begin{pmatrix} 1 & 1 & 1 \\ 1 & a^2 & a \\ 1 & a & a^2 \end{pmatrix} \cdot \begin{pmatrix} I_{a0} \\ I_{a1} \\ I_{a2} \end{pmatrix} \qquad (5.40)$$

$$I_a = I_{a0} + I_{a1} + I_{a2} = 0 \qquad (5.41)$$

$$I_b = I_{a0} + a^2 \cdot I_{a1} + a \cdot I_{a2} \qquad \varepsilon := 2.7183 \qquad a := \varepsilon^{j \cdot \frac{2\pi}{3}}$$

$$I_c = I_{a0} + a \cdot I_{a1} + a^2 \cdot I_{a2} \qquad a^2 + a = -1 \qquad a + a^2 = -1 \qquad \text{See Table 3-1}$$

$$I_a + I_b + I_c = 3 \cdot I_{a0} \qquad (5.42)$$

$$I_a = 0 \qquad \text{See Eq. (5.26)}$$

$$I_b + I_c = 3 \cdot I_{a0} \qquad I_n = I_b + I_c \qquad \text{Short circuit current} \qquad I_n = 3 \cdot I_{a0} \qquad (5.43)$$

Likewise for the voltage phasors, see section 3-2, and Eqns. (3.1) to (3.2)

$$\begin{pmatrix} V_a \\ V_b \\ V_c \end{pmatrix} = \begin{pmatrix} 1 & 1 & 1 \\ 1 & a^2 & a \\ 1 & a & a^2 \end{pmatrix} \cdot \begin{pmatrix} V_{a0} \\ V_{a1} \\ V_{a2} \end{pmatrix} \qquad (5.44)$$

Phasors that belong to same sequence have the same magnitude. But in this specific case, see Eq. (5.28) and Fig. (5-5), all the phasors are equal in magnitude regardless their sequence.

$$\left|V_{a0}\right| = \left|V_{a1}\right| = \left|V_{a2}\right| = \left|\frac{V_a}{3}\right|$$

$$\left|V_a\right| = \left|V_{a0}\right| + \left|V_{a1}\right| + \left|V_{a2}\right| = 3 \cdot \left|V_{a1}\right| \tag{5.45}$$

Figure 5-6 illustrates the connection diagram of the sequence networks for the case of two lines fault to ground.

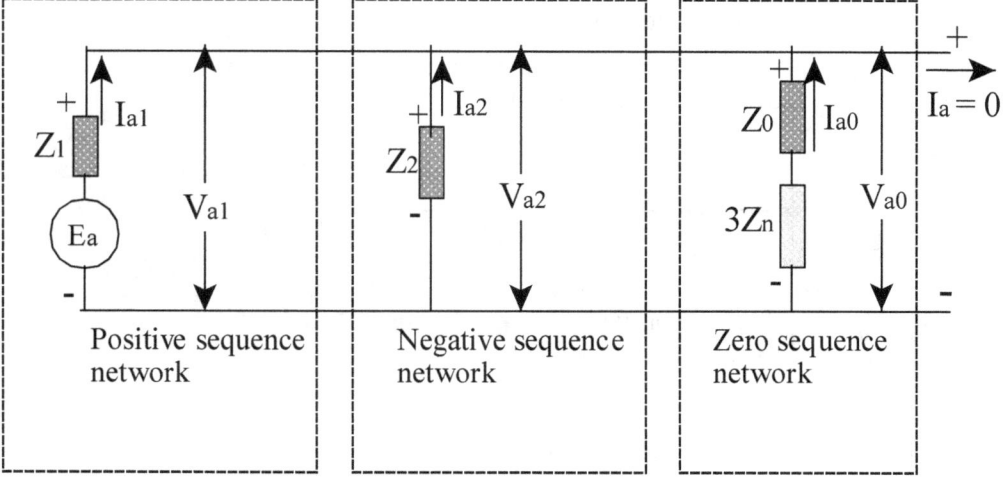

Figure 5-6 Sequence networks of a generator at no load condition experiencing a double line to ground short circuit.

5.4 Asymmetrical Faults on Electrical Power Systems

The simple electrical power system illustrated in Fig. (5.7) is general enough to demonstrate the application of symmetrical components to solve short circuit faults of the types considered in the previous section. In the analysis of short-circuit events occurring in three-phase power systems, the assumptions listed below are usually made.

- The power system is balanced before the short circuit occurrence.

- The resistance component of generators and transformers impedances are neglected

- Normal operating load is neglected.

- Transformers shunt admittances and transmission line capacitances are neglected

- The separation between transmission lines are large enough that the mutual impedance between lines due to the zero-sequence currents can be neglected.

- The mechanical power input to each generator remains constant.

- The classical model could be used for each generator, that is, constant emf behind constant subtransient reactance. Both are assumed to remain constant through the short circuit event.

- The analysis is conducted assuming that the fault occurs in phase **a**. This is not a restriction since any phase can be designated as phase **a**.

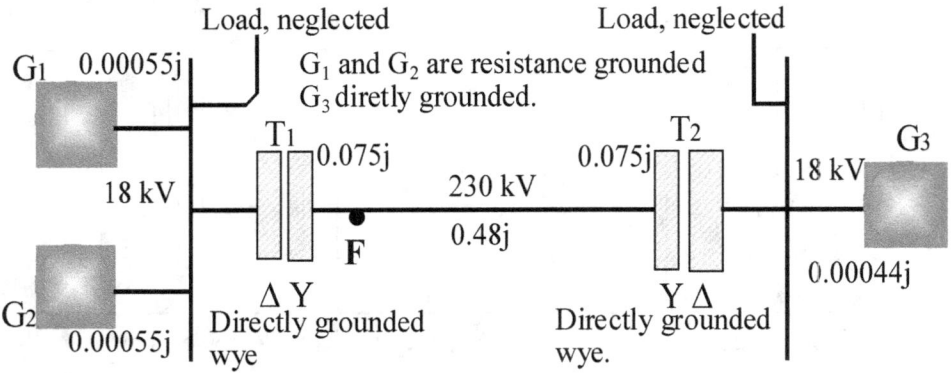

Figure 5-7 Single-line diagram of a balanced three-phase power system. All positive sequence per unit reactance values based on 230kV, 300 MVA.

New per unit base: 300 MVA and 230 kV

$$\frac{500}{300} = 1.6666667 \qquad \frac{18}{230} = 0.0782609$$

Applying Eq. (4.1) we obtain:

$$0.09 \cdot \left(\frac{18}{230}\right)^2 = 0.00055 \qquad 0.12 \cdot \left(\frac{18}{230}\right)^2 \cdot \frac{300}{500} = 0.00044$$

$$0.016 \cdot \left(\frac{18}{230}\right)^2 = 0.0001 \qquad 0.03 \cdot \left(\frac{18}{230}\right)^2 \cdot \frac{300}{500} = 0.00011$$

See Table 5-2

	G1	G2	G3
Type	Combustion turbine	Combustion turbine	Steam turbine
MVA	300	300	500
kV	18	18	18
Poles RPM	2 3600	2 3600	2 3600
Subtransient reactance	0.09 pu	0.09 pu	0.12 pu
Negative Seq reactance	0.09 pu	0.09 pu	0.12 pu
Zero sequence reactance	0.016 pu	0.016 pu	0.03 pu

Table 5-1 Generator Factory Ratings. Per-unit values based on 18 kV and generator MVA

	G1	G2	G3
Type	Combustion turbine	Combustion turbine	Steam turbine
MVA	1	1	1.67
kV	0.078	0.078	0.078
Poles RPM	2 3600	2 3600	2 3600
Subtransient reactance	0.00055 pu	0.00055 pu	0.00044 pu
Negative Seq reactance	0.00055 pu	0.00055 pu	0.00044 pu
Zero sequence reactance	0.0001 pu	0.0001 pu	0.00011 pu

Table 5-2 Generator Per-Unit Values Converted to 230 kV, 300 MVA

Transformers T1 and T2 of Fig.5-7 are three-phase, 600 MVA, and have a per-unit positive sequence reactance of 0.15 based on 600 MVA and 230 kV. Their per-unit value converted to 300 MVA base is 0.075.

$$0.15 \cdot \frac{300}{600} = 0.075$$

The transmission line per-unit positive and negative reactances are: 0.48j each based on 230 kV, 300 MVA. And the zero sequence per-unit reactance is: 0.96j

Let us assume that the generator grounding impedance (resistance connected in the secondary side of a grounding transformer and reflected to the primary plus ground transformer primary reactance) is 1.5 times the generator subtransient reactance. Symbolically:

For G1 and G2: $1.5 \cdot 0.09 = 0.135$ G3 is directly grounded

Converting these values to 230 kV and 300 MVA base we obtain:

$$0.135 \cdot \left(\frac{18}{230} \right)^2 = 0.000827 \quad \text{pu}$$

Or directly from Table 5-2:

$$1.5 \cdot 0.00055 = 0.000825 \qquad 3 \cdot 0.000827 = 0.0025$$

5.6 Single Line to Ground Short-Circuit on the Power System Illustrated on Fig. 5-7

The conditions imposed by the short-circuit to ground are:

$$\vec{I}_b = 0 \qquad\qquad \vec{I}_c = 0 \qquad\qquad \vec{V}_a = 0 \tag{5.46}$$

The analysis is based on results obtained in section 5.1, except that Vf replaces Ea, the line to neutral emf, in Eq. (5.8). Symbolically,

$$I_{a1} = \frac{V_f}{Z_0 + Z_1 + Z_2} \tag{5.47}$$

From Eq. (5.5) we know that:

$$I_{a1} = I_{a2} = I_{a0} \tag{5.48}$$

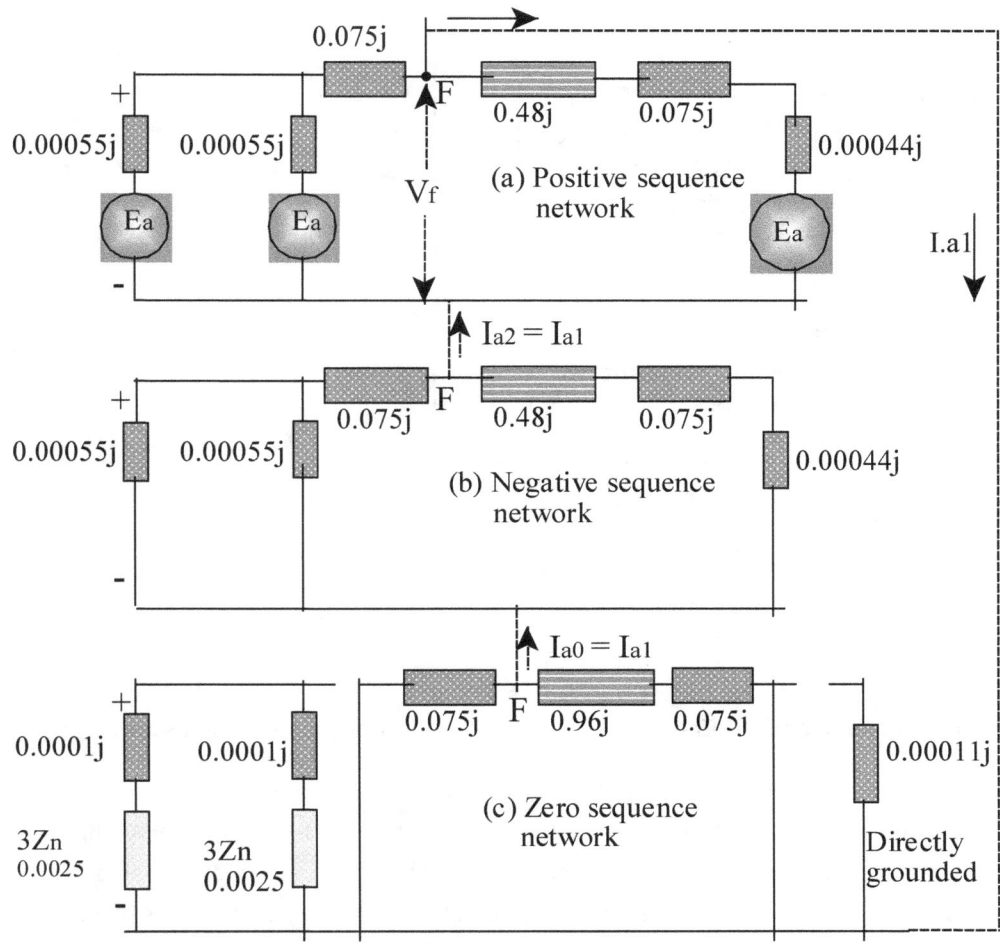

Figure 5-8 The three sequence networks of the balanced three-phase power system illustrated in Fig. 5-7

Equations (5.39) and (5.42) are the foundation to represent the single line to ground fault by connecting the three sequence networks in series via the point at which the fault occurs as indicated on Fig. (5-8). The Thévenin equivalent of the sequence networks are:

$$\frac{\left(\dfrac{0.00055j}{2} + 0.075j\right) \cdot (0.048j + 0.075j + 0.00044j)}{\left(\dfrac{0.00055j}{2} + 0.075j\right) + (0.048j + 0.075j + 0.00044j)} \quad \text{simplify} \ \rightarrow 0.0467602j$$

Positive sequence network

$$\frac{(0.075j) \cdot (0.96j + 0.075j)}{(0.075j) + (0.96j + 0.075j)} \quad \text{simplify} \ \rightarrow 0.069932j$$

Zero sequence network

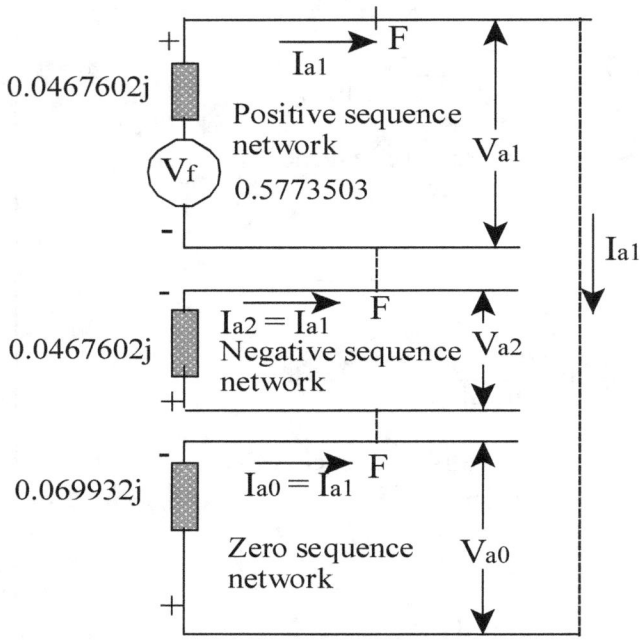

Figure 5-9 Connection of the Thevenin equivalent of the three-sequence networks

Vf in Fig. 5-9 is the voltage between point F and ground before the occurrence of a short circuit to ground at point F. Which is equal to the generated line to neutral voltage because the load has been neglected. The reader should remember that in a three-phase system the generated emfs are 120 degree apart in time phase. In fact, the generated emfs in a three-phase system have specific relative polarities and angular position with respect to each other, and once you arbitrarily assign the angular position of one of them the angular position of the others become fixed. Assuming that the generators are operating at rated voltage, and because the load current has been neglected, then the pre fault voltage at the point F is 132.8 kV or 0.5773 per-unit.

$$V_f = \frac{230}{\sqrt{3} \cdot 230} = 0.5773503$$

Per unit voltage at point F before short circuit

$$\vec{V_f} = 0.57735 + 0j$$

Taking Vf as reference

Applying Eq. (5.47) we obtain the per unit short circuit currents circulating in the sequence networks.

$$I_{a1} = \frac{0.57735}{2 \cdot 0.0467602j + 0.069932j} = -3.532221j$$

Per unit amperes

$$\frac{I_{a1}}{I_{base}} = -3.5322j$$

pu

$$I_{base} = \frac{MVA_b}{\sqrt{3} \cdot kV_b} = \frac{300}{\sqrt{3} \cdot 230} = 0.7530656$$

base kiloamperes

$$-3.5322j \cdot 0.7530656 = -2.660j \qquad \vec{I_{a1}} = -2.660j$$

kiloamperes

$$\vec{I_a} = 3 \cdot \vec{I_{a1}} = 3 \cdot (-2.660j) = -7.980j$$

kiloamperes

Total short circuit current = 7980 amperes

Once the short circuit current is determined, the problem is in general considered as solved because you have most of the required information for circuit protection or equipment selection. You might need to know the actual voltage at the point of fault if you are concerned with possible flash-over or if the voltage is large enough to sustain and arc inside a specific equipment. The short circuit current lags 90 degrees the voltage at point F. The symmetrical components of the voltage at point F are:

$$\vec{V_{a0}} = 0.0699j \cdot (-3.5322j) = 0.2469$$

Per-unit voltage drop

$$\vec{V_{a2}} = 0.0468j \cdot (-3.5322j) = 0.1653$$

Per-unit voltage drop

$$\vec{V_{a1}} = 0.57735 - 0.0468j \cdot (-3.5322j) = 0.4120$$

Per-unit voltage raise

The matrix equations that provides the symmetrical components of the voltage at the point of fault is obtained applying Eq. (4.5), where Vf replaces Ea. As shown in Eq. (5.49).

$$\begin{pmatrix} V_{a0} \\ V_{a1} \\ V_{a2} \end{pmatrix} = \begin{pmatrix} 0 \\ V_f \\ 0 \end{pmatrix} - \begin{pmatrix} Z_0 & 0 & 0 \\ 0 & Z_1 & 0 \\ 0 & 0 & Z_2 \end{pmatrix} \cdot \begin{pmatrix} I_{a0} \\ I_{a1} \\ I_{a2} \end{pmatrix}$$

(5.49)

Substituting in Eq. (5.49) the per unit values already obtained we get (5.50)

$$\begin{pmatrix} V_{a0} \\ V_{a1} \\ V_{a2} \end{pmatrix} = \begin{pmatrix} 0 \\ 0.57735 \\ 0 \end{pmatrix} - \begin{pmatrix} 0.0699j & 0 & 0 \\ 0 & 0.0468j & 0 \\ 0 & 0 & 0.0468j \end{pmatrix} \cdot \begin{pmatrix} I_{a1} \\ I_{a1} \\ I_{a1} \end{pmatrix} \qquad (5.50)$$

$$\begin{pmatrix} 0.0699j & 0 & 0 \\ 0 & 0.0468j & 0 \\ 0 & 0 & 0.0468j \end{pmatrix} \cdot \begin{pmatrix} -3.5322j \\ -3.5322j \\ -3.5322j \end{pmatrix} = \begin{pmatrix} 0.2469008 \\ 0.165307 \\ 0.165307 \end{pmatrix} \qquad (5.51)$$

$$\begin{pmatrix} V_{a0} \\ V_{a1} \\ V_{a2} \end{pmatrix} = \begin{pmatrix} 0 \\ 0.57735 \\ 0 \end{pmatrix} - \begin{pmatrix} 0.2469 \\ 0.1653 \\ 0.1653 \end{pmatrix} = \begin{pmatrix} -0.2469 \\ 0.41205 \\ -0.1653 \end{pmatrix} \quad \text{Per-unit kV} \qquad (5.52)$$

In Eq. (5.52) the voltage drop are considered negative and the voltage raise positive.

$-0.2469 \cdot 230 = -56.787$ kilovolts   drop

$0.41205 \cdot 230 = 94.7715$ kilovolts   raise

$-0.1653 \cdot 230 = -38.019$ kilovolts   drop

$-56.787 - 38.019 = -94.806$ drop   94.7715 raise

$-0.2469 - 0.1653 = -0.4122$ pu Drop = pu Raise   See Eq. (5.52)   (5.53)

The analysis of asymmetrical faults using symmetrical components is very cumbersome and mistake prone. Apparently the advocates of this method avoided the conventional method using Kirchhoff's law which is applicable to any unbalanced polyphase power system.

Σ voltage around a close loop = 0   And   Σ i into a node = Σ i leaving same node

The symmetrical components method requires that the three-phase system be balanced and linear, in this way it can be converted into a two-terminal linear system that can be replaced by a Thévenin equivalent. The reader should keep in mind that the sequence networks, like the symmetrical components phasor diagrams, are not really connected to each other. Although in electrical computations is helpful, as a computation guide, to inter-connect the sequence networks in accordance with the specific application. In electrical engineering short circuit studies, faults are always considered as "bolted." No impedance in the short circuit itself, it could really happen.

5.7 Line-to-Line Short Circuit on the Three-Phase System Illustrated in Fig. 5-7

The conditions imposed by the line-to-line short circuit in this section are identical to the ones imposed in Sec. 5.2. Furthermore, the results of the analysis in Sec 5.2 are also valid.

$$\vec{V_b} = \vec{V_c} \qquad \vec{I_a} = 0 \qquad \vec{I_b} = -\vec{I_c} \qquad \vec{I_b} + \vec{I_c} = 0 \qquad (5.54)$$

The results are:

$$\vec{V_{a2}} = \vec{V_{a1}} \qquad (5.55)$$

$$\vec{I_{a2}} = -\vec{I_{a1}} \qquad (5.56)$$

$$\vec{I_{a1}} = \frac{\vec{V_f}}{\vec{Z_1} + \vec{Z_2}} \qquad \text{Where Vf replaced Ea.} \qquad (5.57)$$

$$\vec{I_{a2}} = -\frac{\vec{V_f}}{\vec{Z_1} + \vec{Z_2}} \qquad (5.58)$$

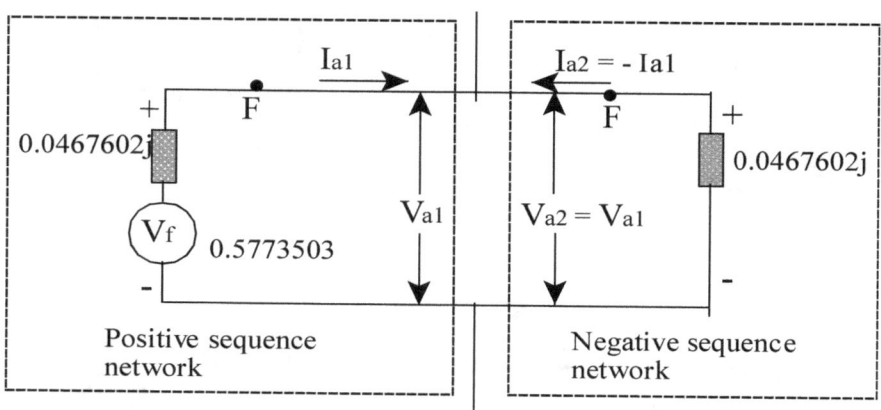

Figure 5-10 Connections of the Thevenin equivalent of the positive and negative sequence networks

Equations (5.54 and 5.55) allow us to conclude that to mimic the line-to-line short circuit, the positive and negative sequence networks should be connected in parallel at the fault point. And because the zero sequence impedance is not present in the above equations, it is not necessary to include a zero sequence network in the model illustrated in Fig. 5-10. As stated before in Sec. 3.2, once you know any symmetrical component of a set, then you can easily find the remaining symmetrical components of the same set using Eqns. (3.4 - 3.5). Fig. 3-1 facilitates the understanding of the sequence components.

$$I_{b1} = a^2 \cdot I_{a1} \qquad I_{c1} = a \cdot I_{a1} \qquad\qquad I_{a0} = I_{b0} = I_{c0} = 0$$

$$I_{b2} = a \cdot I_{a2} = a \cdot \left(-I_{a1}\right) \qquad\qquad I_{c2} = a^2 \cdot I_{a2} = a^2 \cdot \left(-I_{a1}\right)$$

To calculate the line short circuit currents we proceed as follows. From Eqns. (5.56) and (5.57) we obtain the per unit symmetrical components of the short circuit current in line **a**.

$$\overrightarrow{I_{a1}} = \frac{0.5773503}{2 \cdot 0.0467602j} = -6.1735226j \quad \text{Per unit} \qquad\qquad \overrightarrow{I_{a2}} = 6.1735226j \quad \text{Per unit}$$

$$I_{base} = \frac{MVA_{base}}{\sqrt{3} \cdot kV_{base}} = \frac{300}{\sqrt{3} \cdot 230} = 0.7530656 \quad \text{Base kiloamperes}$$

$$\overrightarrow{I_{a1}} = -6.1735j \cdot 0.7531 = -4.6493j \quad \text{Kiloamperes, lags Vf by 90 degrees}$$

$$\overrightarrow{I_{a2}} = 4.6493j \quad \text{Kiloamperes, leads Vf by 90 degrees}$$

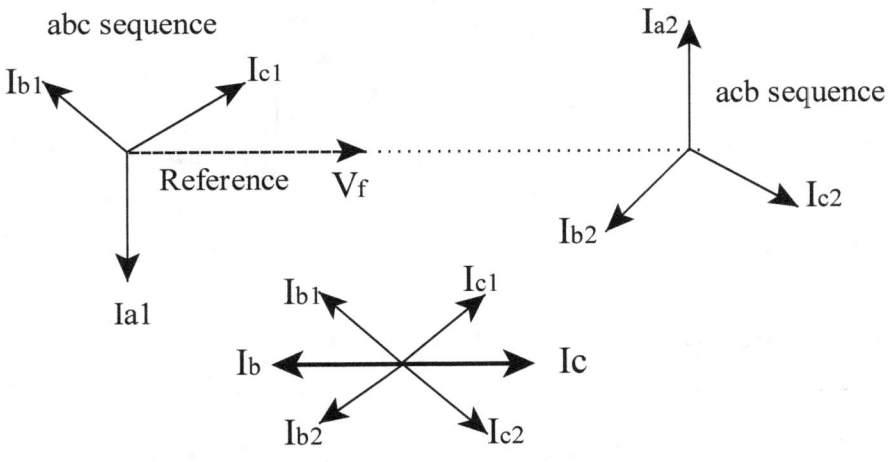

Figure 5-11 Phasor diagrams illustrating line b and c currents.

To calculate the short circuit currents in lines b and c we use Eqns. (3.28) to (3.32).

$$\vec{I_b} = \vec{I_{b1}} + \vec{I_{b2}} = a^2 \cdot \vec{I_{a1}} + a \cdot \vec{I_{a2}} = (-0.5 - 0.866j)(-4.6493j) + (-0.5 + 0.866j) \cdot 4.6493j$$

$$\vec{I_b} = -8.0526 \quad \text{kiloamperes}$$

$$\vec{I_c} = 8.0526 \quad \text{kiloamperes}$$

Because all the symmetrical components phasors have the same magnitude we could have simplified the computations as follows:

$$\vec{I_b} = -2\left(4.6493 \cdot \cos\left(\frac{\pi}{6}\right)\right) = -8.0528 \quad \text{kiloamperes}$$

$$\vec{I_c} = 8.0528 \quad \text{kiloamperes}$$

From Eq. (3.34) we know that: $\quad 3 \cdot \vec{I_{a1}} = a \cdot \vec{I_b} + a^2 \cdot \vec{I_c}$

$$(-0.5 + 0.866j) \cdot (-8.0528) + (-0.5 - 0.866j) \cdot 8.0528 = -13.9474496j$$

$$\vec{I_{a1}} = \frac{-13.9474496j}{3} = -4.6491j$$

5.8 Double Line-To-Ground Short Circuit on the System Illustrated on Fig. 5-7

The conditions imposed by the line-to-line short circuit in this section are identical to the ones imposed in Sec. 5.3. Furthermore, the results of the analysis in Sec 5.3 are also valid.

$$\vec{V_b} = \vec{V_c} = 0 \qquad \vec{I_a} = 0$$

In Eq. (5.28) we concluded that

$$\vec{V_{a0}} = \vec{V_{a1}} = \vec{V_{a2}} = \frac{\vec{V_a}}{3} \tag{5.59}$$

In Eqns. (5.36) to (5.39) we obtained the following results

$$\overrightarrow{I_{a1}} = \frac{\overrightarrow{V_f}}{\left(Z_1 + \dfrac{Z_0 \cdot Z_2}{Z_0 + Z_2} \right)}$$ Vf replaced Ea   (5.60)

$$\overrightarrow{I_{a2}} = \frac{-\overrightarrow{V_{a2}}}{Z_2}$$ (5.61)

$$\overrightarrow{I_{a0}} = \frac{-\overrightarrow{V_{a0}}}{Z_0}$$ (5.62)

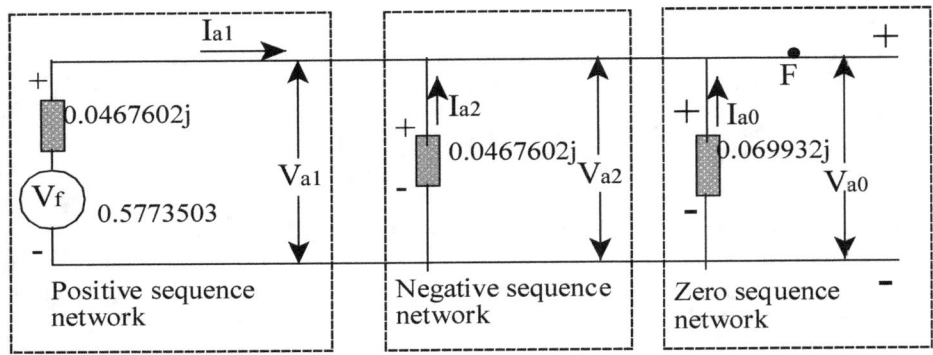

Figure 5-12 Connection diagram of the Thevenin equivalent of the sequence networks.

From Eq. (5.60) we obtain the per unit positive symmetrical component of the short circuit current in line **a**.

$$\overrightarrow{I_{a1}} = \frac{0.5773503}{0.0467602j + \dfrac{(0.069932j) \cdot (0.0467602j)}{0.069932j + 0.0467602j}} \quad \text{simplify} \rightarrow I_{a1} = -7.7203485j$$

pu

$$I_{base} = 0.7530656 \quad \text{Base kiloamperes}$$

$$\overrightarrow{I_{a1}} = -7.7203485j \cdot 0.7530656 = -5.8139j \quad \text{kiloamperes}$$

Equations (3.30 - 3.32) express the line currents as function of operator **a,** and the symmetrical components of phase a. See Eqns. (5.63 - 5.65).

$$\vec{I_a} = I_{a0} + I_{a1} + I_{a2} = 0 \tag{5.63}$$

$$\vec{I_b} = I_{a0} + a^2 \cdot I_{a1} + a \cdot I_{a2} \tag{5.64}$$

$$\vec{I_c} = I_{a0} + a \cdot I_{a1} + a^2 \cdot I_{a2} \tag{5.65}$$

To calculate the short circuit currents in lines b and c we first use Eq. (5.29) to get the symmetrical components of the line voltage in phase a.

$$\begin{pmatrix} V_{a0} \\ V_{a1} \\ V_{a2} \end{pmatrix} = \begin{pmatrix} 0 \\ V_f \\ 0 \end{pmatrix} - \begin{pmatrix} Z_0 & 0 & 0 \\ 0 & Z_1 & 0 \\ 0 & 0 & Z_2 \end{pmatrix} \cdot \begin{pmatrix} I_{a0} \\ I_{a1} \\ I_{a2} \end{pmatrix} \qquad \text{All phasors}$$

$$\vec{V_{a0}} = \overrightarrow{\left(-Z_0 \cdot I_{a0}\right)} \tag{5.66}$$

$$\vec{V_{a1}} = \overrightarrow{\left(V_f - Z_1 \cdot I_{a1}\right)} \tag{5.67}$$

$$\vec{V_{a2}} = \overrightarrow{\left(-Z_2 \cdot I_{a2}\right)} \tag{5.68}$$

$$\vec{I_{a0}} = \frac{-\overrightarrow{V_{a0}}}{Z_0} \tag{5.69}$$

$$\vec{I_{a2}} = \frac{-\overrightarrow{V_{a2}}}{Z_2} \tag{5.70}$$

From Eq. (5.28) we know that

$$\vec{V_{a0}} = \vec{V_{a1}} = \vec{V_{a2}} = \frac{\vec{V_a}}{3} \tag{5.71}$$

Using the data provided in Fig. 5-12 and Eqns. (5.67 and 5.71) we obtain

$$\vec{V_{a2}} = \vec{V_{a1}} = 0.5773503 - \left(0.0467602j \cdot \left(-7.7203485j\right)\right) = 0.2163453$$

$$\vec{V_a} = 3 \cdot 0.2163453 = 0.65 \qquad \text{pu}$$

$$\vec{V}_a = 0.65 \cdot 230 = 149.50 \qquad \text{kV}$$

$$\vec{I}_{a2} = \frac{-V_{a2}}{Z_2} = \frac{-(0.2163453)}{0.0467602j} = 4.6266975j \qquad \text{pu} \tag{5.72}$$

$$\vec{I}_{a2} = 4.6266975j \cdot 0.7530656 = 3.4842067j \qquad \text{kiloamperes}$$

From Eq. (5.69) we obtain

$$\vec{I}_{a0} = -\frac{0.2163453}{0.069932j} = 3.0936524j \qquad \text{pu} \tag{5.73}$$

$$\vec{I}_{a0} = 3.0936524j \cdot 0.7530656 = 2.3297232j \qquad \text{kiloamperes}$$

From Eqs. (3.30 to 3.32) we obtain the short circuit currents I_b and I_c

$$\begin{pmatrix} I_a \\ I_b \\ I_c \end{pmatrix} = \begin{pmatrix} 1 & 1 & 1 \\ 1 & a^2 & a \\ 1 & a & a^2 \end{pmatrix} \cdot \begin{pmatrix} I_{a0} \\ I_{a1} \\ I_{a2} \end{pmatrix} \qquad \text{All phasors}$$

$$\vec{I}_a = \overline{(I_{a0} + I_{a1} + I_{a2})} = 0$$

$$\vec{I}_b = \overline{\left(I_{a0} + a^2 \cdot I_{a1} + a \cdot I_{a2}\right)}$$

$$\vec{I}_c = \overline{\left(I_{a0} + a \cdot I_{a1} + a^2 \cdot I_{a2}\right)} \qquad \varepsilon := 2.7183 \qquad a := \varepsilon^{j \cdot \frac{2\pi}{3}}$$

$$\vec{I}_n = \overline{(I_b + I_c)} \quad \text{Or} \qquad \vec{I}_n = 3 \cdot \vec{I}_{a0}$$

$$3.0936524j + (-7.7203485) + 4.6266975j = 0.0000014j \qquad \vec{I}_a = 0$$

$$I_b = 3.0936524j + (-0.5 - 0.866j) \cdot (-7.7203485) + (-0.5 + 0.866j) \cdot (4.6266975)$$

$$\vec{I}_b = -10.69 + 4.64 \qquad \text{pu}$$

$$\vec{I}_b = (-10.69 + 4.64j) \cdot 0.7530656 = -8.05 + 3.49 \qquad \text{kiloamperes}$$

$$|-8.05 + 3.49j| = 8.774 \qquad \text{kA}$$

$$\vec{I}_c = 3.0936524j + (-0.5 + 0.866j) \cdot (-7.7203485) + (-0.5 - 0.866j) \cdot 4.6266975$$

$$\vec{I}_c = 10.69 + 4.64 \qquad \text{pu}$$

$$\vec{I}_c = (10.69 + 4.64j) \cdot 0.7530656 = 8.05 + 3.49 \qquad \text{kiloamperes}$$

$$|8.05 + 3.49j| = 8.774 \qquad \text{kA}$$

Table 5-3 provides a comparison between the severity of different types of short circuits at point F in Fig. 5-7. Some line to neutral voltages were not provided, because during short circuit conditions the system is saturated and non-linear. And the wave shape of the affected voltages are non-sinusoidal and contains time-variable DC bias or offsets. Sequentially the voltage produced by one of the involved generator phases becomes increasingly larger than the other. This instantaneous voltage difference between the two phases drives the short circuit current.

TYPES OF SHORT CIRCUIT AT POINT F			
	Line to ground	Line to line	Double line to ground
I_a	-7.980j kA	0	0
I_b	0	-8.0526 kA	8.774 kA
I_c	0	8.0526 kA	8.774 kA
V_a	0	132.79 kV	149.50 kV
V_b	....	....	0
V_c	....	....	0

Table 5-3 Short circuit currents for three different types of short circuits

Chapter 6

Large AC Generators

Design Elements

6-1 Introduction

AC electrical generators are based in electromagnetic induction by which mechanical energy is converted into AC electrical energy. The mechanical energy is usually delivered to the generator's rotor by turbines driven by steam, compressed hot air or water. The steam is produced in coal, oil, or nuclear fired boilers. In the case of combustion turbines the compressed air is heated up in gas or oil burners (combustor) before delivering it to the turbine. In the case of water driven turbines the required energized water is obtained from natural water falls or manmade dams and water ways (penstocks).

The AC electrical energy is produced by synchronous generators; a typical one of them consists of:

- A magnetic field winding mounted in the generator's rotor and carrying a direct current delivered to the rotating field winding via slip rings and stationary carbon brushes.
- A three-phase electrical winding mounted inside slots in the generator's stator or armature.

Concisely, a synchronous generator consists of a pedestal supported rotor inside a copper and steel cylindrical cage. The generated electrical power is delivered at the generators output terminals that usually consists of three hot-terminals and a neutral terminal. The efficiency of large power AC generators and their driving turbines increases with the generator rating, the target efficiency is around 99% for very large generators in the order of 1200 or 1500 MW. The main limitation for this very large machines is the required length of the rotor. Because introduces complicated manufacturing problems.

Efficiency is the ratio of electrical power output of the AC generator to the mechanical power input plus the power expend in auxiliary devices, such as rectifiers, deionized water and lubrication systems, plus the power wasted in current conductors and steel laminations. Is customary to express the efficiency ratio as a percentage. The American National Standard Institute (ANSI) and the National Electrical Manufacturers Association (NEMA) as well as the International Electro-technical Commission (IEC) provide standards for calculating AC generators efficiency. These standards include indirect methods of efficiency measuring. Instead of measuring the power in and out of the generator, it add all the losses computed following specific and separated tests. These losses are classified into two categories: fixed losses which are independent of load, and variable losses which are load dependent.

Fixed losses: Friction & Windage losses, Core losses, Lubrication losses.

Variable losses: Copper losses, Stray load losses (eddy currents and harmonics), Exciter losses,

Deionizer losses.

With some isolated exceptions the electrical frequency used in United States is 60 cycles per second (60 hertz). The frequency delivered by any generator is a function of the number of magnetic poles and its rotor's revolutions per second. Symbolically:

$$f = \frac{P \cdot (RPM)}{2 \times 60} \frac{cycle}{sec}$$

(6.1)

P = number of magnetic poles

f = frequency in cycles per second

RPM = revolutions per minute

$$\frac{P}{2} = \frac{60 \times f}{(RPM)} \qquad (6.2)$$

For any specific case the frequency is always known, therefore according with Eq. (6.2) the number of magnetic poles depend of the RPM's that the rotor's driver could provide. So the driving method determines the number of poles. Steam or combustion turbines could provide high RPM and consequently the number of poles would be relatively low, somewhere from 2 to 8. In contrast water driven turbines provide only relatively low RPM to the generator's rotor and therefore for the generator to generate 60 hertz power, the number of magnetic poles required in the rotor is high, as high as 120. This high number of poles requires a rotor with very large diameter. Applying Eq. (6.2) we obtain the necessary number of pair of poles to generate an output voltage of 60 cycles per second.

$$\frac{P}{2} = \frac{3600}{RPM} \qquad (6.3)$$

If the rotor driving machine is a high speed turbine the number of pair of poles is:

RPM = 3600   3600/3600 = 1 PAIR

RPM = 1800   3600/1800 = 2 PAIRS

RPM = 1200   3600/1200 = 3 PAIRS

RPM = 900   3600/900 = 4 PAIRS

If the rotor driving machine is a low speed "water wheel" then the number of pair of poles is:

RPM = 600   3600/600 = 6 PAIRS

RPM = 400   3600/400 = 9 PAIRS

RPM = 200   3600/200 = 18 PAIRS

RPM = 100   3600/100 = 36 PAIRS

RPM = 90   3600/90 = 40 PAIRS

RPM = 80   3600/80 = 45 PAIRS

RPM = 60   3600/60 = 60 PAIRS

Coal-fired steam turbines typically spin at 3600 RPM because they use superheated steam at very high pressure and temperature and carrying high value enthalpy (heat content) in the order of 1522 BTU per lb at 400 Lb/Sq In. and 1000 degrees Fahrenheit. Coal-fired steam turbine typically drive two-pole generators, whose rotor magnetic field has one winding and two poles. Their stator also contains two poles three-phase distributed windings.

Combustion turbines typically drive two poles generators, but sometimes they spin so fast that they need to use a gear reduction before connecting to the generator's rotor. In conclusion most combustion turbines spin at 3600 RPM

6.2 Generators Preliminary Size

Combustion turbines are quickly becoming the favorite machine to generate power in industrial and utilities applications. The configuration and size of the rotor depends of the generator' manufacture experience and each one of them, in the entire world there are only a few manufactures of large AC generators, has its own secret procedure and computation method, to quickly estimate the dimensions of the rotor core. So the procedure given below for a two pole, cylindrical rotor generator is only tentative and do not contemplate many factors involved in the manufacturing process. Manufactures sort their accumulated data per MW rating, RPM, number of magnetic poles, type of rotor (cylindrical or salient,) electrical and mechanical losses and type of cooling system.

Let us consider the following data to estimate the rotor size

Generator rating = 500MW   P = 2   3600 RPM   26kV

Losses = 0.015 p.u. Electrical and mechanical losses

The megawatts of power losses to be removed are 500x0.015 = 7.5 MW

In general, buyers request a compact design. This demands a deionized water cooling system capable of maintaining the water conductivity at smaller than 2x10-7 mhos and designed to remove 5.5 MW of power losses. Plus a 98% pure hydrogen at 45 psig designed to remove 2 MW of power losses from the insulation wrapping the conductors and covering the insides of the slots. The insulation temperature must be maintained below 155 degree C.

The armature or stator is a cylindrical cage that surrounds the rotor cylindrical core and contains the windings where the AC emf are generated. The stator's windings carry the load currents and is the place in which most of the losses are generated. The winding's conductors consists of long cooper bars that are wrapped with insulation and placed into longitudinal slots running in the inside diameter the stator core. These conductors are held in place by wedges that slide into the top of each slot. The magnetic core of the stator is made of grain oriented silicon steel laminations which are electrically isolated from each other to reduce core losses.

Besides, the laminations are staked and separated in groups to allow force ventilation of the stator's core.

Most generators with cylindrical rotors are mounted inside a single, on-site fabricated enclosure which is tested for vibrations at all frequencies of concern. The enclosure contains: the stator or armature, the rotor's pedestals, the rotor's shaft complete with slip rings and brushes or anything else associated with the excitation system. Besides it contains oil and hydrogen pipes, electrical conduits, lightning arresters, surge capacitors. This enclosure covers the entire generator. For a preliminary design the manufacture experience, indicates the following:

Rotor core length = 2 feet per megawatt of total power losses.

2 x 7.5 = 15 feet   **Estimated rotor core length = 4.572 meter**

15 + 12 = 27 feet **Estimated total rotor length = 8.23 meter**

3600/60 = 60 rev/sec   1/60 = 0.0167 sec/rev

7.5 x 0.0167 = 0.12525 MWsec/rev   Generator total energy loss per revolution

5.5 x 0.0167 = 0.09185 MWsec/rev   Energy per revolution to be removed by deionized water cooling system.

2 x 0.0167 = 0.0334 MWsec/rev   Energy to be removed by hydrogen cooling system per revolution.

The next important dimension is the diameter of the rotor core. In order not to exceed the steel elastic limit due to the centrifugal forces acting on the rotor core, the diameter of the core cannot be larger than 1.2 meter for 3600 RPM machines. This will avoid causing irreversible plastic deformation to the rotor core.

Definitions:

D =   diameter of rotor core, m

L =   rotor core length, m

RPS = revolutions per second

ξ =   generator volumetric efficiency, MVA / m3 x RPS

The generator volume depends of the rotor total length and core diameter, and because the radial dimension of the air gap between the stator and the rotor is insignificant compared to the rotor core diameter. The generator volumetric efficiency is directly related to the generator MVA output per rotor core unit volume, per rotor revolution. Symbolically:

$$\xi = \frac{MVA}{\frac{\pi \cdot D^2}{4} \cdot L \cdot RPS}$$

(6.4)

From Eq. (6.4) it is easy to conclude that the greater is ξ, the more efficient is the generator. Because it generates more MVA per revolution, per unit of rotor core volume. Now days there is emphasis in compact design, especially for combustion turbines.

Substituting in Eq. (6.4):

$D := 1.2\,m$ or $D = 3.94\,ft$ $L = 4.572$ meter

$\xi = 4 \times 500/3.14 \times 1.44 \times 4.572 \times 60 = 1.61$ MVA / m3 x RPS

The value obtained for volumetric efficiency, given that the diameter of the rotor core was defined to be 1.2 meter, is $\xi = 1.61$ MVA per cubic meter per revolution per second. The larger the core diameter the more space is for the stator windings and also improves the generator cooling.

6-3 Cylindrical Rotors and DC Field Windings

The rotor and field windings are submitted to a uniform circular motion in which the vector representing the rotor velocity at any point is constant in magnitude but is constantly changing in direction. Symbolically:

$v = \omega \times r$ $F = m \cdot a$ $|a| = \frac{v^2}{r}$

$$|F_{cen}| = \frac{m \times v^2}{r}$$ Centrifugal force magnitude

$$|F_{cen}| = \frac{m \times \omega^2 \times r^2}{r} = m \times \omega^2 \times r$$

(6.5)

The magnitude of the centrifugal force acting on any part of the rotor's surface is given by Eq. (6.5) and is proportional to the mass of the rotor's shaft and core, to the square of its angular velocity and the distant to the axis of rotation. The generator's designer needs to live with the centrifugal force magnitude provided by Eq. (6.5). Because the following reasons:

If the centrifugal force magnitude is larger than the value provided by Eq. (6.5) the generator would be severally damage or destroyed. He could reach this condition by selecting a rotor radius larger than 0.6 meter.

The inside diameter of the stator cannot be larger than 1.21 meter, approximately. Actually, the air gap between the stator and rotor is uniform and very small, in the order of less than one centimeter. With distributed windings, both the inside and outside diameters of the stator are determined by the number of conductors per pole. And the number of poles required is fixed by the RPM delivered by the turbine. *Nothing is more important than the data accumulated by the generator's manufacturer which is a decisive factor in the generator's physical design.*

The angular velocity, ω, is determined by the RPM that the rotor's driver could provide. It is possible, but not easy, to vary or adjust the RPM by changing the heat content (enthalpy) of the steam supplied to the turbine. Is even harder in combustion turbines. It would be necessary to change the amount and temperature of the fuel injected into the combustor, but the danger here is to increase the temperature of the hot air so much that would melt the surrounded metal.

In conclusion, the rotor assembly must be a high quality piece of equipment capable of resisting without harmful effects the centrifugal forces created by its high rate of rotation and the stresses produced by thermal expansion. A typical rotor assembly consists of a single piece of forged steel, cylindrical shaped and having a middle core where the field winding is installed in rectangular radial slots running length wise along the core.

The rotor is mounted on pedestals fitted with tilting pad bearings at both ends. Each pole winding consists of insulated copper bars that are inserted into insulated slots and then immobilized by metal wedges that slide into the top of the slots. The reader must realize that each turn of the coil (very few) must be insulated from the next turn. Distributed field windings are preferred because they reduce the harmonics content of the voltage they induced in the stator windings. At both ends of the rotor cylindrical core the windings are held in place by retaining rings. These retaining rings must be corrosion resistance and nonmagnetic. They are installed by shrinking them onto the rotor body and securing them in place with snap rings.

6.4 Salient Pole Rotors

The study of rotating rigid bodies that are symmetrical about a fixed rotational axis requires the application of the mechanical formulas listed in Table 6-1, which also includes the corresponding formulas for similar concepts in linear motion.

Linear motion: $F = m \cdot \dfrac{d^2}{dt^2} S$ Mass times linear acceleration

Rotational motion: $T = I \cdot \dfrac{d^2}{dt^2} \theta$ Moment of inertia times angular acceleration.

The rotor of hydroelectric generators are driven by water driven turbines which can only provide relatively low RPM, so for the generator to deliver the required 60 hertz power its rotor must have a large number of poles, anywhere from 12 to 120 poles. This type of rotor are called *Salient Pole Rotors* and are characterized by their large diameter because of the large space required to mount the large number of poles required and that are mounted on a large circular steel frame fixed to the rotor and where adjacent poles are of opposite polarities. Due to its large mass, moment of inertia, and angular velocity the rotational inertia of this type of rotor is very large, see Eq. (6.8) which contribute to maintain the frequency of the power output constant. Symbolically:

$$KE = \frac{I \cdot \omega^2}{2} \tag{6.6}$$

$$M = I \cdot \omega \tag{6.7}$$

$$KE = \frac{M \cdot \omega}{2} \qquad \text{Generators stored rotational energy in megawatt-sec} \tag{6.8}$$

KE = rotational kinetic energy, megajoules or megawatt-sec

I = moment of inertia, megajoules x sec2 / rad2

ω = angular velocity, rad / sec

M = angular momentum, megajoules x sec / rad   or   megawatt x sec2 /rad

Damper winding are often installed in the rotor pole faces. Each winding consist of one or two turns forming a closed loop. In steady state condition the rotor is rotating at constant synchronous speed, ωs, and the damper windings carry not current. However, when sudden changes in the load produces fluctuations on the angular speed of the rotor around the nominal synchronous speed, then the magnetic flux linking the damper winding changes and generates an emf that in accordance with the Lenz law creates a large circulating current in the damper winding of such direction that the mmf it produces opposes the cause that created the oscillations of the angular speed. In the magnetic circuit, the interaction between the rotor field and the stator field reaction, produced by the oscillating current circulating in the stator conductors, results in an oscillating net magnetic field that creates the rotor torque that damps the oscillation. Symbolically:

$$\vec{B}_{net} = \vec{B}_R + \vec{B}_S \qquad \vec{B}_R = \vec{B}_{net} - \vec{B}_S \qquad \vec{B}_S = \vec{B}_{net} - \vec{B}_R$$

The mmf created in the damper winding affect the flux density created by the stator in such direction that the resulting torque induced in the rotor changes in the direction that reduces the amplitude of the oscillations. The torque induced in the rotor is proportional to the vector product of the vectorised flux densities of the rotor and stator. Symbolically:

$$\overrightarrow{T_{ind}} = k \cdot \left(\overrightarrow{B_R} \times \overrightarrow{B_S} \right) = k \cdot \left[\overrightarrow{B_R} \times \left(\overrightarrow{B_{net}} - \overrightarrow{B_R} \right) \right]$$

$$\overrightarrow{T_{ind}} = k \cdot \left(\overrightarrow{B_R} \times B_{net} \right) - k \cdot \left(\overrightarrow{B_R} \times \overrightarrow{B_R} \right) \qquad \left(\overrightarrow{B_R} \times \overrightarrow{B_R} \right) = 0$$

$$T_{ind} = k \cdot \left(\overrightarrow{B_R} \times \overrightarrow{B_{net}} \right) \qquad \text{k = proportionality constant} \qquad (6.9)$$

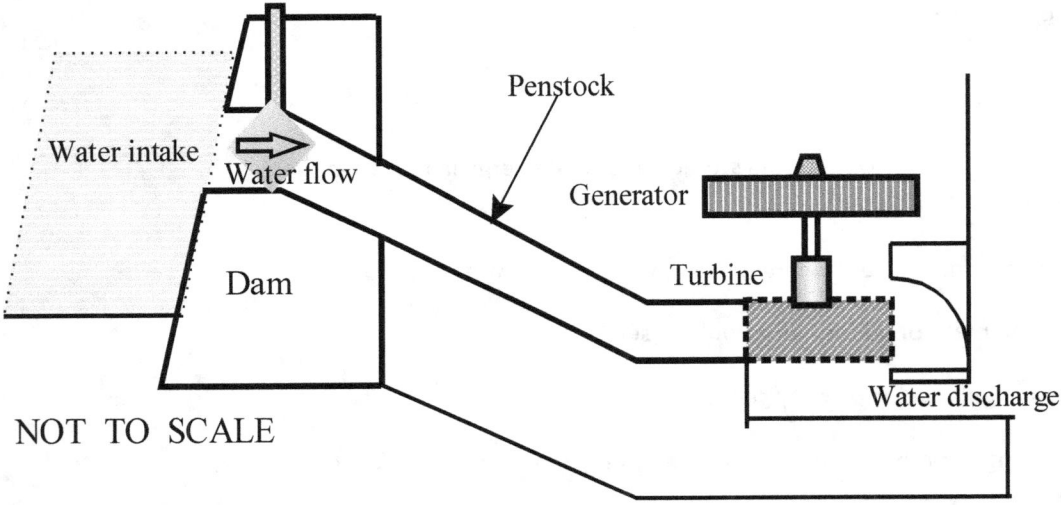

Figure 6-1 Illustration of a hydro-electric plant containing water intake, dam, penstock, turbine, generator, and water discharge.

Generators inertial constant, H, is defined as the stored rotational energy of the generator at synchronous speed in megawatt-sec per unit of generator rating in megavolt-amperes (MVA). Applying Eq. (6.8) we obtain:

$$H = \frac{M \cdot \omega_s}{2 \cdot G} \qquad \text{G is the generator rating in MVA} \qquad (6.10)$$

$$G \cdot H = \frac{M \cdot \omega_s}{2} \qquad \text{Stored rotational energy at synchronous speed, megawatt seconds}$$

$$(6.11)$$

This energy plus the mechanical and electrical losses must be supplied by the water turbine to the rotor. Let us now find the amount of energy that the water dam / penstock system can deliver to a turbine when the dam water level is at the design (normal) level. See Fig. 6-1.

The Bernoulli equation, see Eq. (6.12), provides the total energy contained in the fluid at any point of the water dam system. Symbolically,

$$E = \frac{w \cdot p}{\gamma} + w \cdot z + \frac{w \cdot v^2}{2 \cdot g} \tag{6.12}$$

Potential energy term = wz due to elevation

Flow energy term   = $\dfrac{w \cdot p}{\gamma}$ due to pressure w/γ = volume   work = pV

Kinetic energy term = $\dfrac{w \cdot v^2}{2 \cdot g}$ due to velocity

The unit of each term of Eq. (6.12) has the dimension of length, which in the English unit system is feet. This is the reason why the terms of the Bernoulli's equation usually are, in our trade, designated as "head". However, because each term represent some kind of energy is better to refer them in _foot-pounds per pound_ which is the same than feet, but it makes clear that each term represent a portion of the energy contained in the fluid per pound of fluid flowing from the lake to the turbine. Let us consider two points of the water system. If we neglect all fluid frictional losses and assuming that no energy is added or lost between points 1and 2, then the total energy at point 1 must be equal to the total energy at point 2. Applying Eq. (6.12) to points 1 and 2 we arrive at Eq. (6.13).

$$\frac{w \cdot p_1}{\gamma} + w \cdot z_1 + \frac{w \cdot v_1^2}{2 \cdot g} = \frac{w \cdot p_2}{\gamma} + w \cdot z_2 + \frac{w \cdot v_2^2}{2 \cdot g} \tag{6.13}$$

Dividing by w in both sides of Eq. (6.13) we obtain Eq. (6.14) which is the most useful form of the Bernoulli equation. Symbolically,

$$\frac{p_1}{\gamma} + z_1 + \frac{v_1^2}{2 \cdot g} = \frac{p_2}{\gamma} + z_2 + \frac{v_2^2}{2 \cdot g} \tag{6.14}$$

Rigorously, Eq. (6.12) is only valid for incompressible fluids, such as water, because the specific weight, or weight per unit volume, of the water was assumed to be the same at the two points. Let us now find the total amount of energy delivered by the water flow to the turbine. For this we apply Bernoulli's equation at two points of our choice.

1--at the surface of the lake near the dam wall.

2-- at the water intake of the turbine.

At the water surface close to the wall p = 0, and the water velocity is near zero and could be neglected. So applying Eq. (6.14) to a point near the wall the only term left in the left side is z1.

γ = specific weight, 62.4 lb / ft3 at 4 degree C

g = gravity acceleration, 32.2 ft. / sec2

z1 = normal elevation of the water surface, 200 ft.

z2 = elevation at turbine intake, 10 ft.   Head to discharge the water down river.

p1= pressure at the water surface, 0

Symbolically,

$$z_1 = \frac{p_2}{\gamma} + z_2 + \frac{v_2^2}{2 \cdot g} \qquad\qquad z_1 - z_2 = \frac{p_2}{\gamma} + \frac{v_2^2}{2 \cdot g}$$

Total head available at the turbine intake disregarding all frictional losses in the water way, such as: water entrance to the penstock, in pipes and in the water exit at turbine nozzles.

$$z_1 - z_2 = 190 \quad \text{ft-lb/lb}$$

If the designed water flow rate is 5000 gallons per second.

$$5000 \cdot 0.1337 = 668.5 \quad \frac{ft^3}{sec} \qquad\qquad 668.5 \cdot 62.4 = 4.17144 \times 10^4 \qquad\qquad \frac{ft^3}{sec} \cdot \frac{lb}{ft^3}$$

41714.4 lb/sec   Weight flow rate

$$190 \cdot 41714.4 = 7.926 \times 10^6 \quad \frac{ft - lb}{sec} \qquad\qquad \frac{ft - lb}{lb} \cdot \frac{lb}{sec}$$
1 kilowatts = 737.8 ft-lb / sec

$$\frac{7.926 \cdot 10^6}{737.8} = 10.74 \times 10^3 \qquad \text{kilowatts} \quad \text{or} \quad 10.74 \text{ megawatts}$$

Let us say that the generator driven by the turbine is a 10 MW salient pole AC generator with vertical rotor and 40 pair of poles that must rotate at 1.5 rev / sec to deliver 60 hertz power. The directly coupled water turbine rotates with a mechanical angular velocity of 9.425 rad/sec.

$$\omega_m = 1.5 \cdot 2\pi = 9.425 \ \frac{\text{rad}}{\text{sec}} \qquad \frac{\text{rev}}{\text{sec}} \cdot \frac{\text{rad}}{\text{cycle}} \cdot \frac{\text{cycle}}{\text{rev}}$$

$$\omega_s = 1.5 \cdot 40 \cdot 2 \cdot \pi = 377 \qquad \text{Electrical radians per sec.}$$

The generator inertial constant H is the stored rotational energy of the generator at synchronous speed in megajoules per unit of generator rating in megavolt-amperes (MVA).

H = stored rotational energy, megajoules / generator MVA rating

1 megajoules = 1 megawatt-second

So H could also be expressed as

H = stored rotational energy, megawatt-second / generator MVA rating. Symbolically,

$$H = \frac{M \cdot \omega_s}{2 \cdot G} \qquad \text{G is the generator rating, MVA} \qquad (6.15)$$

M is the angular momentum in megajoule-second per radian

$$G \cdot H = \frac{M \cdot \omega_s}{2} \qquad \text{megawatt} - \text{seconds}$$

$$(6.16)$$

GH = Generator stored rotational energy at synchronous speed, megawatt-seconds   (6.17)

The inertial constant H of synchronous AC generators has a small range of values for each type of generator independent of its MVA and angular speed ratings. In fact, for hydroelectric salient pole AC generators the value of H ranges from 2 to 4. In addition, H can be calculated provided that the weight and radius of gyration of all rotating parts of the generator and prime mover are known.

6.5 Stator Winding Conductors

The working of an AC generator could be summarized with the following initial statement. The magnetic flux density in the stator depends of the amount of DC current flowing in the rotor field winding. The larger is the magnetic flux density in the stator, the greater are the emfs induced in the stator three-phase windings and the smaller are the AC currents flowing in the winding conductors for a given load. So the voltage induced and the current circulating in the stator windings can be controlled by varying the DC current flowing in the rotor winding. With very few exceptions, if any, the stator windings are connected in wye and the neutral is connected to ground, directly or through a grounding resistor, reactance, or impedance. Stator windings are stationary and therefore they could be directly connected to the load. And unlike the rotor windings their insulation is not submitted to centrifugal forces. In a wye connected stator three-phase winding the voltage per winding phase is 58% of the line to line voltage. In contrast a delta winding would see the entire line to line voltage. Therefore the thickness of the insulation required for a wye connected three phase winding is smaller than for a delta connected winding. The saved insulation space could be occupied by a larger cross section conductor which would result in smaller copper losses and hence will reduce the cooling required. Figure 6-2 Illustrates the stator windings of a wye connected three-phase generator and the line voltages at the output terminals.

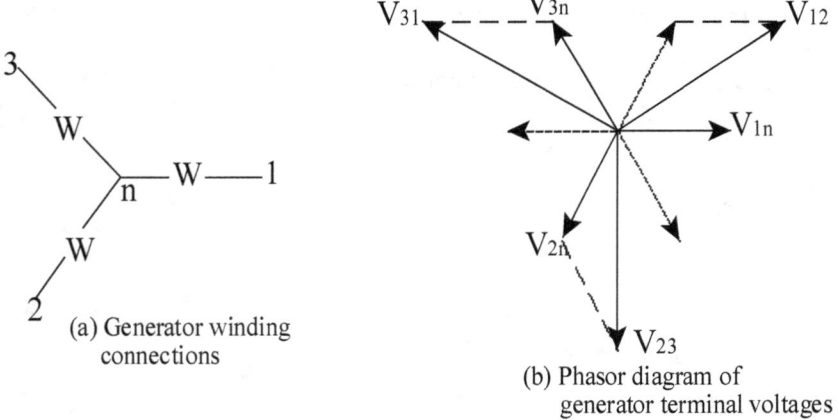

(a) Generator winding connections

(b) Phasor diagram of generator terminal voltages

Figure 6-2 Generator winding connections and terminal voltages

$$\overrightarrow{V_{12}} = \overrightarrow{V_{1n}} + \overrightarrow{V_{n2}} = \overrightarrow{V_{1n}} - \overrightarrow{V_{2n}}$$ (6.18)

$$\overrightarrow{V_{23}} = \overrightarrow{V_{2n}} + \overrightarrow{V_{n3}} = \overrightarrow{V_{2n}} - \overrightarrow{V_{3n}}$$

$$\overrightarrow{V_{31}} = \overrightarrow{V_{3n}} + \overrightarrow{V_{n1}} = \overrightarrow{V_{3n}} - \overrightarrow{V_{1n}} \qquad\qquad \varepsilon := 2.7183 \qquad j := \sqrt{-1}$$

The generator line to neutral voltages are:

$$\overrightarrow{V_{1n}} = V \cdot \varepsilon^{j \cdot 0} \qquad\qquad\qquad \overrightarrow{V_{1n}} := V \cdot (\cos(0) + j \cdot \sin(0))$$

$$\overrightarrow{V_{2n}} = V \cdot \varepsilon^{-j \cdot 120^0} \qquad \overrightarrow{V_{2n}} := V \cdot \left(\cos\left(\frac{2 \cdot \pi}{3}\right) - j \cdot \sin\left(\frac{2 \cdot \pi}{3}\right) \right) = (-0.5 - 0.866j) \ V$$

$$\overrightarrow{V_{3n}} = V \cdot \varepsilon^{j \cdot 120^0} \qquad \overrightarrow{V_{3n}} := V \cdot \left(\cos\left(\frac{2 \cdot \pi}{3}\right) + j \cdot \sin\left(\frac{2 \cdot \pi}{3}\right) \right) = (-0.5 + 0.866j) \ V$$

The generator line to line terminal voltages are:

$$\overrightarrow{V_{12}} = V \cdot (\cos(0) + j \cdot \sin(0)) - (-0.5 - 0.866j) \ V = (1 + 0 \cdot j) \cdot V - (-0.5 - 0.866j) \ V$$

$$\overrightarrow{V_{12}} = V + 0.5 \cdot V + 0.866 \cdot j \cdot V = 1.5 \cdot V + j \cdot 0.866 \cdot V = (1.5 + 0.866 \cdot j) \cdot V \qquad (6.19)$$

$$|(1.5 + 0.866 \cdot j)| = 2 \qquad\qquad \text{atan}(0.866/1.5) = 0.523561 \text{ rad} = 30 \text{ degrees}$$

$$\overrightarrow{V_{12}} = \sqrt{3} \cdot V \cdot \varepsilon^{j \cdot 30^0} \qquad\qquad\qquad\qquad\qquad\qquad\qquad (6.20)$$

$$\overrightarrow{V_{23}} = \sqrt{3} \cdot V \cdot \varepsilon^{-90^0 \cdot j} \qquad\qquad \overrightarrow{V_{31}} = \sqrt{3} \cdot V \cdot \varepsilon^{j \cdot 150^0} \qquad\qquad (6.21)$$

Waves formed by symmetrical positive and negative loops do not contain *even* harmonics. Equations (6.18) express that the line to line voltage between any pair of output terminals is equal to phasor difference of corresponding neutral voltages. And because the third harmonic in all the phases are in phase (same argument) they cancel each other in Eqns. (6.18). The same is true for all the multiples of the third (9, 15, 21...). However, the third harmonic and its multiples are present in the line to neutral voltages. Therefore if the stator windings were connected in delta, which never are, the third harmonic in all the three windings will add up and produce a large circulating current within the delta that would produce large copper losses. In wave analysis Eq. (6.18) is applied, separately, to each harmonic present in the original wave. The stator coils are usually distributed windings which occupied several slots per pole per phase. The load current, which usually is very large, flows in every conductor of these distributed windings. However, in anyone of the phases the voltage generated in conductors situated at different slots are not in phase. For the fundamental harmonic the difference in phase is equal to the angle, in electrical degrees, between adjacent slots. For any other harmonic the out of phase angle is the corresponding multiple of the fundamental difference. For instant for the seventh harmonic the difference in phase is seven times the difference for the fundamentals. Usually the stator coils conductors are copper bars of the required cross section. One very important guide line in the design and manufacturing of AC generators is to maintain the stator copper losses (RI2) as low as possible, and because the copper bars used as winding conductors have very low resistance, the main emphasis is to reduce the load current, to do this, the generator output voltage is selected as high as practically possible having all the other factors in account. Higher output voltage means higher line to neutral voltage and thicker insulation, which requires larger slots. Because, there is a limitation in how high the generator output voltage could be, the output current and therefore the copper losses of large generators are very large. At the present time, 2013, the design of large and very large AC generators in the range of 300 to 1500 MW the solution is to use de-ionized water or high pressure hydrogen cooling systems or even a combination of both. In fact, the stator copper bars perform a dual service as current conductors and heat sink at the same time. The deionized water cooling system is by far the best, twelve times better than hydrogen at high pressure, 45 psig. Fig. (6-3) is a diagram of the network of hollow copper conductors, carrying de-ionized water, and illustrating how they are water-interconnected as part of the generator cooling system. Both sides of each coil winding are independently cooled. The deionized water enters the copper bars at the electrical connections side. And exit the hollow copper bars at the turbine side of the stator. See Fig. 6-3. Is beyond the scope of this book to describe the entire deionized water cooling system. However, the main components found in practically every deionized water cooling system are:

-- Deionized water storage tank

-- Cooling water pumps

-- At least two temperature compensated conductivity meters

-- Mixed bed deionizer tank

-- Water flow measuring devices or orifices

-- Heat exchangers

-- Filters

-- Temperature control valves

-- Teflon hoses as required

Deionized water must circulate with turbulent flow (NR > 4000) inside a hole along the center of each copper bar and must maintain a resistivity larger than 2000000 ohm-meters or a conductivity smaller than 0.0000005 mho-meters. The simple example given below would provide an idea of the expected leakage current via cooling water.

Example 6-1. Calculate the stator leakage current via the cooling water system. In reality the cooling water forms a very high impedance wye connected in parallel with the very low impedance copper wye. Assuming zero inductance in the water branches and disregarding the skin and proximity effects because deionized water is not magnetic, we have:

$$R = \rho \cdot \frac{l}{A}$$ Resistance of cooling water per phase

ρ = resistivity of deionized water, ohm-m $\rho := 2 \cdot 10^6$

l = half a coil length by water, m $l := 6$

A = cross sectional area of hole, m2 $A := 1.267 \cdot 10^{-4}$

We are assuming only one copper bar, with one round hole, 1.267 cm in diameter in the center of the bar cross section per pole, per phase. There are two conductor (1-coil) per phase.

Generated **line to line** voltage = 26,000 volts   Line to neutral voltage = 15011 volts

Cooling water resistance per phase in ohms $2 \cdot 10^6 \cdot \dfrac{6 \cdot 2}{1.267 \cdot 10^{-4}} = 189.424 \times 10^9$

Leakage current per phase in amperes $\dfrac{15011}{189.424 \cdot 10^9} = 79.246 \times 10^{-9}$

$3 \cdot 79.246 \times 10^{-9} = 237.738 \times 10^{-9}$ Stator leakage current in amperes

Example 6-2 Calculate the energy lost in a standard annealed 8x3 cm copper bar with a hole in the middle of the cross section running along the full length of the bar which is 100 cm long. When the 60 Hz current flowing in the bar is 1000 amperes.

Data: Operating temperature = 155 °C   Hole diameter = 1.27 cm

$\rho := 1.724 \cdot 10^{-6}$ Ohm − cm   Copper resistivity at 20 °C

$\alpha := 0.0039$ $\dfrac{1}{°C}$ Temperature Coefficient of Resistance at 20 °C

$\dfrac{\pi \cdot 1.27^2}{4} = 1.267$ cm^2 Hole area

$(8 \cdot 3) - 1.267 = 23$ cm^2 Bar net cross section   Slot = 12 cm deep

Copper losses are proportional to the length of the conductor. The DC resistance of the cooper bar at $20\,°C$ is:

$$R_1 = 1.724 \cdot 10^{-6} \cdot \frac{100}{22.733} = 7.584 \times 10^{-6} \qquad \text{ohms}$$

$$R_{155} = R_1 \left(1 + \alpha \left(t_2 - t_1\right)\right)$$

$$7.584 \times 10^{-6}[1 + 0.00393 \cdot (155 - 20)] = 11.608 \times 10^{-6} \quad \text{ohms}$$

Copper bar DC resistance at $155\,°C$, 100 cm long and with a cross section of 22.733 cm2.

The ac resistance is increased by the skin and proximity effects. To minimize these effects the copper bars should be tall and narrow and separated as much as possible. Let us assume that the ac resistance is 12% larger than the DC resistance.

$$R_{ac} = 1.12 \cdot 11.608 \cdot 10^{-6} = 13.001 \cdot 10^{-6} \quad \frac{\text{ohms}}{\text{m}}$$

AC resistance at 155 $°C$ per meter

The power loss per thousand amperes, per meter length of the copper bar at $155\,°C$ is:

$$R_{155} \cdot I^2 = 13.001 \cdot 10^{-6} \cdot 1000^2 = 13.001 \quad \frac{\text{watts}}{\text{m}}$$

For a 500 MVA two poles, three-phase generator with 7 conductors per stator pole, 7 meters long each including rear and front ties, the copper losses in the stator distributed windings are:

$$3 \cdot 7 \cdot 2 \cdot 7 = 294 \quad \text{meters} \qquad \text{Total length of conductors in the stator}$$

$$500 \cdot 10^6 = \sqrt{3} \cdot 26000 \cdot I$$

$$I = \frac{500 \cdot 10^6}{\sqrt{3} \cdot 26 \cdot 10^3} = 11103 \quad \text{Amperes}$$

$$13.001 \cdot 10^{-6} \cdot 7 = 91.007 \times 10^{-6} \quad \text{ohms} \qquad \text{AC resistance of each winding conductor}$$

$$91.007 \times 10^{-6} \cdot 11103^2 = 11.219 \times 10^3 \quad \text{watts} \qquad \text{Power loss in each winding conductor}$$

$$294 \cdot \left(13.001 \times 10^{-6}\right) = 3.8223 \times 10^{-3} \quad \text{ohms}$$

Total copper bar AC resistance at $155\,°C$

$$R \cdot I^2 = 3.822 \times 10^{-3} \cdot 11103^2 \qquad 3.822 \times 10^{-3} \cdot 11103^2 = 471.163 \times 10^3 \text{ watts}$$

Copper losses in the stator windings = 471.163 kW

Total number of conductor in the stator 3x2x7=42

Copper losses in each stator winding conductor $\quad \dfrac{471.163}{42} = 11 \quad$ kW

The reader must realize that the alternating magnetic flux existing in the stator magnetic material induce hysteresis losses and eddy current losses. The hysteresis losses are proportional to the frequency (60Hz) and the area within the hysteresis loop. The eddy current losses could be considered negligible because the stator is fabricated with grain oriented steel laminations which make the eddy current losses insignificant when compared to the copper and core losses.

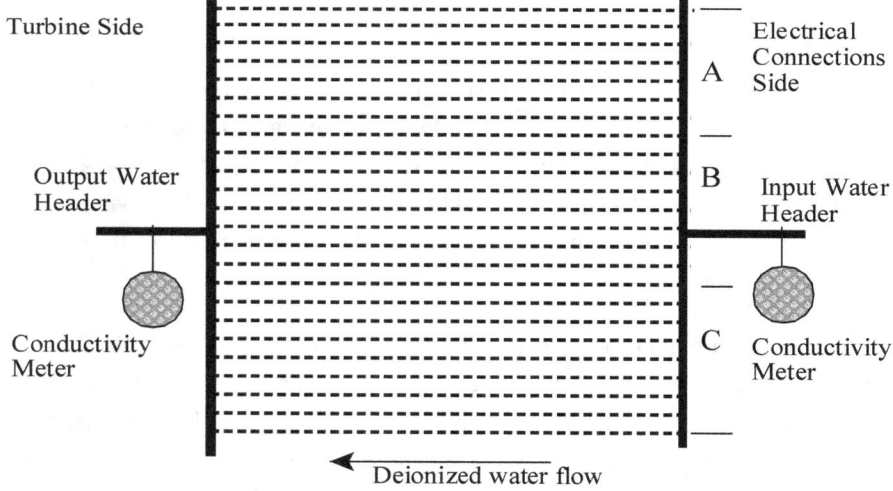

Figure 6-3 Deionized water stator cooling diagram

The heat flow from the winding conductors to the deionized water is a complicated combination of heat conduction and convection processes. Figure 6-4 provides an illustration of the water cooled hollow copper busses.

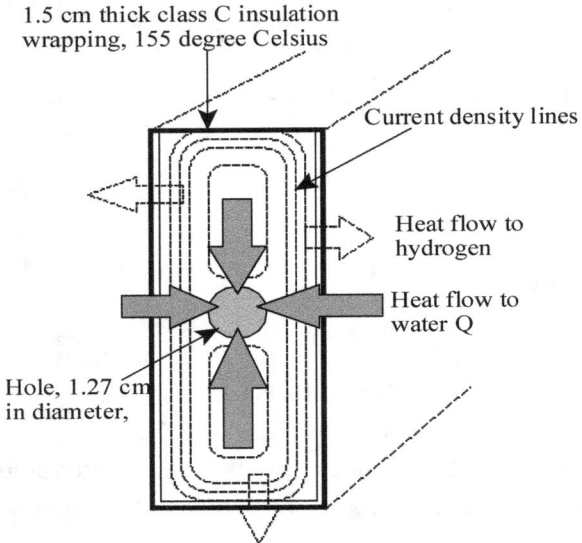

1.5 cm thick class C insulation
wrapping, 155 degree Celsius

Current density lines

Heat flow to
hydrogen

Heat flow to
water Q

Hole, 1.27 cm
in diameter,

Figure 6-4 Illustration of the cross section of a 8x3 cm
insulation wrapped copper bus with a hole in the
middle where the deionizer water flows.

The reader should realize that most of the heat is conducted away from the copper busses by
the deionized water flowing inside the busses, and as indicated in Fig. 6.4 heat generation
occurs throughout the heat conduction material, cooper. Due to the skin effect, the AC current
density distribution is larger toward the periphery of the conductor, hence this is where most
of the heat is generated. Furthermore, the proximity effect also impact the current
distribution in the cross section

of the conductor and makes it even less uniform. The important thing is to maintain the
temperature in the periphery of the conductor below 155 degree Celsius which is the rating
for mica paper based insulation material. Class C insulation. This objective is helped by cooling
the outside of the insulation material with a flow of hydrogen gas at 45 psig. The hydrogen
operating purity should be greater than 98 percent. And most always be maintained above 75
percent purity, which is the explosive limit, to this effect a gas analyzer should continuously
monitor the hydrogen flow. The hydrogen gas must be contained inside a gas tight enclosure
built around the stator and rotor and that unfortunately is penetrated, at both ends by the
driving shaft. The enclosure consists of a cleaver design of shields and rings, and the Seal Oil
system which must prevent escape of the gas through the penetrating openings for the rotor
driving shaft. The sealing oil is carried to the shaft zone between sealing rings via a grove in
the seal housing. The oil flows in both shaft directions and gets into the constricted zones
between the shaft and the sealing rings forming a film that prevent the hydrogen from leaking
along the shaft. To determine how many gpm of deionized water are required to cool the
copper busses, where the heat is generated, we will disregard whatever heat is removed by
the hydrogen gas and use Eq. (6.22). The reader should understand than to determine the

cooling effect of the hydrogen gas it would be required to know the generator design in great detail. Furthermore some heat is carried away from the generator enclosure by conduction via the cables connecting the generator to the rest of the power system.

$$Q = C_v \cdot \Delta T \qquad \text{1 joule = 1 watt-sec} \qquad (6.22)$$

Q = heat transferred, watt-sec

Δ T = change in temperature, degree Kelvin or Celsius

C.v = volumetric heat capacity of water, joule/cm3 x K degree = 4.1796. Heat required to change the temperature of one cm3 of water one degree Kelvin or Celsius $|°K| = |°C|$

From the results of example 6-2

$$13.001 \cdot 10^{-6} \cdot 7 = 91.007 \times 10^{-6} \qquad \text{ohms per conductor}$$

$$91.007 \times 10^{-6} \cdot 11103^2 = 11.219 \times 10^3 \qquad \text{watts-sec per conductor}$$

$$Q := 11.219 \times 10^3 \quad \text{joules}$$

$$\Delta T = T_{out} - T_{in}$$

The incoming water maximum temperature is 40 degree C and the final temperature, to avoid pressurizing the hoses interconnecting the conductors, should not be more than 100 degree C.

Data:

$$D := 1.27 \quad cm \qquad \text{Diameter of the hole inside the copper bar.}$$

$$1 gallon = 3785 \quad cm^3$$

Volumetric heat capacity of liquid water at 25 $°C$ $C_v = 4.1796 \quad \dfrac{watt - sec}{cm^3 \cdot °C}$

Volumetric heat capacity of liquid water at 100 $°C$ $C_v = 4.2160 \quad \dfrac{watt - sec}{cm^3 \cdot °C}$

Volumetric heat capacity of liquid water at 40 $°C$

$$C_V = \left(\frac{4.2160 - 4.1796}{75}\right) \cdot 15 + 4.1796 = 4.1869 \qquad \frac{watt - sec}{cm^3 \cdot °C}$$

The heat transferred from a conductor to the water flowing inside the buss is the same anywhere along the seven meters length of the bus. Actually, you can assume that the water is static during the time it takes to travel the entire length of the bus and then calculate the temperature rise of the water due to the heat removed from the bus.

From Eq. (6.22) we obtain:

$$\Delta T = \frac{Q}{C_V} = \frac{11219}{4.1869} = 2679.5 \qquad cm^3 \cdot °C \qquad \frac{watt - sec}{\left(\dfrac{watt - sec}{cm^3 \cdot °C}\right)}$$

$$W_V = \left(\frac{\pi \cdot D^2}{4}\right) \cdot 700 = 886.738 \qquad cm^3 \qquad \text{Volume of water inside conductor}$$

$$\frac{\Delta T}{W_V} = \frac{2679.5}{886.738} = 3.0217 \qquad \frac{cm^3 \cdot °C}{cm^3} = °C \qquad (6.23)$$

Equation (6.23) gives the cm^3 temperature rise of the water contained inside the bus. Assuming a total volumetric water flow rate of 590.376 gallons per minute for the 42 conductors or 0.2342757 gallons per second per conductor. The water temperature rise could also be calculated as follows:

$$\text{Volumetric water flow rate} = 0.2342757 \qquad \frac{gallons}{sec \cdot conductor}$$

Total number of conductors = 3 x 14 = 42

$$\text{Total water flow rate} = 42 \times 0.2342757 = 9.8396 \qquad \frac{gallons}{sec}$$

$$\text{Total water flow rate per minute} = 9.8396 \times 60 = 590.376 \qquad \frac{gallons}{minutes}$$

$$\psi = A_o \cdot v$$ Continuity equation for steady flow of incompressible fluids

ψ water flow rate, cm^3/sec $A_o := 1.266760$ cm^2

A_o cross section of hole, cm^2

v average velocity of flow, cm / sec

$$\psi = 0.2342757 \cdot 3785 = 886.738 \quad \frac{cm^3}{sec}$$

$$\frac{\psi}{A_o} = 700 \quad \frac{cm}{sec} \qquad v := 700 \quad \frac{cm}{sec} \qquad \text{conductor length} = 700 \text{ cm}$$

Therefore, we can conclude that the water travels from one end to the other end in one second.

$$\frac{\Delta T}{\psi} = \frac{2679.5}{886.738} = 3.0217 \quad °C \cdot sec \qquad \frac{cm^3 \cdot °C}{\frac{cm^3}{sec}} = °C \cdot sec$$

$\Delta T = 3.0217 \, °C$ Water temperature rise across one conductor. Hydraulically, all conductors are in parallel.

The heat transferred from the copper bars to the cooling water is done by a process called forced convective heat transfer. It takes place between the surface of a solid and a moving fluid. Specifically, between the inside surface of the copper bars and the flowing deionized water. See Eq. (6.24). A thin film of slowly moving flow is formed adjacent to the copper bar surface. This film is of a complex nature, very close to the solid surface the water flow is laminar (very low Reynolds' number). A transition zone is formed immediately after the laminar flow zone, and before the turbulent flow zone. Most of the Δ T between the copper bar surface, and mid-stream water flow occurs in this film in which the heat is transferred, essentially by conduction.

$$q = h_c \cdot A \cdot \Delta T_c \quad \text{watt} \qquad \textbf{Heat transferred by convection} \qquad (6.24)$$

| h_c | **Convective heat transfer coefficient** | $\dfrac{\text{watt}}{\text{cm}^2 \cdot {}^\circ\text{C}}$ |

This coefficient is a function of many things, and is very hard to obtain a value that you could apply with confidence.

A   Area of the heat transfer surface, cm^2

ΔT_c Temperature of the inside surface of the copper bar minus temperature of mid stream water flow, in degree Celsius.

$h_c := 0.05$ Assumed value for the flow of 0.2342757 gallons per second per conductor.

$q := 11219$ watt

$A := \pi \cdot D \cdot 700 = 3 \times 10^3 \ \text{cm}^2$

$$\Delta T_c = \frac{q}{A \cdot h_c} = 80.34 \ {}^\circ\text{C} \qquad\qquad \frac{\text{watt}}{\text{cm}^2 \cdot \dfrac{\text{watt}}{\text{cm}^2 \cdot {}^\circ\text{C}}} = {}^\circ\text{C}$$

The heat released by the copper bar is uniform, so the mid-stream average temperature is

40+ (43 - 40) / 2 = 41.5

$\Delta T_c = T_{bus} - 41.5$ $\qquad\qquad$ $T_{bus} = 80.34 + 41.5 = 121.84 \ {}^\circ\text{C}$

This is the temperature of the inside surface of the copper bus at full load current. It will be less at normal operating current.

Reducing by one half the flow rate or to 0.1171 gallons per second per conductor

Total water flow rate = 42 x 0.1171 = 4.92 gallons per second

Total water flow rate in gallons per minute = 4.92 x 60 = 295.2 <<<<

$h_a := 0.04$ Assumed value for the flow of 0.1171 gallons per second per conductor.

$$\Delta T_c = \frac{q}{A \cdot h_c} = 100.43 \ °C$$

$$\psi = A \cdot v$$

$$v = \frac{\psi}{A_o} = \frac{0.1171 \cdot 3785}{1.266769} = 349.885 \quad \frac{cm}{sec}$$

At this speed it takes 2 seconds to travel the entire conductor length

$$\Delta T = \frac{Q}{C_v} = \frac{11219}{4.1869} = 2679.5 \quad cm^3 \cdot °C$$

$$\psi = 0.1171 \cdot 3785 = 443.224 \quad \frac{cm^3}{sec}$$

$$\frac{\Delta T}{\psi} = \frac{2679.5}{443.224} = 6.045 \quad °C \cdot sec$$

$$2 \cdot 6.045 = 12 \quad °C \quad \text{Water temperature rise across one conductor}$$

$$40 + \frac{(52 - 40)}{2} = 46 \quad °C \quad \text{Mid-stream average temperature}$$

$$T_{bus} = 100.43 + 46 = 146.43 \ °C$$

When the water flow was reduced to 295 gpm, the inside surface of the copper bus temperature increased from 122 to 146 degree Celsius

6.6 Rotor Windings

The rotor winding is always distributed in a manner that produce approximately a sinusoidal distribution of the rotor mmf, and magnetic flux, in the air gap. Field windings are usually connected in spiral. The field winding illustrated in Fig. 6-5 is connected as illustrated in Fig. 6-6. In a generator with cylindrical rotor the armature reaction causes only a small distortion of the magnetic field in the air gap. Actually, the magnetic field intensity or flux density existing in the air gap is the field resultant from the vectorial addition of the mmf's produce by rotor and stator mmf's at the air gap. See Fig. 6-15. In fact, at any specific field current the armature reaction changes the shift angle between the resultant magnetic field axis and the rotor magnetic axis. The rotor magnetic field rotates inside the hollow space in the stator cage and induces a three-phase set of emf's in the stator three-phase windings which are function of time. The magnetic field produced by the rotor is controlled by the exciter system. Which provides a simple way of controlling the AC generator output voltage by changing the DC current delivered to the rotor winding. Also the excitation system is responsible for the AC system stability when the generator is submitted to sudden load changes, including short circuits. In this regards the quick response time of the excitation system is of vital importance. In fact, faster solid state static exciters are now part of the solution to improve the stability of AC generators. Fig. 6-5 illustrates a cross section of a two pole AC generator with cylindrical rotor. All of the magnetic flux lines produced are closed and most of them stay entirely within the rotor-stator magnetic circuit, a very small amount of them spread into the surrounding air in part of their paths. The magnetic potential drop in the rotor and stator (steel laminations) are so small in comparison with the drop in the air gap that they could be neglected in flux density computations.

Compute the ampere-turns required to create a magnetic flux density of 15000 gausses in the rotor-stator magnetic circuit depicted in Fig. 6-5 disregarding the armature reaction.

The applicable formula in the cgs system is:

$$B = \mu \cdot H \qquad\qquad \mu_a = 1 \quad \text{Air gap permeability}$$

$$B = H \quad \text{Where B is in gausses and H is in oersted.}$$

$$\text{oersted} = \frac{\text{gilbert}}{\text{cm}} \qquad \text{1 ampere-turn = 1.257 Gilbert} \qquad \text{0.4 X 3.1416 = 1.257}$$

$$B := 15000 \quad \text{gausses} \qquad H = 15000 \quad \text{oersted}$$

$$F = N \cdot I \quad \text{Magnetomotive force in ampere-turns}$$

$$F = 0.4 \cdot \pi \cdot N \cdot I = 1.257 \cdot N \cdot I \quad \text{Magnetomotive force in Gilbert}$$

$$F = H \cdot L_a \qquad\qquad \text{La = 0.5 cm or 5 mm}$$

Where H is the magnetic circuit magnetizing force or field intensity and La is the air gap length. We neglected the magnetic potential drop in the iron part of the circuit.

$$0.4 \cdot \pi \cdot N \cdot I = 15000 \cdot 0.5$$

$$N \cdot I = \frac{15000 \cdot 0.5}{0.4 \cdot \pi} = 5.968 \cdot 10^3 \quad \text{ampere} - \text{turns} \qquad\qquad 5968 \sim 6000$$

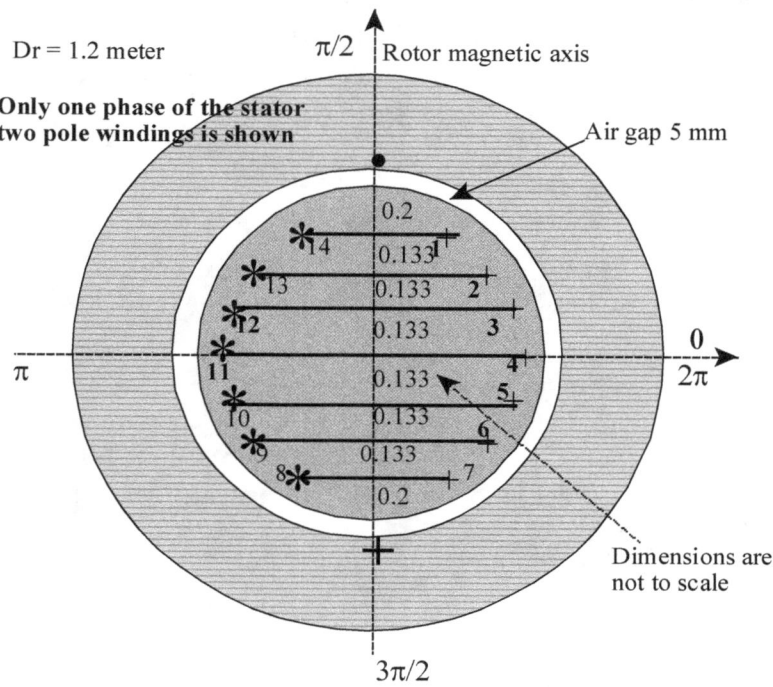

Dr = 1.2 meter

Only one phase of the stator two pole windings is shown

π/2 | Rotor magnetic axis

Air gap 5 mm

0.2
0.133
0.133
0.133
0.133
0.133
0.133
0.2

π

0

2π

Dimensions are not to scale

3π/2

Figure 6-5 Generator illustration, not to scale, showing the front end connections of the rotor spiral winding.

158

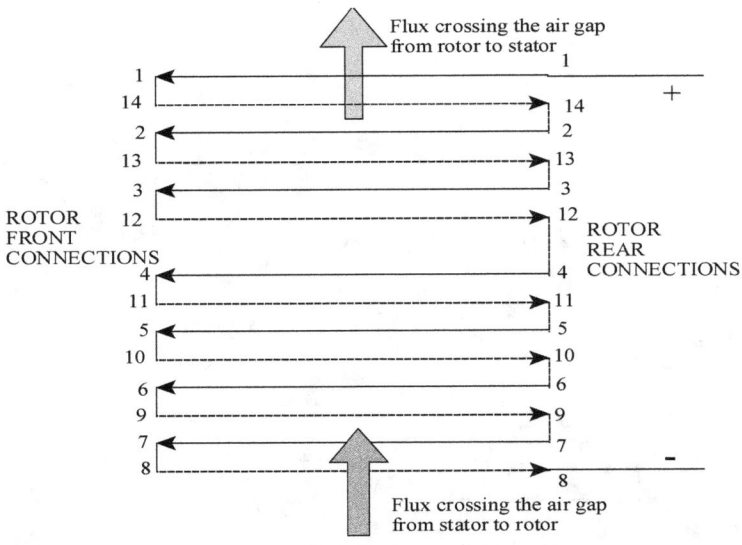

Figure 6-6 Diagram illustrating the rotor spiral winding connections

Sinusoidal steps determination

$$F = 6200 \cdot \sin(\theta)$$

$\theta = 15° = 0.262 = \dfrac{\pi}{12}\text{rad}$ $6200 \cdot \sin(15°) = 1605$

$\theta = 30° = 0.524 = \dfrac{\pi}{6}\text{rad}$ $6200 \cdot \sin(30°) = 3100$

$\theta = 45° = 0.785 = \dfrac{\pi}{4}\text{rad}$ $6200 \cdot \sin(45°) = 4384$

$\theta = 60° = 1.047 = \dfrac{\pi}{3}\text{rad}$ $6200 \cdot \sin(60°) = 5369$

$\theta = 90° = 1.571 = \dfrac{\pi}{2}\text{rad}$ $6200 \cdot \sin(90°) = 6200$

$\theta := 0, 0.175 .. 3.16$

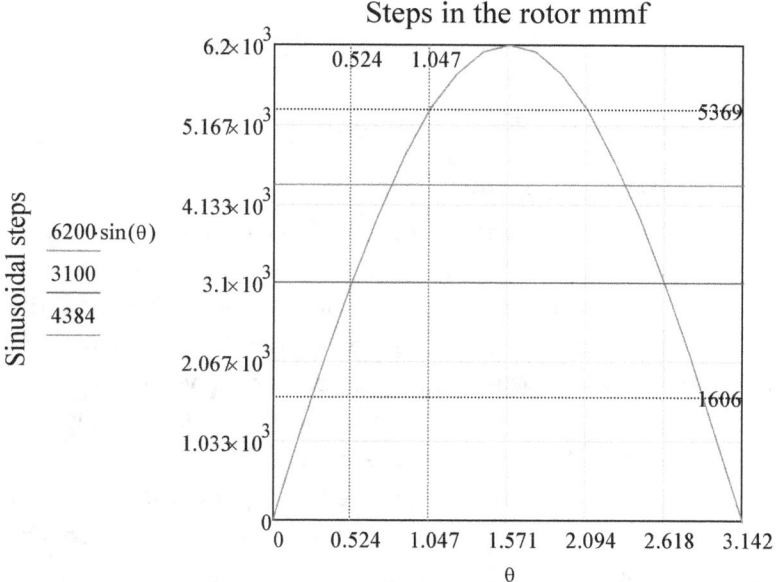

Figure 6-7 Value of sinusoidal steps.

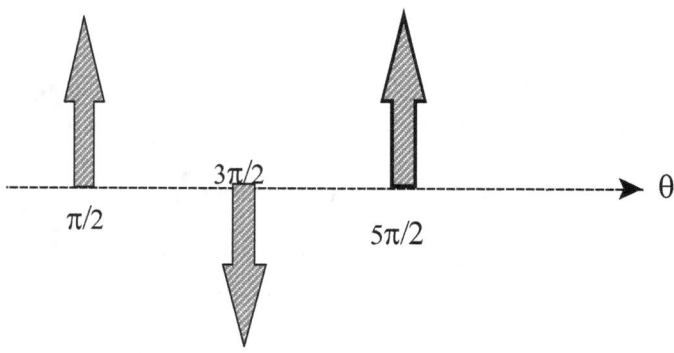

Figure 6-8 Illustration showing the change in direction of the mmf acting on the upper conductor of one of the stator coils produced by the rotation of the rotor.

The rotor in Fig. 6-5 contains a total of 14 slots, 12 cm deep each. And the number of conductor per slot is:

slots 1,14,7,8 = 2   slots 2,13,6,9,3,12,5,10 = 3   slots 4,11 = 6   conductors per slot

Rotor total number of conductors = 44

Total number of turns or coils = 22

Current levels = 2 I_1 = 428 amps I_2 = 500 amps

The manufacturer has several solutions available to produce a generator output with sinusoidal wave shape. However, many consequences of his final decision depends of the system he decides to use as exciter for the AC generator. Frequently the excitation system consists of a DC generator feeding DC current to the rotor of the AC generator via carbon brushes and slip rings. These conventional system requires frequently maintenance which represent a burdensome and costly task. Westinghouse corporation developed a brush less excitation system that consists of an auxiliary three-phase generator whose field is stationary but its revolving armature, rides together with a three-phase rectifier on the same shaft than drive the rotor of the main AC generator and therefore the DC output of the rectifier could be directly connected to the rotor of the synchronous generator. There are several possible versions of the brush less excitation system. The reader must realize that all the rotor coils are wound in the same direction, that the exciter provides DC current, and that is impossible to have a fraction of a turn. For the 500 MW synchronous generator under consideration, we would use a brush-less excitation system because it allow us to obtain two different levels of DC current, much easier than with the conventional type of exciter.

Amp-turns desired	Coil current amps	Coils	Turns per coil	Amp-turns obtained
$5369 - 4384 = 985$	500 amps	$(1 - 14), (7 - 8)$	2 turns each coil	500*2 = 1000
$4384 - 3100 = 1284$	428 amps	$(2 - 13), (6 - 9)$	3 turns each coil	428*3 = 1284
$3100 - 1605 = 1495$	500 amps	$(3 - 12), (5 - 10)$	3 turns each coil	500*3 = 1500
$1605 + 1605 = 3210$	500 amps	$(4 - 11)$	6 turns	500*6 = 3000

With respect to the rotor, the magnetic flux is always of the same direction as illustrated in Fig. (6-6). However with respect to the stator, the flux alternate as illustrated in Figs. (6-8 and 6-9). In fact, an observer situated on the stator will detect that the flux changes direction at regular intervals of time. The losses produced by the rotor winding are only due to copper losses. There are not hysteresis losses because the rotor winding current is DC. Furthermore there are not eddy current losses in the rotor either, because the magnetic field does not move with reference to the rotor. However, the magnetic field created by the rotor alternate with reference to the stator and therefore it would produce hysteresis and eddy current losses in the stator. The eddy current losses in the stator are very small because the stator magnetic core is built with grain oriented steel laminations.

Rotor copper losses

Maximum operating temperature = 155 degree Celsius.

$\rho := 1.724 \cdot 10^{-6}$ ohm $-$ cm   Copper resistivity at $20°C$

$\alpha := 0.00393 \dfrac{\text{ohm}}{\Delta°C}$ Temperature coefficient of resistance

Copper bar cross section = 2 x 1   cm2

Copper bars length = 700 cm   including half length of the ends connections.

$$R = \rho \cdot \frac{L}{A}$$

$$R_{20} = 1.724 \cdot 10^{-6} \cdot \frac{700}{2} = 603.4 \times 10^{-6} \quad \text{ohms}$$

$$R_{155} = R_{20}(1 + \alpha\,(155 - 20))$$

$$R_{155} = 603.4 \times 10^{-6}[1 + 0.00393 \cdot (155 - 20)] = 923.534 \times 10^{-6} \quad \text{ohms}$$

$$R_{155} \cdot I^2 = 923.534 \times 10^{-6} \cdot 500^2 = 230.883 \quad \frac{\text{watt}}{\text{bar}} \qquad \text{16 bars}$$

$$R_{155} \cdot I^2 = 923.534 \times 10^{-6} \cdot 428^2 = 169.177 \quad \frac{\text{watt}}{\text{bar}} \qquad \text{6 bars}$$

$230.883 \cdot 16 + 169.177 \cdot 6 = 4709.19$ watt   Total copper losses in rotor winding

Figure 6-9 Diagram illustrates the mmf in the air gap of Fig. 6-5, as seen from the stator, and produced by the distributed rotor winding shown in Fig. 6-6. The magnetomotive force produced by the rotor have a fixed value of ampere turns, and it always point in the same direction with respect to the rotor. However, due to the rotation of the rotor, an observer placed in the upper (or lower) conductor of a stator coil, of one of the phases (there are three-phases) would see the mmf successively changing in value and direction. A helpful way of understanding Fig. 6-9 is replacing the rotor with a rotating bar magnet. The reader surely knows that one end of the bar is the North Pole (N) and the other end is the South Pole (S). The magnetic flux always enters the bar by the S pole and exits the bar by the N pole. When a stator coil conductor is successively swept by the flux leaving the N pole (flux direction is from rotor to stator) followed by the flux entering the S pole (flux direction is from the stator to rotor) of the rotating bar magnet, an alternating emf would be induced in the stator coil. Because, when an N pole sweeps across a stator coil conductor it induces a positive voltage and similarly a negative voltage will be induced in the same conductor when an S pole sweeps the conductor.

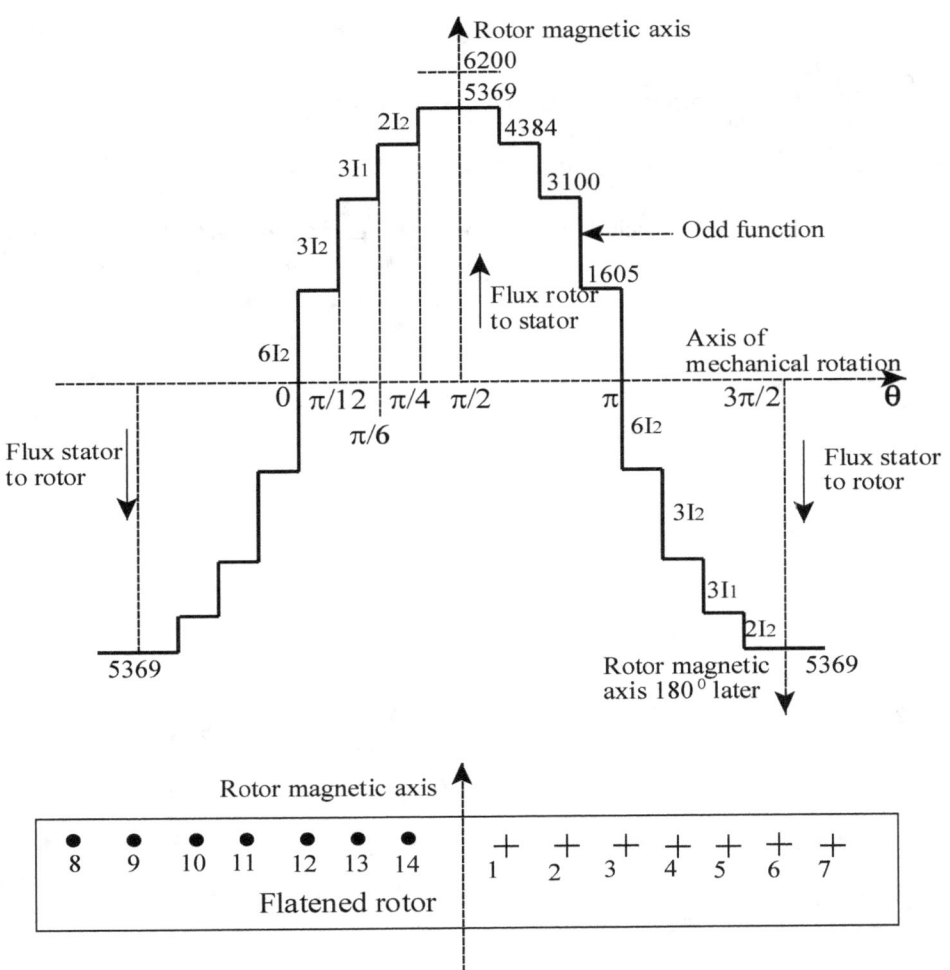

Figure 6-9 Illustration of the rotor mmf steps

6.7 Rotor Magnetomotive Force Wave Analysis

The wave shown in Fig. 6-9 is a non-sinusoidal, periodic odd function that could be represented by the Fourier Series. The reader is referred to Section 1.6. Because the wave consists of symmetrical positive and negative loops it cannot contain even harmonics nor DC bias. Therefore Eq. 1.26 is reduced to:

$$f(\theta) = A_1 \cdot \sin(\theta) + B_1 \cdot \cos(\theta) + A_3 \cdot \sin(3\theta) + B_3 \cdot \cos(3\theta) + A_5 \cdot \sin(5\theta) + \blacksquare$$

$$B_5 \cdot \cos(5\theta) + \ldots \tag{6.25}$$

In the Fourier expansion the A and B coefficients are mutually exclusive of each other and only one of them is necessary. In this case, as explained in section 1.6, the B coefficients are all zero and could be eliminated because the original function is an odd function consisting of symmetrical positive and negative loops. Eliminating the cosine terms from Eq. (6.25) it becomes Eq. (6.26). Where to simplify we did not consider harmonics higher than the thirteenth.

$$f(\theta) = A_1 \cdot \sin(\theta) + A_3 \cdot \sin(3\theta) + A_5 \cdot \sin(5\theta) + A_7 \cdot \sin(7\theta) + A_9 \cdot \sin(9\theta) + \blacksquare$$

$$A_{11} \cdot \sin(11\theta) + A_{13} \cdot \sin(13\theta) + \ldots \tag{6.26}$$

The original wave illustrated in Fig. 6-9 poses quarter-wave symmetry and this allows us to reduce the wave analysis to a quarter of a cycle or $\pi/2$ radians. Applying Eq. (1.35) we have

$$A_n = \frac{1}{\pi} \cdot \int_0^{2\pi} f(\theta) \cdot \sin(n \cdot \theta) \, d\theta$$

$$A_n = \frac{1}{\pi} \cdot \int_0^{\frac{\pi}{12}} 1605 \cdot \sin(n \cdot \theta) \, d\theta + \frac{1}{\pi} \cdot \int_{\frac{\pi}{12}}^{\frac{\pi}{6}} 3100 \cdot \sin(n \cdot \theta) \, d\theta + \blacksquare$$

$$\frac{1}{\pi} \cdot \int_{\frac{\pi}{6}}^{\frac{\pi}{4}} 4384 \cdot \sin(n \cdot \theta) \, d\theta + \frac{1}{\pi} \cdot \int_{\frac{\pi}{4}}^{\frac{\pi}{2}} 5369 \cdot \sin(n \cdot \theta) \, d\theta$$

$$\frac{1}{\pi} \cdot \int_0^{\frac{\pi}{12}} 1605 \cdot \sin(n \cdot \theta) \, d\theta = 17.408 \qquad \frac{1}{\pi} \cdot \int_{\frac{\pi}{12}}^{\frac{\pi}{6}} 3100 \cdot \sin(n \cdot \theta) \, d\theta = 98.578$$

$$\frac{1}{\pi} \cdot \int_{\frac{\pi}{6}}^{\frac{\pi}{4}} 4384 \cdot \sin(n \cdot \theta) \, d\theta = 221.766 \qquad \frac{1}{\pi} \cdot \int_{\frac{\pi}{4}}^{\frac{\pi}{2}} 5369 \cdot \sin(n \cdot \theta) \, d\theta = 1.208 \times 10^3$$

$$17.408 + 98.578 + 221.766 + 1208 = 1546 \qquad n := 1$$

$$A_1 = 1546 + 3 \cdot 1546 = 6184 \qquad \text{Fundamental coefficient}$$

$$n := 3$$

$$\frac{1}{\pi} \cdot \int_{0}^{\frac{\pi}{12}} 1605 \cdot \sin(n \cdot \theta) \, d\theta = 49.878 \qquad \frac{1}{\pi} \cdot \int_{\frac{\pi}{12}}^{\frac{\pi}{6}} 3100 \cdot \sin(n \cdot \theta) \, d\theta = 232.582$$

$$\frac{1}{\pi} \cdot \int_{\frac{\pi}{6}}^{\frac{\pi}{4}} 4384 \cdot \sin(n \cdot \theta) \, d\theta = 328.916 \qquad \frac{1}{\pi} \cdot \int_{\frac{\pi}{4}}^{\frac{\pi}{2}} 5369 \cdot \sin(n \cdot \theta) \, d\theta = -402.817$$

$$49.878 + 232.582 + 328.916 - 402.817 = 208.559$$

$$A_3 = 4 \cdot 208.559 = 834.236 \quad \text{Third harmonic coefficient} \qquad \frac{834.236}{6184} = 0.135 \qquad 13.5\%$$

$$75.732 + 221.99 - 44.353 - 241.69 = 11.679$$

$$A_5 = 4 \cdot 11.679 = 46.716 \quad \text{Fifth harmonic coefficient} \qquad \frac{46.716}{6184} = 0.0076 \qquad 0.76\%$$

$$91.874 + 85.595 - 313.609 + 172.636 = 36.496$$

$A_7 = 4 \cdot 36.496 = 145.984$ Seventh harmonic coefficient $\dfrac{145.984}{6184} = 0.024$ 2.4%

$$96.904 - 77.527 - 109.639 + 134.272 = 44.010$$

$A_9 = 4 \cdot 44.01 = 176.040$ Ninth harmonic coefficient $\dfrac{176.04}{6184} = 0.028$ 2.8%

$$91.306 - 164.336 + 199.569 - 109.859 = 16.680 \qquad 4 \cdot 16.68 = 66.720$$

$A_{11} = 66.72$ Eleventh harmonic coefficient $\dfrac{66.72}{6184} = 0.011$ 1.1%

$$77.259 - 139.054 + 168.866 - 92.958 = 14.113 \qquad 4 \cdot 14.113 = 56.452$$

$A_{13} = 56.452$ Thirteenth harmonic coefficient $\dfrac{56.452}{6184} = 0.009$ 0.9%

The Fourier series expansion of the mmf in the air gap depicted in Fig. 6-10 and Figure 6-11 depicts the summation of the fundamental, third, fifth, seventh, ninth, eleventh and thirteenth harmonics.

$$\theta := 0, \frac{\pi}{10} .. 2\pi$$

$F_1 = 6184 \cdot \sin(\theta)$ $F_3 = 834.236 \cdot \sin(3\theta)$ $F_5 = 46.716 \cdot \sin(5\theta)$

$F_7 = 145.984 \cdot \sin(7\theta)$ $F_9 = 176.04 \cdot \sin(9\theta)$ $F_{11} = 66.72 \cdot \sin(11\theta)$

$F_{13} = 56.452 \cdot \sin(13\theta)$

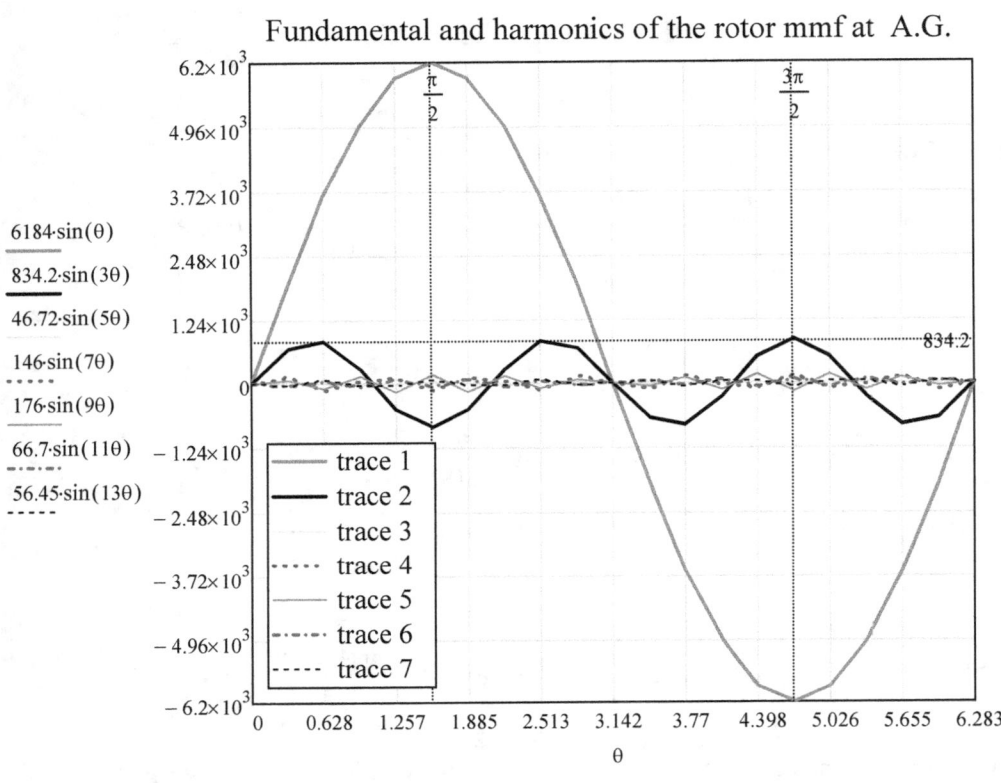

Figure 6-10 Furier series expansion of the mmf in the air gap

The truncation of the Furier equivalent of the original function as shown in Figure 6-11 occurs at exactly the same value than the maximum of the original function. See Fig. 6-9.

MMF1= 6184sin(θ) + 834.236sin(3θ) + 46.716sin(5θ) + 145.984sin(7θ) + 176.04sin(9θ)+

66.72sin(11θ) + 56.452sin(13θ)

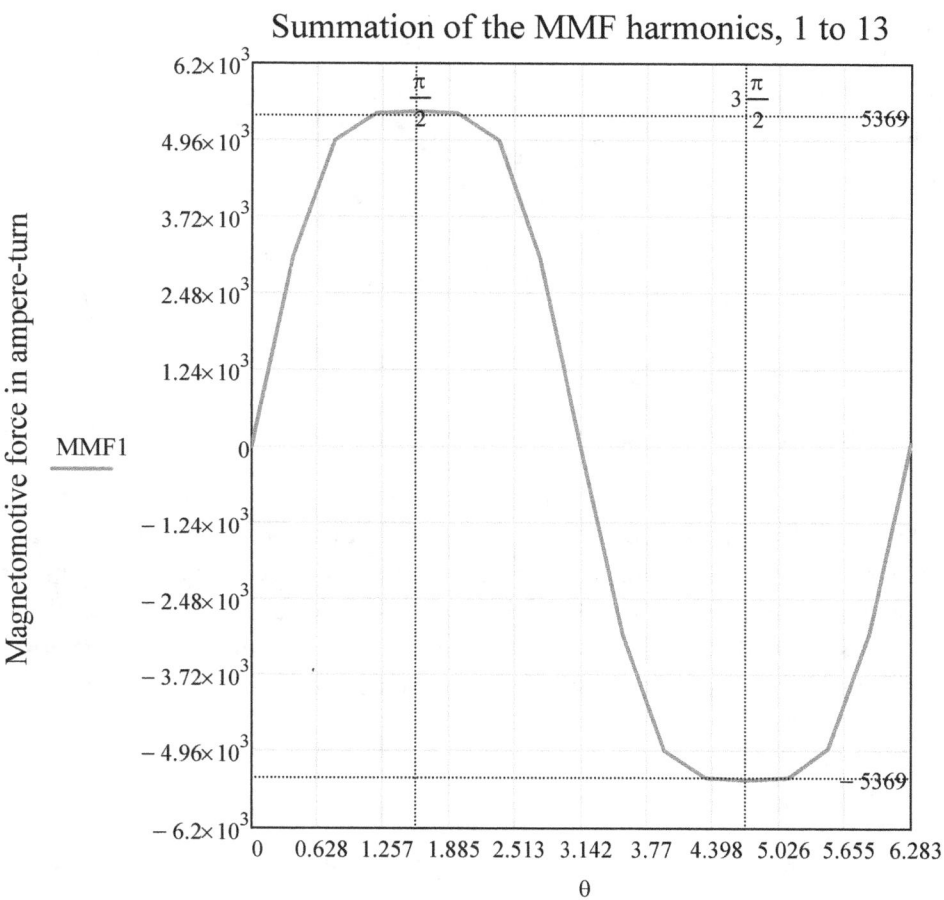

Figure 6-11 Summation of the fundamental, third, fifth, seventh, ninth, eleven, and thirteen harmonics.

Now we will determine the wave shape of the line to line voltage, which is the voltage impressed on the generator load and it is also the wave shape of the load current or generator output current. See Fig. (1-1) which is an illustration of the magnetic induction law. From this illustration it is obvious than the electromotive force induced in the wire has the same wave shape than the changing flux that induced it, but lagging it by 90 degrees.

$$F = H \cdot L_{mc}$$

Where Lmc is the magnetic circuit length. So H is the magnetic potential drop per unit length of the magnetic circuit and is expressed in ampere-turns per meter. Furthermore the ratio B/H is defined as μ, the permeability of the medium, and for air, in the cgs system μ = 1. So in the cgs system of units, in air, B = H and therefore:

$$\frac{\phi}{A_{mc}} = \frac{F}{L_{mc}}$$

$$\phi = \frac{A_{mc}}{L_{mc}} \cdot F \qquad \frac{d}{dt}\phi = \frac{A_{mc}}{L_{mc}} \cdot \left(\frac{d}{dt}F\right)$$

$$e = -N \cdot \left(\frac{d}{dt}\phi\right) \qquad e = -N \cdot \left[\frac{A_{mc}}{L_{mc}} \cdot \left(\frac{d}{dt}F\right)\right] \qquad e = -\left(N \cdot \frac{A_{mc}}{L_{mc}}\right) \cdot \left(\frac{d}{dt}F\right)$$

Amc = Cross sectional area of magnetic circuit

Lmc = Length of the magnetic circuit

The above equations disregard mathematical rigor, and they are written for average values of the flux density and field intensity. However they indicate the logic of the reasoning path. So we can conclude than the wave shape of the generator output voltage is the same than the wave shape of the mmf, but lagging 90 degrees. Because sine and cosine waves have the same shape, but phase-shifted 90 degrees. Below is a plot of the summation of the mmf harmonics without the third and the ninth harmonics, which is almost perfect when compared to a pure sinusoidal function.▢

$$\theta := 0.000\,,0.393 \,..\, 6.283$$

$$MMF1 = 6184 \cdot \sin(\theta) + 46.716 \cdot \sin(5\theta) + 145.984 \cdot \sin(7\theta) + 66.72 \cdot \sin(11\theta) + \blacksquare$$

$$56.452 \cdot \sin(13\theta)$$

The generator line to line output voltage has the same wave shape that the mmf shown in Fig. (6-12) but lagging 90 degrees from the mmf plot. See figure (6-13) which is a plot of the output voltage at no load condition, when there is no armature reaction.

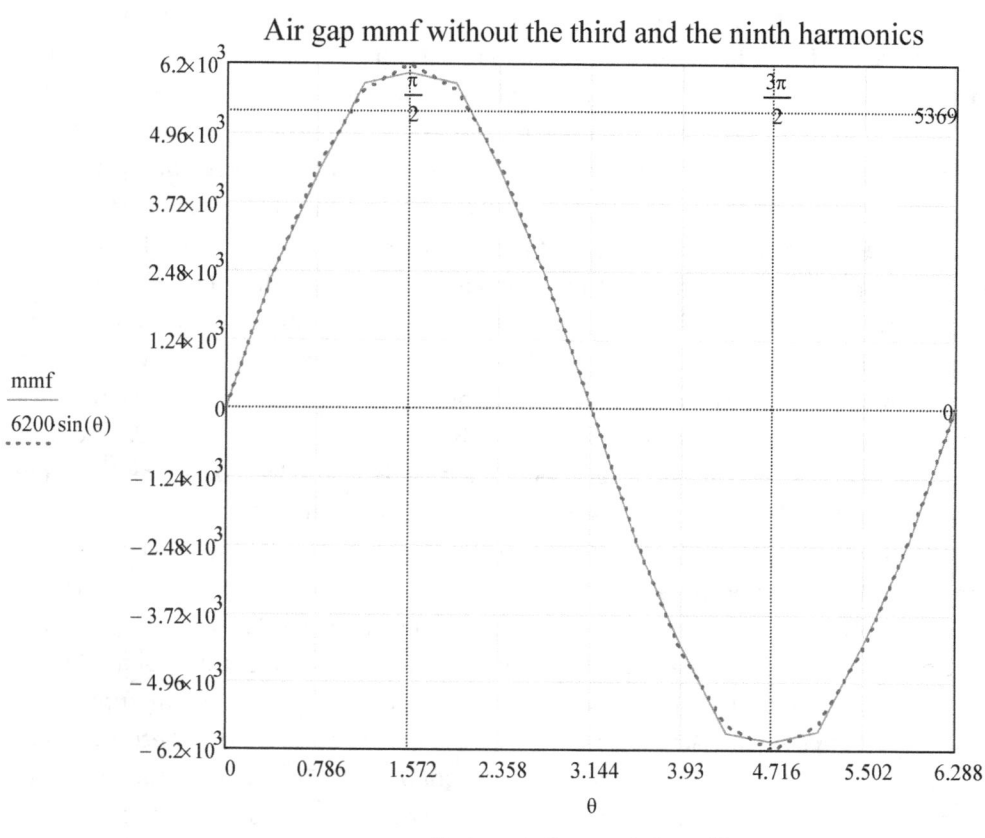

Figure 6-12 Summation of the fundamental, fifth, seventh, eleventh, and thirteenth harmonics versus a pure sinusoidal wave shape.

6.8 Resultant Magnetomotive Force in the Air Gap

The magnetic flux created in the generator magnetic circuit is three-dimensional and encompass the rotor, the air gap, and the stator and runs the entire length of the rotor core. The reader must remember that any one of the flux lines forms a close loop. As we know from previous sections the line to line voltage at the generator output terminals does not contains third harmonics nor any of its multiples. Therefore the current delivered to the load does not in general contains third harmonics nor any of its multiples either, provided that the load is linear. However, in some transient conditions the magnetic's of some devices, such as transformers, could saturate and become non-linear and introduce third harmonics components in the line currents. However, the reader most realize that the line and transformer reactances attenuate the high frequency harmonics the most. Besides cores' saturation in the power system might introduce line currents unbalance during transient conditions, such as short circuits to ground, and introduce unexpected harmonics.

In steady state conditions the mmf existing in the air gap is the phasor addition (component by component) of the mmfs produced by the rotor and the stator. The reader should realize that both mmf are rotating at the same synchronous speed or 60 rev/sec. (for a 2-pole generator). In fact, the rotor mmf rotates because the rotor is driven by an external machine and the stator mmf rotates because it is induced by a set of three-phase currents, each of equal magnitude and separated from the other two by 120 degrees, flowing in a three-phase winding. Consequently, these two rotating mmf are stationary with respect to each other. However, they are not in phase. In fact the mmf (and also the flux) produced by armature reaction lags behind the rotor mmf. Actually these two magnetic fields tend to align with each other and therefore the rotor magnetic field pulls on the armature magnetic field while doing work the mechanical torque so created opposes the rotation of the rotor shaft which usually is driven by a turbine. The larger the electrical load, the larger is the armature reaction and the opposing torque, and the more mechanical energy must be delivered by the turbine to sustain rotation. This is a wonderful and efficient way of converting mechanical to electrical energy. Figure 6-14 illustrates the phase difference between the fundamentals of the rotor mmf and the fundamental of the mmf produced by the phase-1 line current, as well as the air gap net mmf. In fact, the voltage induced in phase one by the changing magnetic flux concatenations, V1n, lags 90 degrees from the air gap net mmf, which is the resultant of the phasor addition (component by component) of all the harmonic components of the rotor and armature reaction magnetomotive forces. *Actually, to find the resultant mmf it would be better to add the waves representing the rotor and the armature reaction mmfs.* This is especially true when the armature winding is distributed. Similarly, the line to line voltage lags the air gap resultant mmf by 60 degrees. And the current in line one and the mmf induced in the air gap by line one, lags the air gap resultant mmf by 60 degrees plus the load power factor angle. See figures 6-2 and 6-14.

$$V_{1n} := 6200\sin(\theta) \qquad\qquad mmf_{net} := 26 \cdot 10^3 \cdot \sin\left(\theta - \frac{\pi}{2}\right)$$

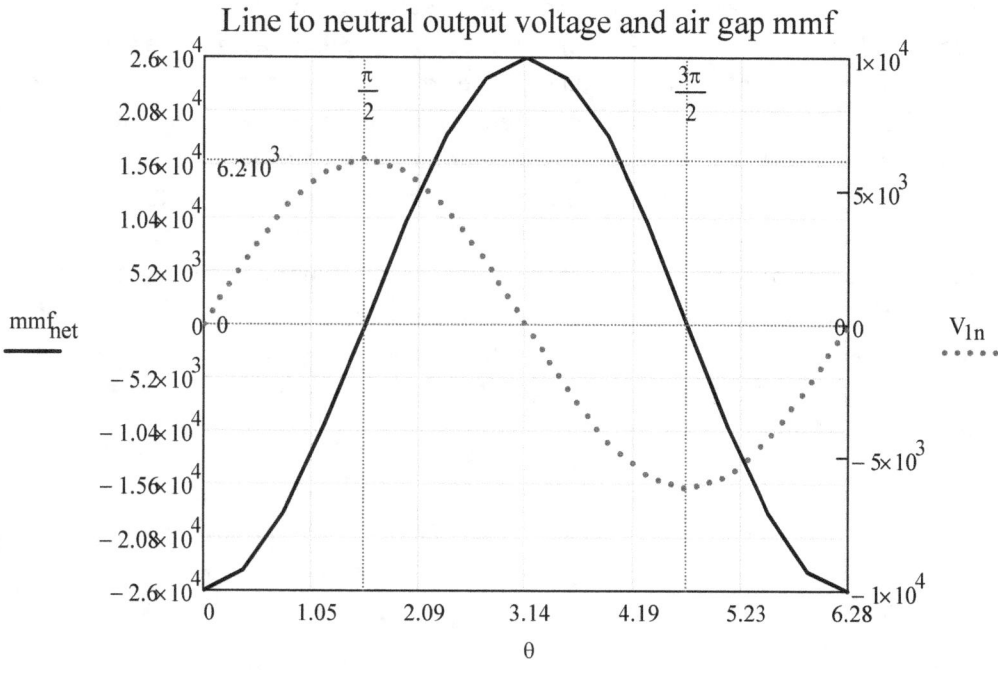

Figure 6-13 Generator line to neutral output voltage at no load condition and the mmf in the air gap.

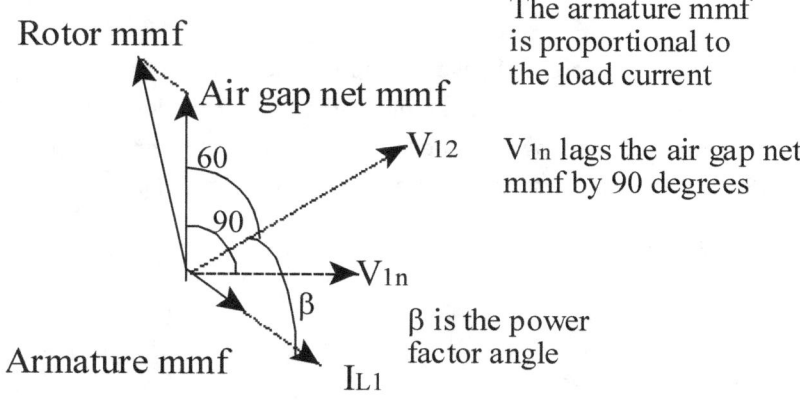

Figure 6-14 Illustration showing the phase difference in the air gap between the fundamentals of the rotor and the armature mmfs.

The stator windings being considered are full pitch distributed windings with seven slots per pole per phase, with each slot containing one insolated bus-bar conductor, which is one of the two actives side of a coil, the other side of the coil is 180 degrees away. In each phase the coils are connected so that the rotating magnetic field they create have the same number of poles than the rotor field winding. In this case 2 poles. Indeed the three-phase set of load currents flowing through the set of three-phase windings in the armature creates a rotating magnetic field containing a north and a south pole and therefore is classified as a two pole stator winding. Invoking the Lenz law we could say that the armature reaction opposes any change in the magnetic field established in the air gap by the rotor and we know that the harmonic's spectrum of the armature reaction must be the same than the spectrum of the field established in the air gap by the rotor, except for the absence of the third and all the multiple of the third harmonics, assuming that the electrical load is linear. So in general the net mmf existing in the air gap would be smaller than the original mmf induced by the rotor. Figure 6-15 illustrate the phasor addition of the flux density in the air gap produced by the fundamentals of both mmf.

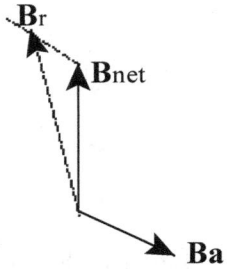

Figure 6-15 Illustration showing the net flux density at the air gap produced by the mmf of the rotor and armature fundamentals.

The purpose of distributed windings is to improve the wave shape of the generator output voltage by making it more sinusoidal, by decreasing the amplitude of the harmonics with respect to the fundamental. For instance the results of the Furier analysis of the wave shape in Fig. 6-9 is provided in Eq. 6.26. Below are the results in percent of the fundamental.

A.1 = 6184   ampere-turns of the fundamental

A.3 = 13.5%   A.5 = 0.76 %   A.7 = 2.4 %

A.9 = 2.8 %   A.11 = 1.1 %   A.13 = 0.9 %

The A.3 and A.9 components, which are the largest, will remain the same because the armature reaction does not contain these harmonics, assuming that the electrical load is linear. And the A.5, A.11, and A.13 are small enough and do not need to be decreased. The only one that appears to be a candidate for improving is the A7 component. However, the frequency of this component is 420 Hz and therefore the ampere-turn due to the 7th harmonic is greatly reduced by the reactances of the generator, transformers, and transmission lines. So there is not a real need for using distributed windings. This is especially true for generators larger than 250 MVA, because distributed windings concentrate its large copper losses in a relatively small section of the generator. From the point of view of heat removal, it would be much better to use a longer concentrated winding or perhaps a distributed winding with no more than three fractional-pitch (5/6) coils to eliminate the 7th harmonic. Of course, the other harmonics, including the fundamental would also be decreased. And therefore the generated voltage would be smaller.

6.9 Full and Fractional Pitch Windings.

In full pitch windings the angular separation between the active sides of a coil is 180 electrical degrees. For a fractional pitch winding the angular separation is smaller than 180 electrical degrees. The magnitude of the fundamental of the emf induced in a full pitch coil is twice the magnitude of the fundamental of the emf induced in one of the active side conductors. The emf induced in the active sides are of such direction that they add up around the coil. Although the phasors that represent the emfs point in opposite direction at any one time. This is because the flux density direction in the north side is from the rotor to the stator, but in the south side is the reverse, that is: from the stator to the rotor. So the fundamental of the total emf induced in the coil is the phasor difference of the fundamentals of the leading and lagging phasors. See figure 6-16.

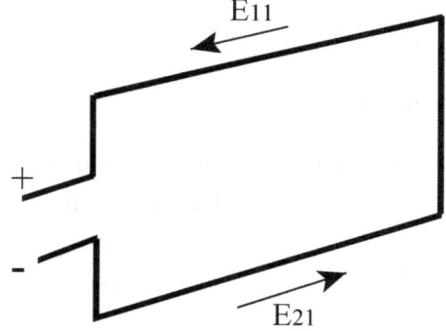

Figure 6-16 Full pitch coil

$$\overrightarrow{E_{11}} = E_1 \cdot \varepsilon^{j \cdot 0^0}$$
Fundamental of the leading phasor selected as reference.

$$\overrightarrow{E_{21}} = E_1 \cdot \varepsilon^{j \cdot 180^0}$$
Fundamental of side two, it is a full pitch coil

$$\overrightarrow{E_{11}} - \overrightarrow{E_{21}} = E_1 \cdot \varepsilon^{j \cdot 0^0} - E_1 \cdot \varepsilon^{j \cdot 180^0}$$

$$\overrightarrow{E_{11}} - \overrightarrow{E_{21}} = E_1 \cdot \varepsilon^{j \cdot 0^0} + E_1 \cdot \varepsilon^{j \cdot 0^0}$$

$$\overrightarrow{E_{11}} - \overrightarrow{E_{21}} = 2E_1 \cdot \varepsilon^{j \cdot 0^0} \qquad \overrightarrow{E_1} = 2E_1 \cdot \varepsilon^{j \cdot 0^0}$$

$$(6.28)$$

Equation (6.28) provides the magnitude of the coil emf fundamental and phase angle with reference to the fundamental of the emf induced in the leading active side of the coil. In a fractional pitch coil the magnitude of the fundamental emf induced is smaller than 2E. In fact in a fractional pitch coil the fundamental of the induced emfs in the active sides are out of phase by an amount of electrical degrees equal to the coil pitch. Furthermore, the difference in phase between the harmonics of the emfs induced in the active sides of a fractional pitch coil is qρ, where q is the order of the harmonic and ρ is the coil pitch expressed in electrical degrees. Symbolically:

$$\Delta\theta = q \cdot \rho \qquad \text{Valid for any harmonic}$$

$$\Delta\theta_1 = 1 \cdot \rho \qquad \text{Difference in phase between the fundamentals}$$

$$\Delta\theta_3 = 3 \cdot \rho \qquad \text{Difference in phase between the third harmonics}$$

$$\Delta\theta_5 = 5 \cdot \rho \qquad \text{Difference in phase between the fifth harmonics}$$

$$\Delta\theta_7 = 7 \cdot \rho \qquad \text{Difference in phase between the seventh harmonics}$$

Figure 6-17 illustrates the phase difference between the q harmonic's of the emf induced in the two active sides of a pitched coil and the total emf induced in the coil for the harmonic considered.

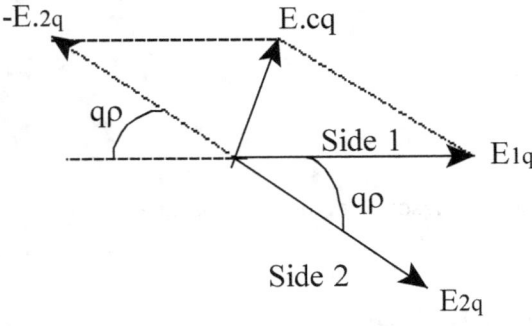

Figure 6-17 Harmonics phase difference in a coil and total induced emf

The q harmonic of the emf induced in the leading side of the coil is taken as reference.

$$\overrightarrow{E_{1q}} = E_{1q} \cdot \varepsilon^{j \cdot 0} \qquad \text{Reference} \qquad \text{Side one}$$

$$\overrightarrow{E_{2q}} = E_{2q} \cdot \varepsilon^{-j \cdot q \cdot \rho} \qquad\qquad \text{Side two}$$

$$-\overrightarrow{E_{2q}} = E_{2q} \cdot \varepsilon^{j(180^\circ - q \cdot \rho)} \qquad \text{Rotated 180}$$

$$\overrightarrow{E_{1q}} - \overrightarrow{E_{2q}} = E_{1q} \cdot \varepsilon^{j \cdot 0} + E_{2q} \cdot \varepsilon^{j(180^\circ - q \cdot \rho)}$$

$$\overrightarrow{E_{1q}} - \overrightarrow{E_{2q}} = E_{1q} + E_{2q} \cdot [(\cos(180^\circ - q\rho) + j \cdot \sin(180^\circ - q\rho)]$$

$$|E_{11}| = |E_{21}| = E_1$$

$$\overrightarrow{E_{1q}} - \overrightarrow{E_{2q}} = E_q[1 + [(\cos(180^\circ - q\rho) + j \cdot \sin(180^\circ - q\rho)]$$

$$\overrightarrow{E_{1q}} - \overrightarrow{E_{2q}} = E_q(1 - \cos(q\rho) + j \cdot \sin(q\rho)) \qquad\qquad (6.29)$$

$$\overrightarrow{E_{cq}} \quad \text{Is the total harmonic q induced on the \textbf{coil}.}$$

$$\overrightarrow{E_{cq}} = E_q \cdot (1 - \cos(q \cdot \rho) + j \cdot \sin(q \cdot \rho)) \qquad\qquad (6.30)$$

Pitch factor $\quad (1 - \cos(q \cdot \rho) + j \cdot \sin(q \cdot \rho)) \qquad\qquad (6.31)$

To eliminate a harmonic from the voltage induced in an armature coil, choose a fractional pitch ρ that makes zero the *pitch factor* for that harmonic. Meaning qρ = 0. Fractional pitch means that the pitch is a fraction of 180 electrical degrees, for instance 4/5, 5/6, 6/7. Actually, this fraction is the ratio of the number of slots spanned by a coil to the number of slots required to be spanned in the case of a full pitch coil.

Evaluating Eq. (6.30) for a 5/6 pitch we obtain:

$$q := 1 \qquad\qquad \rho := \frac{5}{6} \cdot 180^\circ \qquad\qquad q \cdot \rho = 150.000 \cdot {}^\circ$$

$$\overrightarrow{E_{c1}} = E_1 \cdot (1 - \cos(q \cdot \rho) + j \cdot \sin(q \cdot \rho))$$

$(1 - \cos(150°) + j \cdot \sin(150°)) = 1.866 + 0.5j$

$\overrightarrow{E_{c1}} = E_1 \cdot (1.866 + 0.5j)$ $|1.866 + 0.5j| = 1.932$ $\operatorname{atan}\left(\dfrac{0.5}{1.866}\right) = 15.000 \cdot °$

$\overrightarrow{E_{c1}} = 1.932 \cdot E_1 \cdot \varepsilon^{j \cdot 15°}$

(6.32)

$q = 3$ $q \cdot \rho = 450 \cdot °$ $(1 - \cos(450°) + j \cdot \sin(450°)) = 1 + j$

$\overrightarrow{E_{c3}} = E_3 \cdot (1 + j)$ $|1 + j| = 1.414$ $\operatorname{atan}\left(\dfrac{1}{1}\right) = 45 \cdot °$

$\overrightarrow{E_{c3}} = 1.414 \cdot E_3 \cdot \varepsilon^{j \cdot 45°}$

(6.33)

$q := 5$ $q \cdot \rho = 750 \cdot °$ $(1 - \cos(q \cdot \rho) + j \cdot \sin(q \cdot \rho)) = 0.134 + 0.5j$

$\overrightarrow{E_{c5}} = E_5 \cdot (0.134 + 0.5j)$ $|0.134 + 0.5j| = 0.518$ $\operatorname{atan}\left(\dfrac{0.5}{0.134}\right) = 75 \cdot °$

$\overrightarrow{E_{c5}} = 0.518 E_5 \cdot \varepsilon^{j \cdot 75°}$

(6.34)

$q := 7$ $q \cdot \rho = 1050 \cdot °$ $q \cdot \rho = 330°$

$(1 - \cos(q \cdot \rho) + j \cdot \sin(q \cdot \rho)) = 0.134 - 0.5j$

$\overrightarrow{E_{c7}} = E_7 \cdot (0.134 - 0.5j)$ $|0.134 - 0.5j| = 0.518$ $\operatorname{atan}\left(\dfrac{-0.5}{0.134}\right) = -75 \cdot °$

$\overrightarrow{E_{c7}} = 0.518 \cdot E_7 \cdot \varepsilon^{-j \cdot 75°}$

(6.35)

$q := 11$ $q \cdot \rho = 1650 \cdot °$ $q \cdot \rho = 210°$

$(1 - \cos(q \cdot \rho) + j \cdot \sin(q \cdot \rho)) = 1.866 - 0.5j$

$\overrightarrow{E_{c11}} = E_{11} \cdot (1.866 - 0.5j)$ $|1.866 - 0.5j| = 1.932$ $\operatorname{atan}\left(\dfrac{-0.5}{1.866}\right) = -15 \cdot °$

$\overrightarrow{E_{c11}} = 1.932 \cdot E_{11} \cdot \varepsilon^{-j \cdot 15°}$

(6.36)

With a full-pitch coil the voltage induced in both active sides of the coil are in phase. But with a fractional pitch coil they are out of phase as shown in Eqns. (6.32 to 6.36). And the magnitudes of the harmonics in fractional-pitch coils (windings) compared with their magnitudes in a full-pitch coil that has the same number of turns are computed from Eqns. (6.32 to 6.36) for a 5/6 pitch as follows:

Magnitude of the voltage generated in fractional-pitch coil / Magnitude of the voltage generated in a full-pitch coil.

For instance for the 5th harmonic the ratio of the magnitudes is:

$$\frac{0.518 \cdot E_5}{2 \cdot E_5} = \frac{0.518}{2} = 0.259$$

$$\rho = \frac{5}{6} \qquad \frac{1.932}{2} = 0.966 \qquad \frac{1.414}{2} = 0.707 \qquad \frac{0.518}{2} = 0.259 \qquad \frac{0.518}{2} = 0.259$$

$$\frac{1.932}{2} = 0.966$$

Evaluating Eq. (6.30) for a 6/7 pitch we obtain:

$$\rho = \frac{6}{7} \cdot 180° \qquad \rho := 154.286°$$

$$q := 1 \qquad q \cdot \rho = 154.286° \qquad \overrightarrow{E_{c1}} = E_1 \cdot (1 - \cos(q \cdot \rho) + j \cdot \sin(q \cdot \rho))$$

$$(1 - \cos(q \cdot \rho) + j \cdot \sin(q \cdot \rho)) = 1.901 + 0.434j$$

$$\overrightarrow{E_{c1}} = E_1 \cdot (1.901 + 0.434j) \qquad |1.901 + 0.434j| = 1.95 \qquad \mathrm{atan}\left(\frac{0.434}{1.901}\right) = 12.860 \cdot °$$

$$\overrightarrow{E_{c1}} = 1.95 \cdot E_1 \cdot \varepsilon^{j \cdot 12.9°} \tag{6.37}$$

$$q := 3 \qquad q \cdot \rho = 463 \cdot ° \qquad \overrightarrow{E_{c3}} = E_3 \cdot (1 - \cos(q \cdot \rho) + j \cdot \sin(q \cdot \rho))$$

$$q \cdot \rho = 103°$$

$$(1 - \cos(103°) + j \cdot \sin(103°)) = 1.225 + 0.974j$$

$$\overrightarrow{E_{c3}} = E_3 \cdot (1.225 + 0.974j) \qquad |1.225 + 0.974j| = 1.565 \qquad \mathrm{atan}\left(\frac{0.974}{1.225}\right) = 38.488 \cdot °$$

$$\overrightarrow{E_{c3}} = 1.565 \cdot E_3 \cdot \varepsilon^{j \cdot 38.5°} \tag{6.38}$$

$$q := 5 \qquad q \cdot \rho = 771 \cdot ° \qquad \overrightarrow{E_{c5}} = E_5 \cdot (1 - \cos(q \cdot \rho) + j \cdot \sin(q \cdot \rho))$$

$q \cdot \rho = 51°$

$(1 - \cos(51°) + j \cdot \sin(51°)) = 0.371 + 0.777j$

$\overrightarrow{E_{c5}} = E_5 \cdot (0.371 + 0.777j)$ $\quad |0.371 + 0.777j| = 0.861$ $\quad \mathrm{atan}\left(\dfrac{0.777}{0.371}\right) = 64.477 \cdot °$

$\overrightarrow{E_{c5}} = 0.861 \cdot E_5 \cdot \varepsilon^{j \cdot 64.5°}$
$\hspace{10cm}$ (6.39)

$q := 7 \quad q \cdot \rho = 1080 \quad \overrightarrow{E_{c7}} = E_7 \cdot (1 - \cos(q \cdot \rho) + j \cdot \sin(q \cdot \rho))$

$q \cdot \rho = 0° \qquad (1 - \cos(0) + j \cdot \sin(0)) = 0$

Pitch factor equal to zero for the 7th harmonic with a pitch of 6/7

$\overrightarrow{E_{c7}} = E_7 \cdot \varepsilon^{j \cdot 0°}$
$\hspace{10cm}$ (6.40)

$q := 11 \qquad q \cdot \rho = 1697 \cdot ° \quad \overrightarrow{E_{c11}} = E_{11} \cdot (1 - \cos(q \cdot \rho) + j \cdot \sin(q \cdot \rho))$

$q \cdot \rho = 257°$

$(1 - \cos(257°) + j \cdot \sin(257°)) = 1.225 - 0.974j$

$\overrightarrow{E_{c11}} = E_{11} \cdot (1.225 - 0.974j)$ $\quad |1.225 - 0.974j| = 1.565$ $\quad \mathrm{atan}\left(\dfrac{-0.974}{1.225}\right) = -38.488 \cdot °$

$\overrightarrow{E_{c11}} = 1.565 \cdot E_{11} \cdot \varepsilon^{-j \cdot 38.5°}$
$\hspace{10cm}$ (6.41)

$\rho = \dfrac{6}{7} \qquad \dfrac{1.95}{2} = 0.975 \qquad \dfrac{1.565}{2} = 0.782 \qquad \dfrac{0.861}{2} = 0.430 \qquad \dfrac{0}{2} = 0 \qquad \dfrac{1.565}{2} = 0.782$

Table 6-1 shows the harmonics magnitudes in fractional pitch coils relative to their magnitudes in full pitch coils having the same number of turns.

Pitch	Harmonics Order				
	1	3	5	7	11
4/5	0.951	0.588	0.000	0.588	0.951
5/6	0.966	0.707	0.259	0.259	0.966
6/7	0.975	0.782	0.430	0.000	0.782

Table 6-1 Harmonics magnitudes in fractional-pitch coils compared with their magnitudes in a full-pitch coil.

Table 6-2 shows the coil voltage harmonics phase angles with reference to the harmonic voltage of the leading side of the coil.

Pitch	Harmonics Order				
	1	3	5	7	11
4/5	18	54	0	-54	18
5/6	15	45	75	-75	-15
6/7	12.9	38.5	64.5	0.000	-38.5

Table 6-2 Phase angles in electrical degrees of the voltage-harmonics with respect to the fundamental of the voltage in the leading side of the coil.

The pitches in Tables 6-1 and 6-2 expressed in electrical degrees are:

$$\frac{4}{5} \cdot 180° = 144 \cdot ° \qquad \frac{5}{6} \cdot 180° = 150 \cdot ° \qquad \frac{6}{7} \cdot 180° = 154.286 \cdot °$$

Assuming that the generator phase-windings are connected in wye and that each winding has a single coil. Then the harmonic angles provided in Table 6-2 are the angles with reference to the fundamental of the generator line-to-neutral voltage.

6.10 Stator Windings

Modern and large, 60 Hz, cylindrical rotor synchronous generators are all three-phase, two poles, and 3600 RPM. And most of them use full-pitch windings.

Figure 6-5 shows a cross section of a three-phase, two poles, AC generator illustrating a concentrated stator winding where only one phase is shown. The winding pitch is 180 degrees which is the arc in degrees between adjacent poles. The winding pitch could be also express in centimeters measured around the inner circumference of the stator.

Figure 6-14 shows the net mmf (NI/2) in the air gap, instead of flux density because the flux density depends of the specific generator design. The phasor addition of the mmf produced by the field winding and the mmf of the stator reaction produces the resultant mmf in the air gap which is the one that induce the emf in the stator windings. As shown in Figs. 6-8 and 6-9 each flux line crosses the air gap twice and therefore the mmf drop in the air gap is half of the resultant total or NI/2. The mmf distribution in the air gap as a function of θ is step-like with an amplitude of NI/2.

The mmf changes direction when the rotor moves from one active side to the other of the stator coil, the upper and the lower conductors. In fact, the mmf jumps NI ampere-turns when the rotor crosses anyone of the coil conductors. This flip in the sign of the mmf occurs in the very short time it takes the rotor to sweep by the conductor. The rotor angular speed is 60 revolutions per second and it takes approximately π / 100 rad or 83 microseconds to sweep by one of the coil conductors. Symbolically,

$$\omega = 60 \cdot 2\pi = 120\pi = 377 \quad \frac{\text{rad}}{\text{sec}} \qquad \Delta\theta = \frac{\pi}{100} = 0.0314$$

Assumed sweep angle in radians

$$\frac{\Delta\theta}{\omega} = \frac{0.0314}{377} = 83.2891 \times 10^{-6} \quad \text{seconds}$$

Approximately sweep time

The voltage drop in a conductor or coil of self inductance L is proportional to the rate of change of the current. The magnitude of the inductive reactance (ω L) of L is proportional to the frequency or angular velocity of the driving voltage. Some helpful and basic equations are given below.

$$v = L \cdot \left(\frac{d}{dt}i\right) \qquad X_L = \omega \cdot L = 2 \cdot \pi \cdot f \cdot L \qquad e = -N \cdot \left(\frac{d}{dt}\phi\right)$$

Lenz law, see Eq. (1.7)

Changing current produces a voltage drop, which in phasor algebra is expressed as follows:

$$\overrightarrow{V_L} = \overrightarrow{Z_L} \cdot \overrightarrow{I_L} = \overrightarrow{(j \cdot X_L)} \cdot \overrightarrow{I_L}$$ Neglecting circuit resistance   (6.42)

The accepted engineering practice, to explain the functioning of synchronous generators using simple models and well known electrical circuit analysis concepts, was to introduce the concept of *synchronous reactance*, which stator winding conductors are supposed to have besides the well know leakage reactance. However, to simplify the treatment it was necessary to make the following assumptions:

- That the entire magnetic circuit has constant permeability. So, the resultant mmf in the air gap could be considered as having two separated components: the rotor component and the armature or stator component which are proportional to their respective DC and AC currents. The constant permeability assumption allows to superpose the effect of the separated components.
- The air gap is uniform, the length of the air gap is constant around the cylindrical rotor.
- The rotor component of the air gap mmf is sinusoidal distributed around the air gap.
- The armature winding conductors are uniformly distributed around the stator and carry balanced sinusoidal currents. This means that all the harmonics are neglected and therefore the mmf of the stator component is proportional to the fundamental of the stator winding current. Concisely, the mmf of the armature reaction is constant in magnitude, sinusoidal distributed around the air gap, and rotates at synchronous speed.

Figure 6-18 illustrates the two components theory of the resultant mmf in the air-gap.

E1 -- is the emf generated at no load condition by the rotor component of the air gap flux.

E2 --is the emf produced by the armature reaction due to the load current.

Er -- is the emf generated by the resultant flux in the air gap.

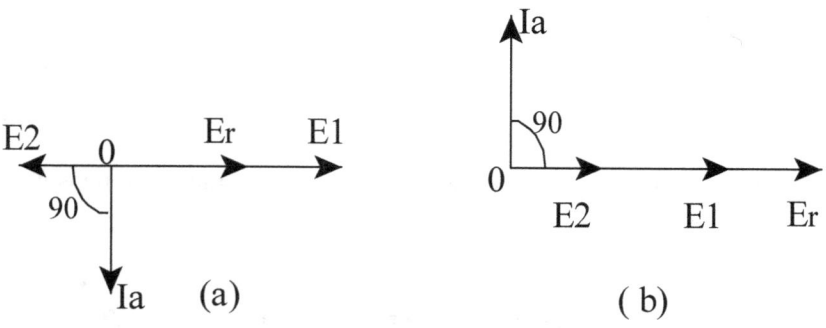

Figure 6-18 Illustration of the air gap two components theory

At no load condition there is not armature reaction and the generated voltage is obviously due only to rotor component or the magnetic flux produced by the DC current in the rotor winding. *As shown in section 1.2 and Fig. 1-1 a sinusoidal current creates a sinusoidal mmf that induces an emf that lags the current by 90 degrees.* Therefore when the fundamental of the armature current lags E1 by 90 degrees, the voltage E2 that it creates, which represent the armature reaction, lags current Ia by 90 degrees and is in direct opposition to E1. See Fig. 6-18a. In the case depicted in Fig. 6-18b the armature current leads E1 by 90 degrees and therefore E2 is in phase with E1. Concisely, if the magnetic circuit has constant permeability, then the resultant flux in the air gap could be considered as having two separated components: the rotor component and the armature or stator component. The armature component of the flux is due to the armature reaction or the flux created by the current circulating in the armature winding, this changing flux induces a voltage in the armature winding that opposes the cause that created it, the combined action of both the DC field excitation and the AC armature current.

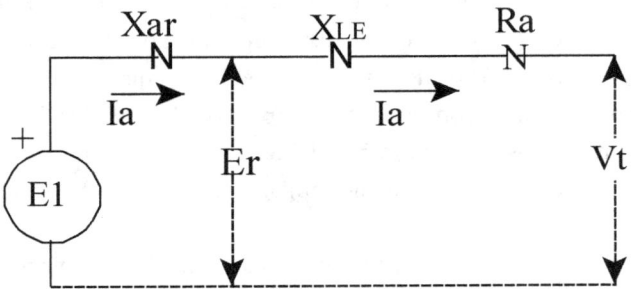

Figure 6-19 Simple model of a synchronous generator.

From Fig. 6-18 we conclude that, the armature reaction, phasor E2, which is always in quadrature and proportional to the armature current, could be mimic by the voltage drop in an inductive reactance Xar as follows:

$$\overrightarrow{E2} = \overrightarrow{(-j \cdot X_{ar})} \cdot \overrightarrow{I_a}$$
(6.43)

From Fig. 6-19 we obtain the resultant voltage generated in one of the generator phases, see Eq. (6.44).

$$\overrightarrow{E_r} = \overrightarrow{E1} + \overrightarrow{E2} = \overrightarrow{E1} - \overrightarrow{(j \cdot X_{ar})} \cdot \overrightarrow{I_a}$$
Generated line to neutral voltage
(6.44)

$$\overrightarrow{V_t} = \overrightarrow{E1} - \overrightarrow{(j \cdot X_{ar})} \cdot \overrightarrow{I_a} - \overrightarrow{(j \cdot X_{LE})} \cdot \overrightarrow{I_a}$$
Line to neutral terminal voltage
(6.45)

$$\vec{V_t} = \vec{E1} - \overrightarrow{j\left(X_{ar} + X_{LE}\right)} \cdot \vec{I_a} = \vec{E1} - \overrightarrow{\left(j \cdot X_s \cdot I_a\right)} \quad \text{Line to neutral terminal voltage} \quad (6.46)$$

$$X_s = X_{ar} + X_{LE} \qquad \text{Denotes synchronous reactance.} \qquad (6.47)$$

Considering the small resistance of the armature as depicted in Fig. 6-19 we obtain:

$$\vec{V_t} = \vec{E1} - \vec{I_a}\left(R_a - j \cdot X_s\right) \tag{6.48}$$

In practice the accepted designation for the generated voltage is Eg not E1, so Eq. (6.46) becomes Eq. (6.49)

$$\vec{V_t} = \vec{Eg} - \vec{I_a} \cdot \left(R_a - j \cdot X_s\right) \tag{6.49}$$

6.11 Types of Stator Windings

This important subject is more an art than a science. Only a brief description without covering all the details is presented here. The reader most keep in mind that the magnetic flux, links only the active sides of the coil windings. It does not links the end connections because they are tangential to the flux path. Figure 6-20 illustrates the simplified version of a *wave* winding showing the conductors routing and end connections. The winding shown in Fig. 6-20 is a *wave winding* because it is wound consecutively from pole to pole. And it is a concentrated winding with all the conductors of any given coil-side located in the same slot. Furthermore, it is a half-coil winding since there is only one-half coil per slot. And it is a full-pitch winding since any of the coils has its two sides 180 degrees apart. Concisely, this three-phase concentrated winding has two open coils per phase. The reader most realize that the phases most be laid out 120 electrical degrees apart.

There are three types of stator windings:

1. Lap windings, they could be used either with concentrated or distributed windings

2. Wave windings, they could be used either with concentrated or distributed windings

3. Spiral windings, they must be used only with distributed windings

Fractional pitch windings should be of the two-layer type. Lap windings were developed to be used with two-layer per slot applications. The conductor "laps" from the bottom of one slot to the top of the other (that would be located at the end of the corded connection). However full-pitch windings could be of the single-layer or two-layer types, and they could be wave or lap windings. The type of winding makes no difference in the voltage generated, assuming the same flux density and distribution. However, because the windings end connections are affected by the type of winding used, the order in which the stator conductors are connected in series between the generator terminals could be different. Furthermore, the insulation required between the end-connection conductors at cross-overs also could be different.

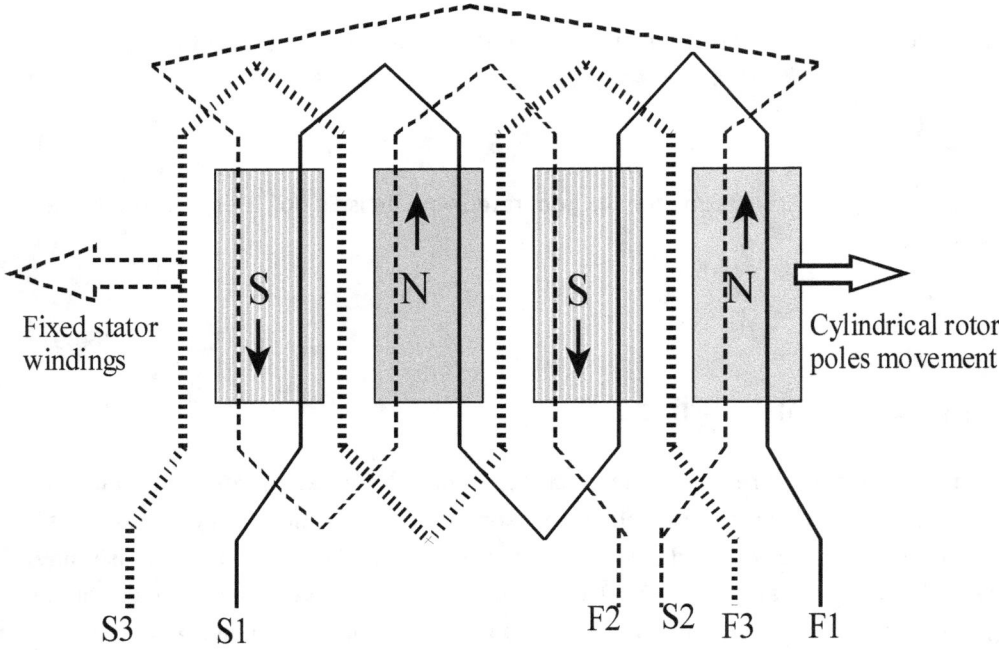

Figure 6-20 Four Poles concentrated wave winding

Chapter 7

Large AC generators Operation

In chapter 5 it was stated that in steady state conditions the magnetomotive force (mmf) existing in the air gap is the phasor addition of the mmfs produced by the rotor and the stator. The reader should realize that both are rotating at the same synchronous speed or 60 rev/sec. (for a 2-pole generator). More explicit, the mmf produced in the air gap by the stator winding AC currents is stationary with reference to the mmf produced also in the air gap by the rotor winding DC current and the resultant mmf in the air gap is practically sinusoidal. Fig. 7-1 shows the single phase equivalent circuit of AC generators with cylindrical rotor (non-salient poles).

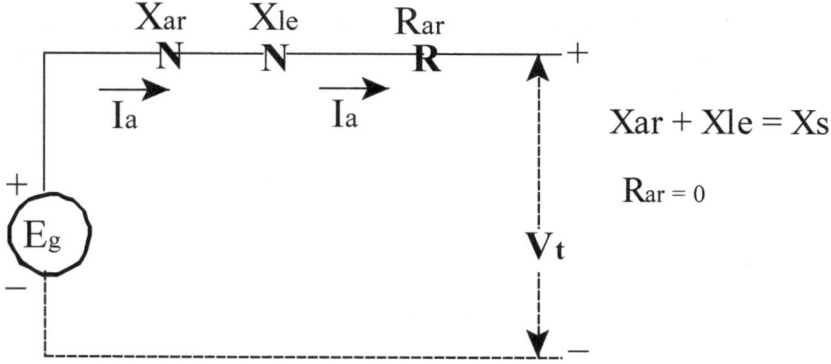

Figure 7-1 AC generator with cylindrical rotor single-phase equivalent circuit for a balanced system.

$$\vec{V}_t = \vec{E}_g - \vec{I}_a\left(R_{ar} + j\left(X_{ar} + X_{le}\right)\right) \tag{7.1}$$

$\vec{V}_t$ Generator output terminal voltage, line to neutral

$\vec{E}_g$ Single phase generated emf or *no load* single phase generated voltage

$\vec{I}_a$ Armature or line current or current delivered to the load

X_{ar} Steady state armature winding single-phase reactance

$X_{l\epsilon}$ Steady state armature winding single-phase leakage reactance

R_{ar} Steady state armature winding single-phase resistance, usually neglected

X_s Steady state overall single-phase reactance of the armature winding.

$$X_s = X_{ar} + X_{l\epsilon}$$

The leakage reactance is due to spreading of the magnetic flux outside the magnetic circuit. This fringing flux increases the magnetic flux density in specific places of the magnetic circuit, such as pole faces tips or the teeth in between slots. And the required correction is attributed to the leakage reactance. Equation (7.1) could be simplified as show in Eq. (7.2).

$$\overrightarrow{V_t} = \overrightarrow{E_g} - \overrightarrow{I_a} \cdot \left(R_{ar} + j \cdot X_s \right) \tag{7.2}$$

$$X_s = X_{ar} + X_{l\epsilon} \quad \textit{is the synchronous reactance.}$$

Neglecting the resistance of the armature winding, we obtain:

$$\overrightarrow{V_t} = \overrightarrow{E_g} - j \cdot \overrightarrow{I_a} \cdot X_s \tag{7.3}$$

The resistance of the armature winding is very small, so in this chapter we always neglect it. However, the reader must realize that the copper losses are very large and can't be neglected.

7.1 Power Flow

Utilities engineers are very concern with the power flow, real and reactive. So let us consider a generator connected to a bus of a very large power system. This system is so large than the bus voltage and frequency will not change due to changes in the excitation or shaft power input to the generator. This bus is called an infinite bus. To maintain constant the power delivered by the generator to the infinity bus sometimes it is necessary to change the magnitude of the generated emf and this could be accomplish by changing the shaft RPM or the generator DC field excitation. Let us refer the discussion to figures 7-1, 7-2, and 7-4.

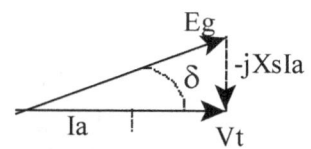

7-2a Normal excited generator

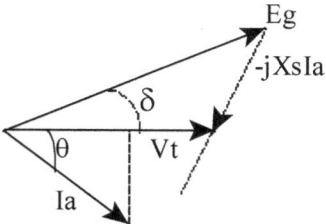

7-2b Over-excited generator

$$Ia = Is = I_R$$

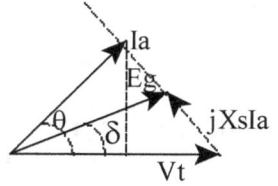

Figure 7-2 shows phasor diagrams
for normally, over, and under
excited generators

7-2c Under-excited generator

The condition for a generator to be considered as normally excited is given by Eq. (7.4)

$$\vec{E_g} \cdot \cos(\delta) = \vec{V_t} \tag{7.4}$$

As could be seen in Fig. 7-2a in a normally excited generator the load current is in phase with the generator terminal voltage. The reader must realize that Fig. 7-2a is provided only as a reference that is not a necessary starting condition. Let us assume that the system, including the load is balanced, then to maintain constant the real power delivered by the generator when the load, or the system changes, it is necessary to maintain constant expression (7.6). Symbolically:

$$\vec{V_t} = \vec{V_b} + \vec{I_a} \cdot \vec{Z_L} \tag{7.5}$$

$\vec{Z_L}$ is the impedance connecting the generator to the infinity bus neglecting the shunt admittances. See Fig. 7-4

$\vec{V_b}$ is the infinity-bus line to neutral voltage, constant

$$\vec{V_t} \cdot \vec{I_a} \cdot \cos(\theta) = \text{constan}^{\cdot} \qquad \text{Real power delivered by the generator} \quad (7.6)$$

θ is the phase angle of the load current with reference to the generator terminal voltage.

$$\left(\vec{V_b} + \vec{I_a} \cdot \vec{Z_L} \right) \cdot \vec{I_a} \cdot \cos(\theta) = \text{constan}^{\cdot} \tag{7.7}$$

Eq. (7.8) is obtained from Eq. (7.3)

$$\vec{E}_g = \vec{V}_t + j \cdot X_s \cdot \vec{I}_a \tag{7.8}$$

$$\vec{E}_g = \vec{V}_t + \overrightarrow{\left(j \cdot X_s\right)} \cdot \vec{I}_a = \vec{V}_b + \vec{I}_a \cdot \vec{Z}_L + \overrightarrow{\left(j \cdot X_s\right)} \cdot \vec{I}_a \tag{7.9}$$

$$\vec{E}_g = \vec{V}_b + \left[\vec{Z}_L + \overrightarrow{\left(j \cdot X_s\right)}\right] \cdot \vec{I}_a \tag{7.10}$$

You should not try, even if you could, to change the generator terminal voltage. This action, for sure would involve changes in the settings and taps of down the line equipment. However, the generator terminal voltage changes in accordance with the current delivered to the infinity bus, due to the change in voltage drop in the connecting line. In Eq. (7.7) the factors you can change are only the load current and the power factor angle, and one change implies the other. *The reader must understand that at the trigger-time all the affected variables start to change and that the phasor equations and figures illustrating the consequences of the trigger event are only valid for the new steady state operating condition.* One way to increase the load current is increasing the generator's rotor DC excitation. This action will increase phasor Eg and therefore the current phasor must change in magnitude and angle to produce the required voltage drop in the line reactance (jXsIa). See Fig. 7-1 and 7-2 and Eq. (7.8). Because the parameters of the circuit shown in Fig. 7-4 and the infinity-bus voltage are constant, an increase in the load current makes the voltage drop in the transmission line connecting the generator to the infinity-bus larger, and therefore the generator terminal voltage should also increase to deliver the corresponding (new Ia) amount of power to the infinity bus.

The over-excited generator illustrated in Fig. (7-2b) provides lagging current with reference to the generator terminal voltage and delivers real and reactive power to the infinity bus. If the generator is under-excited as illustrated in Fig. (7-2c) then the generator provides leading current with reference to its terminal voltage and delivers real power to the infinity bus, but takes reactive power from the system. So, in the three cases shown in Fig. 7-2 the generator provides real power to the system. *However, over-excited generators supply reactive power to the system and under-excited generators take reactive power from the system.*

In all cases for a generator to comply with Eq. (7.3), phasor Eg must always lead phasor Vt. In general a leading current through a reactance produces a rise in voltage. And a lagging current through a reactance produces a decrease voltage

7.2 Power Angle

Assuming that a generator, whose model is shown in Fig 7-1, is operating under the conditions established in section 7.1. The angle δ between Eg and Vt, shown in Fig. 7-2, is the power angle in which Eg always lead Vt. The power balance is always maintained no matter what. So when the input mechanical power delivered to the generator shaft increases by increasing the mechanical torque supplied by the turbine, then the generator rotor angular speed increases, hence the angular speed of the mmf created by the DC winding of the rotor also increases with respect to the rotating mmf created by the stator. Resulting in a new steady state operating condition with a greater power angle. Fig. 7-3 illustrates this change when compared to Fig. 7-2b. Increasing δ without changing the DC excitation, results in a larger Ia

and smaller θ. The reader must realize that $j \cdot X_s \cdot \overrightarrow{I_a}$ is the only term that can change to comply with Eqns. (7.8 and 7.9). And phasor Ia is the only variable within this term. *Succinctly, as a consequence of the increase in δ, the generator has increased the current and the power delivered to the network.*

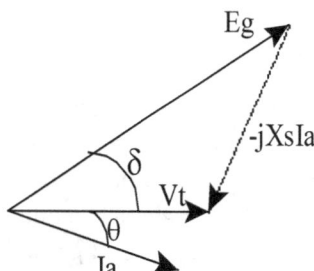

Figure 7-3 As a consequence of increasing the mechanical power delivered to the shaft Ia, cos θ and δ all increase. And the generator supply more power to the network.

Although |Eg| remain the same after increasing the torque delivered to the generator's shaft; the magnitude of the line current, cos θ, and δ all increased. Hence the generator supplies more power to the infinity bus. And now the mechanical power delivered by the turbine to the generator shaft is equal to the electrical power delivered by the generator to the electrical network plus losses. In conclusion the power balance is reestablished at a new operating condition. In the condition depicted in Fig. 7-3 the rotor field excitation is the same than in the previous operating condition depicted in Fig. 7-2, and therefore |Eg| is the same, but the real electrical power output given by |Vt||Ia|cos θ is larger for the operating condition depicted in Fig. 7-3 than the one depicted in Fig.7-2. The increment in the electrical power delivered to the infinity bus as consequence of increasing the mechanical shaft power could be easily determined by using the synchronizing power coefficient. *Succinctly the increment of the power angle times the synchronizing power coefficient provides the increment in electrical power delivered by the generator.* Symbolically:

$$P_{e\Delta} = P_{sy} \cdot \delta_{\Delta} \tag{7.11}$$

$$P_{sy} = P_{emax} \cdot \cos\left(\delta_0\right) \quad \text{Synchronizing power coefficient} \tag{7.12}$$

$$P_{emax} = \frac{|E_g| \cdot |V_t|}{X_s} \quad \text{See Eq. (6.31)} \tag{7.13}$$

$P_{e\Delta}$ Change in electrical output power of the generator

P_{sy} Synchronizing power coefficient

δ_{Δ} Change in power angle

δ_0 Power angle before changing the torque delivered to the generator's shaft.

7.3 Power Flow from Generator to Infinity Bus.

The power flow in the network illustrated in Fig. 7-4 is given in terms of the generalized circuit constant A, B, C, D, see Fig. 7-5, and they include the generator synchronous reactance as well as the impedance of the transmission line connecting the generator to the infinity bus. The network is considered symmetrical and balanced and therefore only a single phase analysis is necessary. Actually, the transmission line is modeled, see Fig. 7-6, as a linear and passive symmetrical π with two pair of terminals. Although in fact the symmetrical π section is a three-terminal network. The reader should keep in mind that the linearity (effects proportional to causes) of the network components is an ideal concept to facilitate the analysis and computations. Furthermore, rigorously speaking, each shunt admittance of the π representation should be shunted by a pure resistance to account for the leakage current between conductors, due to air humidity or imperfect insulation. In addition, the always necessary transformers are nonlinear devices owing to iron core saturation produced by even harmonics, or DC offset during short circuit conditions. However, the fact that the analysis is for steady-state conditions makes the assumption that all the components of the network are linear acceptable.

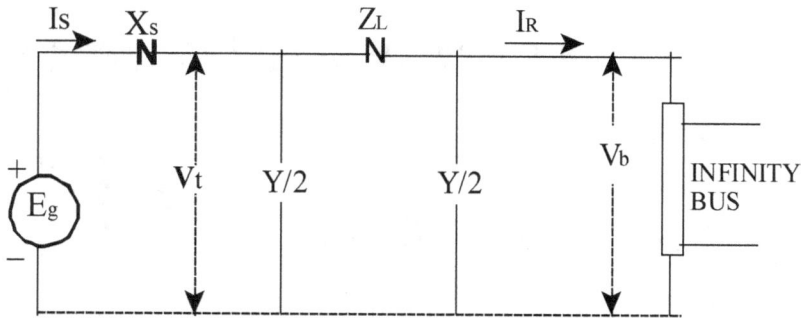

Figure 7-4 Single-phase version of a cylindrical rotor AC generator feeding an infinity bus

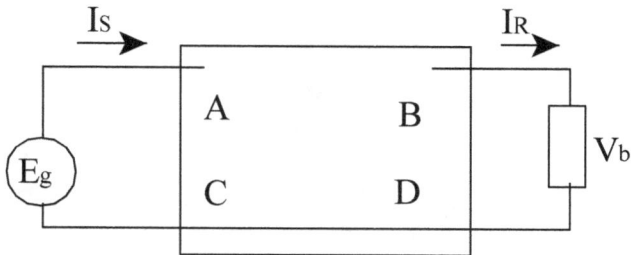

Figure 7-5 Generalized circuit constants of two-terminal pair network.

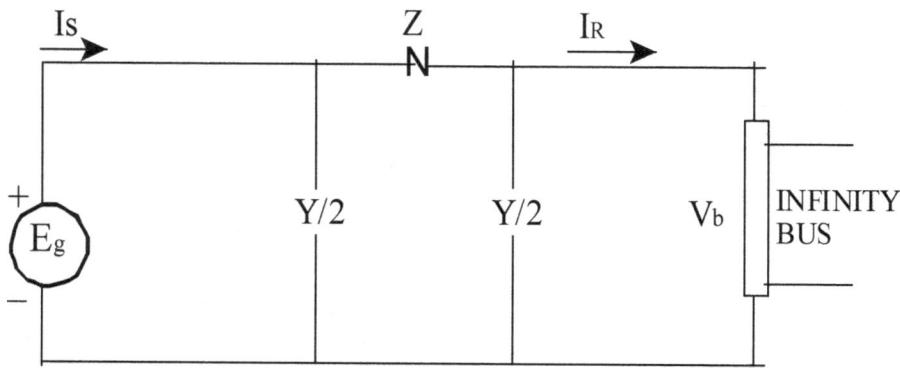

Figure 7-6 Network equivalent symmetrical π

Expressing the voltage and current at the sending-end of the network in terms of the voltage and current at the receiving-end, we have:

$$E_g = \left(I_R + \frac{Y}{2} \cdot V_b \right) \cdot Z + V_b$$

The arrows on top of the phasors have been neglected

$$E_g = \left(\frac{Z \cdot Y}{2} + 1 \right) \cdot V_b + Z \cdot I_R \tag{7.14}$$

$$E_g = A \cdot V_b + B \cdot I_R \qquad \text{Sending voltage in term of A and B} \tag{7.15}$$

Equating (7.14) to (7.15) we obtain:

$$A = \frac{Z \cdot Y}{2} + 1 \quad \text{Unit less} \tag{7.16}$$

$$B = Z \quad \text{ohms} \tag{7.17}$$

$$I_S = \frac{Y}{2} \cdot E_g + \frac{Y}{2} \cdot V_b + I_R$$

$$I_S = \left[\left(\frac{Z \cdot Y}{2} + 1 \right) \cdot V_b + Z \cdot I_R \right] \cdot \frac{Y}{2} + V_b \cdot \frac{Y}{2} + I_R$$

$$I_S = \frac{Y}{2} \left(\frac{Z \cdot Y}{2} + 1 \right) \cdot V_b + \frac{Y}{2} \cdot V_b + \left(\frac{Z \cdot Y}{2} + 1 \right) \cdot I_R$$

$$I_S = \left(\frac{Z \cdot Y^2}{4} + \frac{Y}{2} + \frac{Y}{2} \right) \cdot V_b + \left(\frac{Z \cdot Y}{2} + 1 \right) \cdot I_R$$

$$I_S = Y \left(\frac{Z \cdot Y}{4} + 1 \right) \cdot V_b + \left(\frac{Z \cdot Y}{2} + 1 \right) \cdot I_R \tag{7.18}$$

$$I_S = C \cdot V_b + D \cdot I_R \qquad \text{Sending current in terms of C and D} \tag{7.19}$$

Equating (7.18) to (7.19) we obtain:

$$C = Y \left(\frac{Z \cdot Y}{4} + 1 \right) \quad \text{mhos} \tag{7.20}$$

$$D = \frac{Z \cdot Y}{2} + 1 \quad \text{unit less} \tag{7.21}$$

Of the four generalized circuit constant only three are independent and they related by the following equation.

$$A \cdot D - B \cdot C = 1 \tag{7.22}$$

Summarizing, the symmetrical π equations in terms of the generalized constant are:

$$E_g = A \cdot V_b + B \cdot I_R$$

$$I_S = C \cdot V_b + D \cdot I_R$$

The receiving-end current is obtained from Eq. (7.15)

$$I_R = \frac{E_g - A \cdot V_b}{B} \tag{7.23}$$

Where the following notation is assigned

$$A = |A| \cdot \varepsilon^{j \cdot \alpha} \quad B = |B| \cdot \varepsilon^{j \cdot \beta} \quad E_g = |E_g| \cdot \varepsilon^{j \cdot \delta} \quad V_b = |V_b| \cdot \varepsilon^{j \cdot 0} \tag{7.24}$$

$$I_R = \frac{|E_g| \cdot \varepsilon^{j \cdot \delta} - |A| \cdot \varepsilon^{j \cdot \alpha} \cdot |V_b|}{|B| \cdot \varepsilon^{j \cdot \beta}} = \left|\frac{E_g}{B}\right| \cdot \varepsilon^{j \cdot (\delta - \beta)} - \left|\frac{A \cdot V_b}{B}\right| \cdot \varepsilon^{j \cdot (\alpha - \beta)} \tag{7.25}$$

$$\overline{I_R} = \left|\frac{E_g}{B}\right| \cdot \varepsilon^{j \cdot (\beta - \delta)} - \left|\frac{A \cdot V_b}{B}\right| \cdot \varepsilon^{j \cdot (\beta - \alpha)} \qquad \text{Conjugate of } I_R \tag{7.26}$$

The complex power at the receiving end is

$$P_b + j \cdot Q_b = |V_b| \cdot \overline{I_R} \tag{7.27}$$

The notation used in Eqns. (7.25 and 7.26) although correct could be misleading. For instance, instead of indicating the magnitude of the _phasor operations_, we will carry out the operations only with the _magnitude of the phasors._ For example:

$$\left|\frac{E_g \cdot V_b}{B}\right| = \frac{|E_g| \cdot |V_b|}{|B|}$$

$$P_b + j \cdot Q_b = \frac{\left|E_g\right| \cdot \left|V_b\right|}{|B|} \cdot \varepsilon^{j \cdot (\beta - \delta)} - \frac{|A| \cdot \left(\left|V_b\right|\right)^2}{|B|} \cdot \varepsilon^{j \cdot (\beta - \alpha)} \tag{7.28}$$

Angles α and β are parameters of the connecting _transmission line_ and the _generator_ and their values depend of the generator synchronous reactance and the series and shunt line-impedances. Therefore in Eq. (7.28) there are not variable.

$$\varepsilon^{j \cdot \theta} = \cos(\theta) + j \cdot \sin(\theta) \qquad\qquad \varepsilon^{-j \cdot \theta} = \cos(\theta) - j \cdot \sin(\theta) \tag{7.29}$$

$$P_b = \frac{\left|E_g\right| \cdot \left|V_b\right|}{|B|} \cdot \cos(\beta - \delta) - \frac{|A| \cdot \left(\left|V_b\right|\right)^2}{|B|} \cdot \cos(\beta - \alpha) \quad \text{Real power} \tag{7.30}$$

$$\text{Reactive power}$$

$$Q_b = \frac{\left|E_g\right| \cdot \left|V_b\right|}{|B|} \cdot \sin(\beta - \delta) - \frac{|A| \cdot \left(\left|V_b\right|\right)^2}{|B|} \sin(\beta - \alpha) \tag{7.31}$$

Equations (7.28), (7.30) and (7.31) provide the complex, real, and reactive power delivered to the infinity bus. By definition you cannot change the voltage of the infinity bus. Therefore the only variables left are the power angle δ and $|Eg|$, but because the output voltage of the generator is regulated $|Eg|$ could be also considered as constant. Therefore the only variable in Eqns (7.28, 7.30 and 7.31) is δ, the power or torque angle. This, for a giving generator, is the angle between a mark in the rotor axis and the stator magnetic field axis. When the load changes suddenly, the generator rotor will accelerate or decelerate with respect to the rotating stator field, changing the power angle. In analytical studies with synchronous generators models, the power angle is the angle between two phasors: the generated emf and the generator's terminal voltage (see Fig. 7.1).

As the load connected to the generator changes, the automatic regulator changes the generator's output power by changing the DC excitation of the rotor winding which changes the magnitude of the Eg phasor and forces the line current and the power angle, δ, to change to comply with the following phasor equation $\vec{E}_g - j \cdot X_s \cdot \vec{I}_a = \vec{V}_t$ where Vt is the generator terminal voltage.

Equation (7.30) shows that the maximum real power that could be delivered to the load occurs when $\delta = \beta$. This is expressed symbolically by Eq. (7.32).

$$P_{b.max} = \frac{\left|E_g\right| \cdot \left|V_b\right|}{|B|} - \frac{|A| \cdot \left(\left|V_b\right|\right)^2}{|B|} \cdot \cos(\beta - \alpha) \tag{7.32}$$

If the power-transferring network contains any significant resistance or shunt admittance, then the maximum power that the load (infinity bus) could receive from the generator is less than the maximum power that the generator could deliver to the network. So, if the power required by the load is greater than the maximum power that could be transferred from the generator to the load, which happens when δ = β, then any synchronous motor connected to the infinity bus would slow down as the additional power required by the load comes from the stored rotational energy in motor-load combinations. As the motors slow down they would reach a point at which δ is no longer equal to β, actually it becomes greater than β, therefore accelerating the decrease in the generated power and accelerating the further increase of δ, and eventually falling out of synchronism with the generator. So Eq. (7.32) also provides the steady-state stability limit of the system shown in Fig. 7-4, which is defined as the maximum electrical power that can be transferred to the receiving end (infinity bus) without loss of synchronism. *To summarize, Eq. (7.32) provides the maximum real power that could be delivered to the load (infinity bus) and the steady-state stability limit; both occur when δ = β.* Neglecting the resistances and shunt admittances in the network depicted in Fig. (7-6) we then arrive at the system portrayed in Fig. (7-7) where β = π /2 and Y = 0 and X includes the per-unit values of the line reactance and the transient reactance of the generator. *The generator's transient reactance was selected instead of the steady-state reactance because the generator's rotor mmf is continuously change position with respect to the magnetomotive force produced by the stator current.*

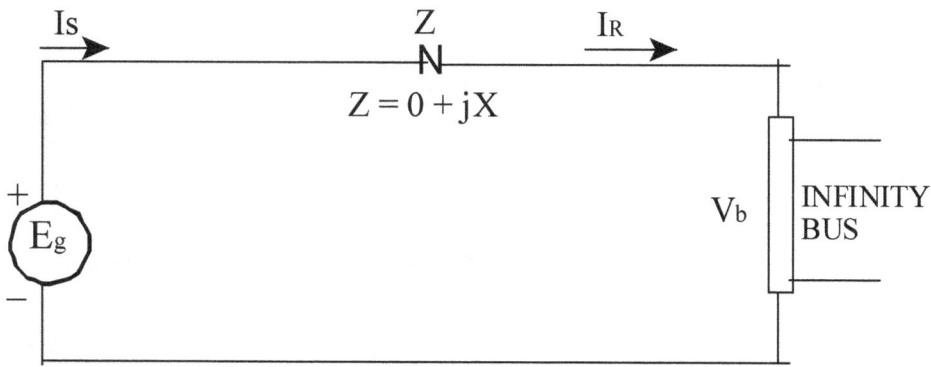

Figure 7-7 Network equivalent symmetrical π where the resistances and shunt admittances where neglected.

The generalized circuit constant for the symmetrical π are given in Eqns. (7.16, 7.17, 7.20, and 7.21).

$$A = \frac{Z \cdot Y}{2} + 1 \qquad B = Z \qquad C = Y \cdot \left(\frac{ZY}{4} + 1 \right) \qquad D = \frac{Z \cdot Y}{2} + 1$$

In Eq. (7.24) the following notation was selected

$$A = |A| \cdot \varepsilon^{j \cdot \alpha} \quad B = |B| \cdot \varepsilon^{j \cdot \beta} \quad E_g = |E_g| \cdot \varepsilon^{j \cdot \delta} \quad V_b = |V_b| \cdot \varepsilon^{j \cdot 0} \qquad \text{Reference}$$

Furthermore, the air capacitive shunt reactance is:

$$X_{Ca} = \frac{1}{\omega \cdot Ca} \quad \text{If} \quad C_a = 0 \quad X_{Ca} = \infty \quad Y = \frac{1}{X_{Ca}} = 0$$

The assumption made (Z = X, Y = 0) and the fact that 1 is a real number, convert the expression of the generalized circuit constant for the symmetrical π to:

$$A = 1 \cdot \varepsilon^{j \cdot \alpha} \quad B = |X| \cdot \varepsilon^{j \cdot \beta} \quad C = 0 \quad D = 1 \cdot \varepsilon^{j \cdot \alpha}$$

$$\alpha = 0 \quad \beta = \frac{\pi}{2} \quad \cos(\beta - \alpha) = 0$$

Therefore the generalized circuit for the symmetrical π depicted in Fig. 7-6 are

$$A = D = 1 \quad B = |X| \cdot \varepsilon^{j \cdot \frac{\pi}{2}} \quad C = 0 \tag{7.33}$$

Substituting the values shown in Eq. (7.33) in Eqns. (7.30) and (7.31) we obtain:

$$P_b = \frac{|E_g| \cdot |V_b|}{|X|} \cdot \sin(\delta) \qquad \text{Real power} \tag{7.34}$$

$$P_{b.max} = \frac{|E_g| \cdot |V_b|}{|X|} \qquad \text{Maximum real power and steady-state stability limit} \tag{7.35}$$

$$P_b = P_{b.max} \cdot \sin(\delta) \tag{7.38}$$
$$\tag{7.36}$$

$$Q_b = \frac{|E_g| \cdot |V_b|}{|X|} \cdot \cos(\delta) - \frac{(|V_b|)^2}{|X|} \cos(\alpha) = \frac{|E_g| \cdot |V_b|}{|X|} \cdot \cos(\delta) - \frac{(|V_b|)^2}{|X|} \tag{7.37}$$

When the infinity bus is receiving the maximum real power, the reactive power it receives is found by substituting $\alpha = \delta = \pi/2$ in Eq. (7.37). The reader should remember that for the generator to remains stable, δ can't be larger than $\pi/2$.

$$Q_b = -\frac{(|V_b|)^2}{|X|} \qquad \text{Reactive power the infinity bus receives while is receiving}$$

maximum real power.

In the operating condition determined by Eqns. (7.35 and 7.38) the minus sign indicates that when the bus receives maximum real power the network is sending back the reactive power indicated by Eq. (7.38). Previously, we covered the power angle and the power flow. And now from Eq. (7.35) we obtain the following conclusions regarding the generator steady-state stability limit.

1. Increasing $|E_g|$ increases the maximum electric power that could be generated and consequently the steady-state stability limit. Remember that Vb is constant. One way of increasing the magnitude of the generated emf is by increasing the generator's excitation. However, connected equipment and insulation ratings limit the amount that the generated voltage could be increased. Furthermore, if the generated real power remains constant when the magnitude of the generated emf is increased, then Eq. (7.34) shows that the power angle, δ, would decrease. And that is a welcome change.

2. Reducing |X| increases the maximum power that could be safely transmitted as well as the steady-state stability limit. This must be carefully considered when selecting the network's equipment (transformers, generators, motors, and conductor's size and separation). However, if the equipment is already installed, then the simplest way to reduce the overall reactance of the network is to install a parallel transmission line capable of carrying the entire load. A parallel line will increase the reliability of the network since one line could carry the entire load in
case of a fault in the other. The additional line will also enhance the transient stability of the system because the more power the system transmits during fault conditions the more stable it will be. Another way of reducing the line voltage drop and increasing the steady stability limit is to connect capacitors in series with the line to decrease the total reactance.

7.4 Cylindrical Rotor Synchronous Generator Load Characteristics.

At constant DC winding excitation and shaft rotation rate, the generator's terminal voltages decline at rapid rate when the armature lagging current (inductive load) increases. The armature reaction increases when the inductive load increases and consequently the net flux density in the air-gap decreases, because the rotor winding excitation is constant. So the emf induced in the armature windings, and the generator terminal voltage most decrease. See, section 6.6 and Fig. 6-13. Figure 7-8 provides an illustration of synchronous generator load characteristic at constant DC excitation and constant shaft RPM. To maintain the terminal output voltage constant at constant shaft RPM, the required DC excitation of the rotor winding is shown in Fig. 7.9 as a function of the load current and power factor. And as indicated all lagging power factor loads and including the unity power factor load require to increase the DC excitation current as the load current increases, to maintain the terminal voltage approximately constant. However, some leading load current at small (0.01PF) power factor requires to reduce the excitation below the no-load excitation current.

198

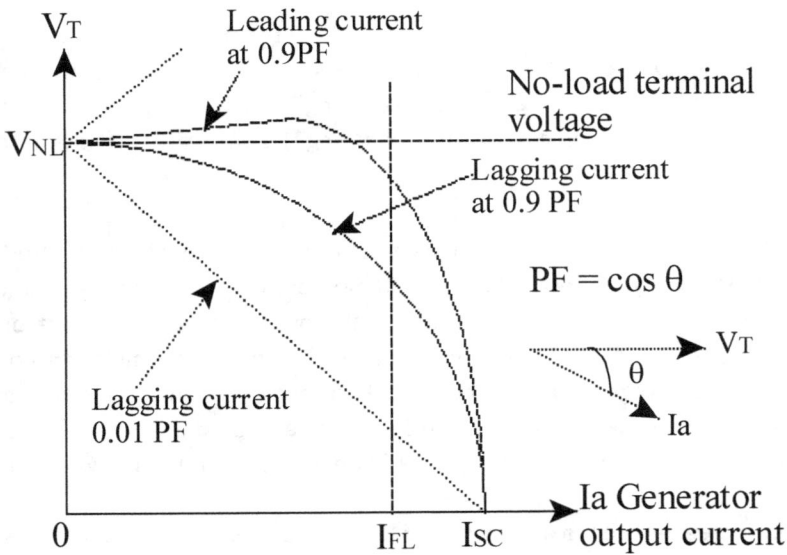

Figure 7-8 Illustration of generator load characteristic at constant excitation and shaft RPM

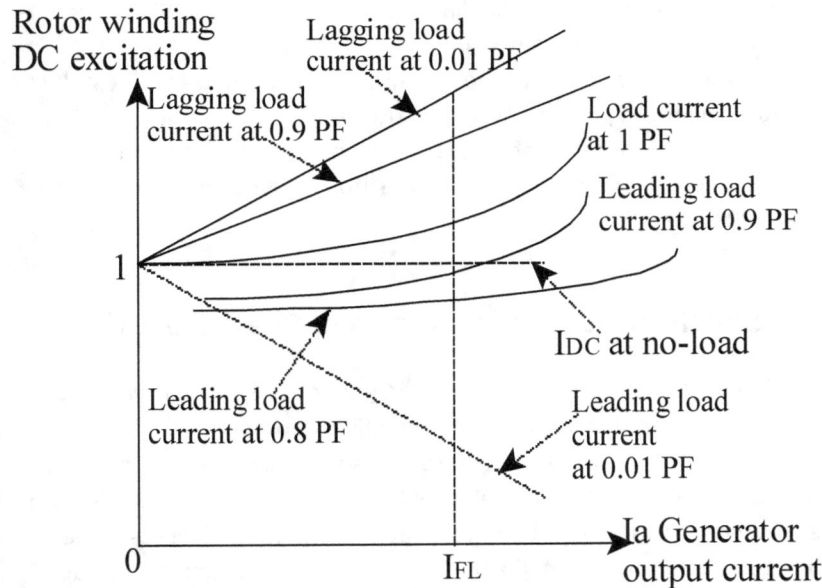

Figure 7-9 Illustration of required DC excitation as a function of the load current.

7.5 Decrease of Prime-Movers Rate of Rotation and of Generator Frequency when Load Increases.

Prime movers of any kind tend to slow down as the load increase. Because, as the load increase the armature reaction mmf increases, the flux density in the air-gap decreases, the rotor angular speed decreases, and these decreases usually are nonlinear. Read section 6-6. However, the generator's automatic control is usually designed to correct these problems by increasing DC excitation current and increasing the amount of steam delivered to the turbine. Furthermore, prime movers (usually turbines) automatically adjust the no-load set-point. Assuming that the generator's automatic control are ON, the prime mover speed drop (SD) is defined as the decrease in revolutions per second as the load changes from no-load to full load in per-unit of revolutions per second at full load. The SD is usually smaller than 0.04 per-unit. Symbolically:

$$SD = \frac{\omega NL - \omega FL}{\omega FL}$$

(7.39)

The generator power output frequency is given in Eq. (6.1) as:

$$f = \frac{Poles \cdot (RPM)}{2 \cdot 60} \frac{cycle}{sec} \qquad f = \frac{Poles \cdot (RPS)}{2} \frac{cycle}{sec}$$

(7.40)

Assuming that the decrease in angular speed is linear, the synchronous generator output terminal characteristic, power vs. angular speed, or power vs frequency are given in Fig. (7-10). The reader should realize that the output terminal characteristics are not applicable to every generator, they are only illustrations. Indeed, every generator has its own specific output terminal characteristic. From Fig. 7-10b we obtain, see Eq. (7.41), the real power output of the generator as a function of the frequency. Symbolically:

$$P = S_f \cdot \left(f_{NL} - f_{FL} \right) \quad megawatts$$

(7.41)

S_f Slope of the dotted line in Fig 7-10b, MW / Hz

$f_{NL} - f_{FL}$ Change in the system frequency, Hz

The reader should realize that a synchronous generator frequency is directly proportional to shaft angular velocity, see Eq. (7.40). And that the shaft mechanical power is directly proportional to the shaft angular velocity or Pm = T^{ω} where, Pm is the incoming mechanical power and T is the shaft torque. Fig. 7-10a shows how **the generator efficiency, η = P / Pm, increases with the power output.**

Angular speed (rev/sec)

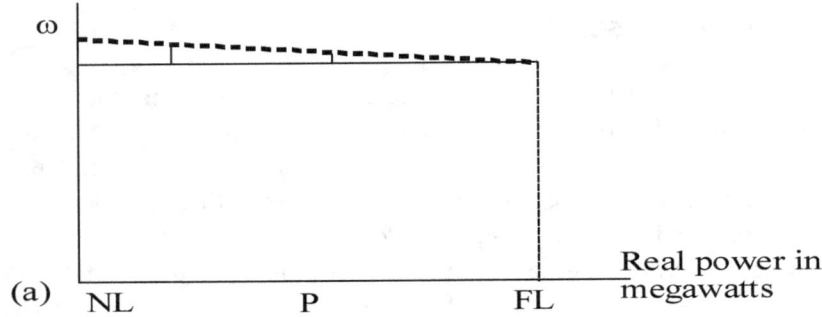

(a)

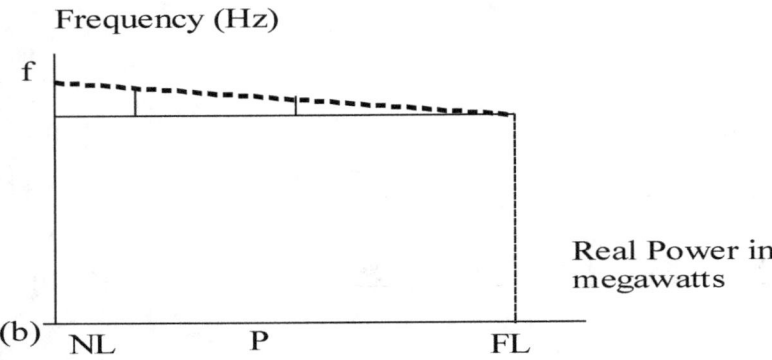

(b)

Figure 7-10 Synchronous generators angular speed and frequency variations with changes in the output of real power

Example 7-1 Consider a synchronous generator supplying real power to load #1 of 2 MW at 0.8 power factor; then load #2 of 1 MW of real power at 0.9 power factor is connected to the generator, which have a no-load operating frequency of 61 Hz and a regulation slope of 0.8 MW per Hz.

Which is the operating frequency before and after connecting load #2, and what could be done to restore the generator output frequency to 60Hz after connecting both loads to the generator? Applying Eq. (7.41) we obtain:

$$P = 0.8\left(61 - f_1\right) \qquad\qquad 2 = 0.8 \cdot 61 - 0.8 \cdot f_1$$

$$f_1 = \frac{0.8 \cdot 61 - 2}{0.8} = 58.5 Hz \qquad\qquad \text{Frequency with load 1 connected}$$

$$P = 0.8\left(61 - f_2\right) \qquad 3 = 0.8 \cdot 61 - 0.8 \cdot f_2$$

$$f_2 = \frac{0.8 \cdot 61 - 3}{0.8} = 57.25 \text{Hz} \qquad \text{Frequency with load 1 and 2 connected}$$

$$60 - 57.25 = 2.7500 \text{ Hz} \qquad 61 + 2.75 = 63.7500$$

To restore the operating frequency to 60 Hz the turbine no-load set point must be increased 2.75 Hz to 63.75 Hz. The reader must keep in mind that the slope of the generator regulation does not change with the load, as long as the regulation curve remains linear. The generator terminal voltage as a function of the generator reactive power output is illustrated in Fig. (7-11). From which we obtain, see Eq. (7.42), the reactive power output of the generator as a function of the terminal voltage.

$$Q = S_V \cdot \left(V_0 - V_{FL}\right) \qquad \text{Reactive megavolt amperes} \qquad (7.42)$$

S_V Is the slope of the dotted line in Fig. 7-11, MVAR/KV

$V_0 - V_{FL}$ Change in the generator terminal voltage, KV

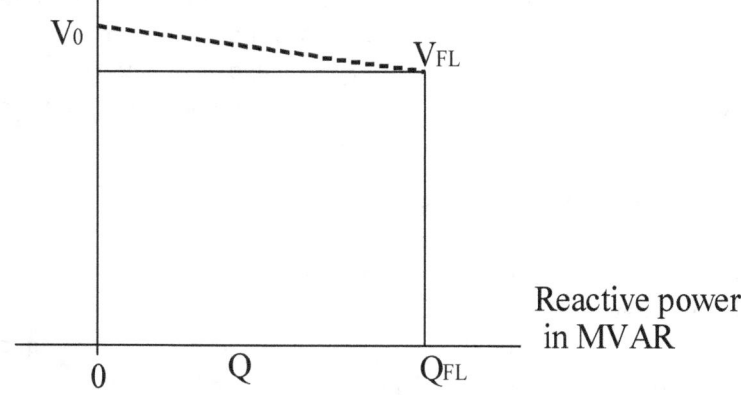

Figure 7-11 Synchronous generator terminal voltage
vs. reactive power

As in the case of frequency, the change in terminal voltage of the generator when the output of reactive power changes is non-linear, but to simplified the analysis it is usually considered as such. It is important to remember that when adding a lagging load (lagging load current), the generator terminal voltage decreases. And if the added load is a leading load (leading load current) then the generator terminal voltage increases. If the synchronous generator is operating alone, isolated from any other power system, then:

-- The amounts of real and reactive powers are as the load demands.

-- The DC field current controls the generator terminal voltage.

-- The turbine governor controls the shaft rotation rate and the operating frequency   of the generator.

7.6 Parallel Operation of Synchronous Generators

There are many reasons and practical needs for operating AC synchronous generators in parallel. Such as:

- Increase the reliability of the power system. Because a failure of one generator does not produce a total power loss of the power system.
- One of the generators could be disconnected from the power system for planned maintenance, while the others keep providing the required power to the system.
- Generators are more efficient operating at full load or near full load. Therefore, at conditions of light load some generators, one or more, could be turned off, or just spinning, while the rest operates in the high efficient zone.

Requirements and procedures for paralleling generators.

Just before the connection, the three-phase set of voltage phasors of the incoming generator output, must be exactly the same than the running generator (or power system) three-phase set of voltage phasors. Same magnitude (plus the very small voltage drop in the connecting line at the small initial current), same phase-angle, and same phase sequence. To be sure the following conditions must be met:

1. The three rms line voltages of the incoming generator must be equal to the existing line voltages of the other (or others) generator.
2. The phase angles of corresponding phases must be equal.
3. Both generators must have the same phase sequence.
4. The frequency of the incoming generator must be a little higher than the frequency of the existing system. When the different in frequencies is small, the phase angles of the incoming generator will change slowly to match the system frequency.

Operation of a generator in parallel with a very large power system or infinity bus.

The infinity-bus voltage and frequency are not affected by changing the load connected to it or drawn from it. Neither is affected by connecting more generators to the infinity-bus. Therefore, the voltages and frequencies of all generators connected to the infinity-bus are the same. This case is a common case in the operation of large utility companies. To simplify let us assume that at no-load condition the three-phase set of voltage phasors of an incoming generator are equal to the set of voltage phasors of the infinity-bus, but just slightly higher in magnitude and frequency. After the connection transient, the incoming generator's three-phase set of voltage phasors and their frequencies will become equal to the infinity-bus voltages and frequencies, and the generator will start supplying a small amount of real power to the infinity-bus. See Fig. 7-12. However if the no-load frequency of the incoming generator is a little lower than the infinity-bus frequency, then after the connection transient, the generator frequency will become equal to the infinity-bus frequency and the generator will start drawing energy from the infinity bus and acting as a synchronous motor. See Fig.7-13.

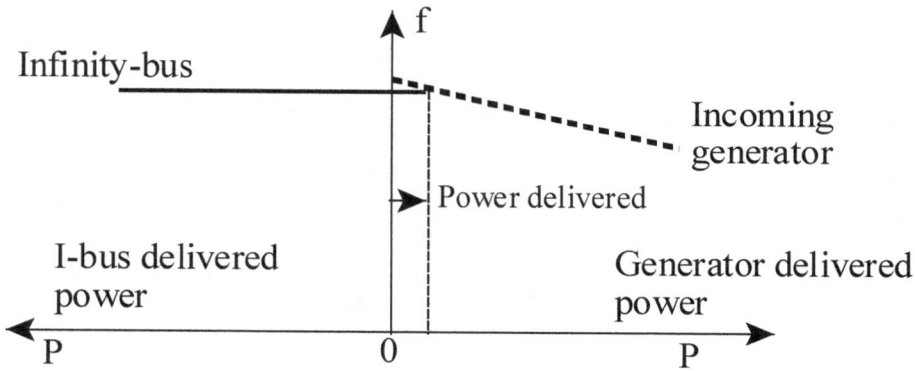

Figure 7-12 Illustration of the power delivered to the infinity-bus when the generator frequency is a little higher.

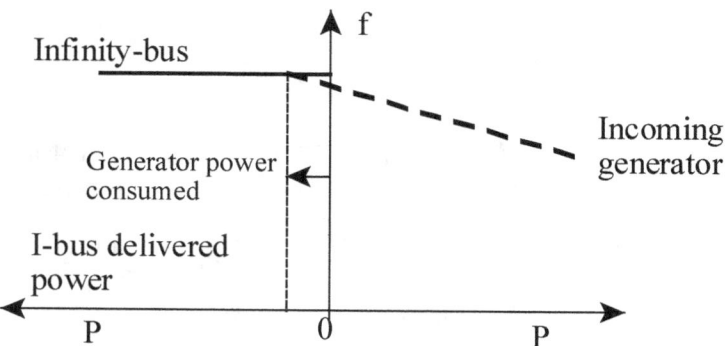

Figure 7-13 Illustration of the power drawn from the infinity-bus when the generator frequency is a little lower

After the incoming generator is connected to the infinity-bus it remains "floating" online delivering or drawing a very small amount of power. To increase the electrical power delivered by the incoming generator *it is necessary to increase the mechanical power delivered by the prime mover*, this requires to increase the shaft rate of rotation. But, as given in Eq. (7.40) the generator frequency is equal to the RPS multiplied by the number of pair of magnetic poles (which are constant), so the frequency (ghost frequency) of the incoming generator increases and the power it delivers also increases. See Figure (7-14). The reader must realize that the frequency of the infinity-bus is constant and would not change and predominates over the frequency of the incoming generator. Meanwhile, because the rotor excitation remains the same, the magnitude of the emf generated by the incoming generator also remains the same, and because the infinity-bus voltage Vb is constant, the power angle δ of the incoming generator must increase to match the fact that the real power delivered by the generator to the infinity bus had increased. See Eq. (7.34). The process is illustrated in Fig. 7-14.

$$P_{B1} + P_{G1} = P_{B2} + P_{G2} = P_L$$

Load connected to the infinity-bus is assumed constant

In conclusion, when the generator operates in parallel to an infinity-bus:

-- The frequency and terminal voltage of the generator are set by the infinity-bus.

-- The mechanical power delivered to the generator's shaft by the prime mover controls the real power delivered by the generator to the infinity-bus.

-- The generator's DC field excitation controls the reactive power delivered by the generator to the infinity-bus provided that the rotor speed remains constant. See section 7-1.

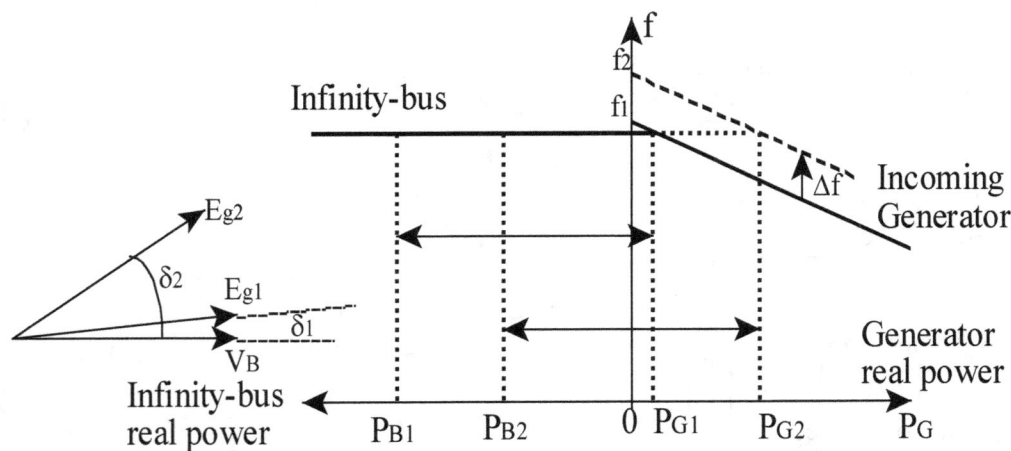

Figure 7-14 Generator share of the load after increasing its frequency

Connecting two generators in parallel. The real and reactive output powers of a generator working alone are determined by the load. And when two generators are connected to a common load, the addition of their real and reactive powers delivered by the two generators is equal to the real and reactive powers demanded by the load. When connecting generators in parallel they should be, preferably, coherent machines. In this way they will swing together responding to system disturbances. Otherwise bad things could happen. Coherent machines could be lumped together for transient analysis even if they have different speed ratings.

Coherency formula for generators G1 and G2

$$\frac{P_{sy1}}{M_1} = \frac{P_{sy2}}{M_2} \quad \frac{\text{rad}}{\text{sec}^2} \tag{7.43}$$

Meaning that in case of a load disturbance, the rotor angular acceleration of both generators must be equal.

Synchronizing power coefficient $\quad P_{sy} = P_{max} \cdot \cos(\delta_0) \, \text{MW} \tag{7.44}$

Rotor angular momentum at synchronous angular speed

$$M = \frac{2 \cdot H}{\omega_s} \cdot G \quad \frac{\text{MW} \cdot \text{sec}^2}{\text{rad}} \qquad \text{See Equation (10-6)} \tag{7.45}$$

$$P_{max} = \frac{|E_g| \cdot |V_b|}{|X|} \qquad \text{Maximum real power and steady state stability limit. See Eq. (7.35)}$$

δ_0 Initial operating power angle

H Stored rotational energy, megawatt-second / generator MVA rating

G Generator rating, MVA

ω_s Generator synchronous angular speed, mechanical radians per second

Consider the case of two generators connected in parallel to a constant load.

Assuming that G1 is already connected to the power system and that G2 is coming online. As stated before the frequency of G2 must be a little higher than the system frequency. Fig. 7-15 illustrates the operating condition described. If the frequency of G2 is increased after the connection, G2 would supplied more power and consequently G1 would supplied less because the total power supplied by the two of them to a fixed load is constant. See Fig. 7-16.

206

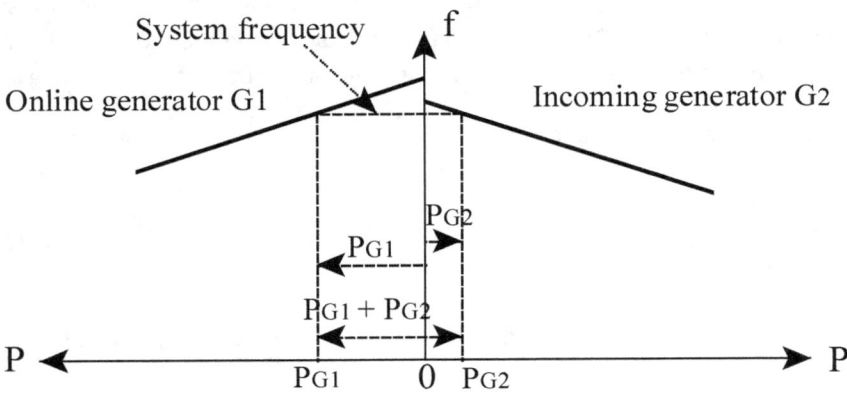

Figure 7-15 Two generators immediately after connecting G2 in parallel with G1.

When two generators are operating in parallel the following operating actions are possible.

1. Increasing the frequency (increasing the shaft rotation rate by letting in more steam) see Fig. 7-16, in one of them produces the following results: (a) Increases the system frequency. (b) Increases the real power delivered by that generator and reduces the real power delivered by the companion generator. Load is assumed constant. Good power sharing between two generators requires that they have significantly sloping frequency-power characteristic, to that effect the shaft speed should drop between 2% to 4% between no-load and full load. See Fig. 7-10b. Otherwise the power sharing will vary widely with the slightly change in rotor speed. **Flat frequency-power characteristic are no good for power sharing.**

2. Increasing the rotor excitation current in one of them, see Fig. 7-17, produces the following results:

    (c) Increases the system operating voltage

    (d) Increases the reactive power delivered by that generator and reduces the reactive power delivered by the other.

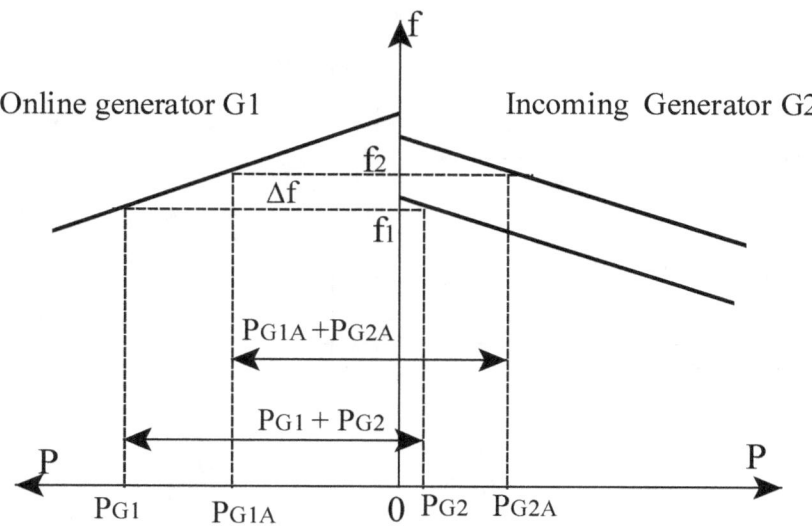

Figure 7-16 Two generators in parallel after increasing the frequency at constant load

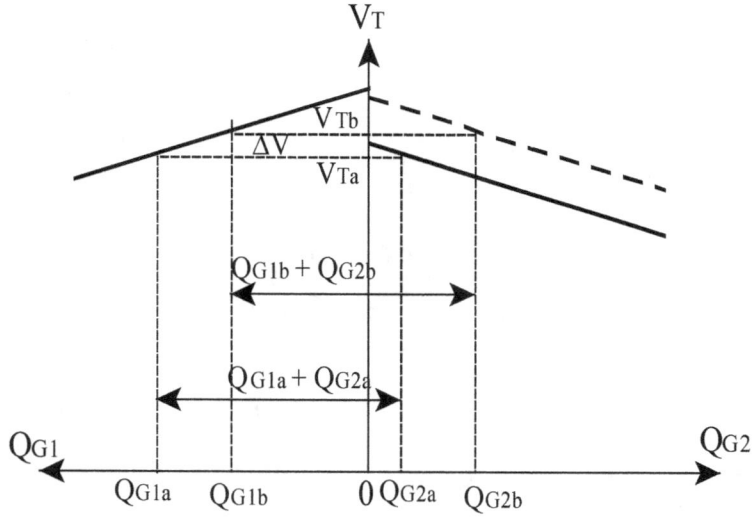

Figure 7-17 Reactive power sharing between two generators in parallel feeding a constant load.

Example 7-2 Two generators of different size, but with the same synchronic speed and frequency slope, are being considered for connection in parallel. They will be feeding a load bus as illustrated in Fig.7-18. The real power provided to the load is 620 MW and the total complex power is 800 MVA. Table 7-1 provides the factory ratings and Table 7-2 provides per unit values based on 300 MVA and 18 kV.

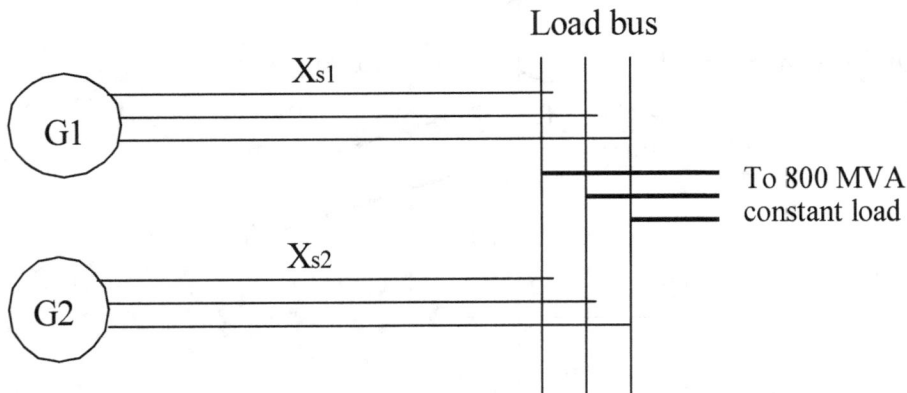

Figure 7-18 Two generators in parallel

Generator	G1	G2	
Type	Steam	Steam	
MVA	500	300	
kV	18	18	
RPM	3600	3600	
H	6	4	MW-sec/MVA
X_s	1.2 pu	0.95 pu	machine based
X_s'	0.15 pu	0.12 pu	machine based

Table 7-1 Generators Factory Ratings

Let us convert the factory ratings to per unit values based on 300 MVA and 18kV.

$$\frac{MVA}{MVA_{base}} = \frac{500}{300} = 1.667 \qquad H_{new} = 6 \cdot 1.667 = 10$$

$$X_{dnew} = 1.2 \cdot \frac{300}{500} = 0.72 \qquad X_{dnew}' = 0.15 \cdot \frac{300}{500} = 0.09$$

Generator	G1	G2
Type	Steam	Steam
MVA	1.667	1
kV	1	1
H	10	4 MW-sec/MVA
X_S	0.72	0.95 pu
X_S'	0.09	0.12 pu

Table 7-2 Generators per unit values based on 300 MVA and 18 kV

Constant Load data:

$$\left(P_L\right)^2 + \left(Q_L\right)^2 = 800^2$$

$$P_L = 620MW \qquad\qquad Q_L = \sqrt{800^2 - 620^2} = 505.57MVAR$$

$$P_1 + P_2 = 620MW \qquad\qquad Q_1 + Q_2 = 505.57MVAR$$

$$800 \cdot \cos(\theta) = 620 \qquad\qquad \theta = acos\left(\frac{620}{800}\right) = 0.684rad = 39.2°$$

Power supplied by both generators:

Real power = 620MW   Reactive power = 505.57 MVAR

Load power factor angle = 39.2 degrees

From Eq. (7.41) we obtain that: $\qquad\qquad P = S_f \cdot \Delta f$

From Eq. (7.40 and 7.41) and the data obtained (assumed) at the installation site for G1 and G2 we have the following sharing of the total real power. See Fig. 7-19

$$f_{NL1} = 61.5Hz \qquad f_{sys} = 60.3Hz \qquad \Delta f_1 = 1.2Hz \qquad S_{f1} = 342\frac{MW}{Hz}$$

$$f_{NL2} = 60.9Hz \qquad f_{sys} = 60.3Hz \qquad \Delta f_2 = 0.6Hz \qquad S_{f2} = 350\frac{MW}{Hz}$$

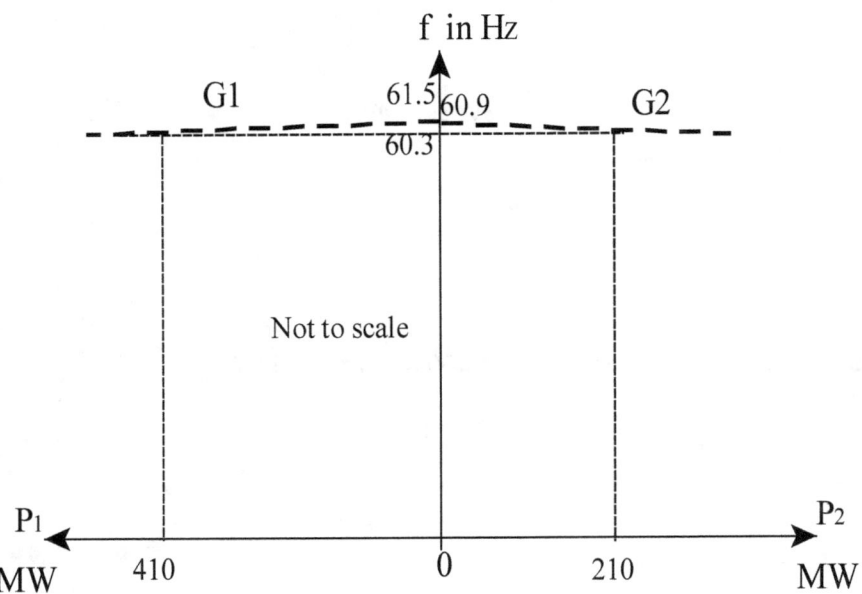

Figure 7-19 Two generators of different size in parallel

Applying Eq. (7.41) we have:

$$P_{load} = P_1 + P_2 = S_{f1}\left(f_{NL1} - f_{sys}\right) + S_{f2}\left(f_{NL2} - f_{sys}\right)$$

$$P_{load} = P_1 + P_2 = S_{f1} \cdot f_{NL1} - S_{f1} \cdot f_{sys} + S_{f2} \cdot f_{NL2} - S_{f2} \cdot f_{sys}$$

$$620 = S_{f1} \cdot f_{NL1} + S_{f2} \cdot f_{NL2} \cdot - f_{sys}\left(S_{f1} + S_{f2}\right)$$

$$f_{sys} = \frac{S_{f1} \cdot f_{NL1} + S_{f2} \cdot f_{NL2} - 620}{S_{f1} + S_{f2}} = \frac{342 \cdot 61.5 + 350 \cdot 60.9 - 620}{342 + 350} = 60.3Hz$$

$$P_1 = 342 \cdot 1.2 = 410.4MW$$

$$P_2 = 350(0.6) = 210MW \qquad \qquad P_1 + P_2 = 410.4 + 210 = 620.4MW$$

Figure 7-19 reflects the data and the results of the computations.

$$P_1{}^2 + Q_1{}^2 = 500^2 \qquad Q_1 = \sqrt{500^2 - 410.4^2} = 286\text{MVAR}$$

Inductive load

$$P_2{}^2 + Q_2{}^2 = 300^2 \qquad Q_2 = \sqrt{300^2 - 210^2} = 214\text{MVAR}$$

Inductive load

$$500 \cdot \cos\left(\theta_1\right) = 410.4 \qquad \theta_1 = \text{acos}\left(\frac{410.4}{500}\right) = 0.608\text{rad} = 34.8°$$

Power factor angle

$$300 \cdot \cos\left(\theta_2\right) = 210 \qquad \theta_2 = \text{acos}\left(\frac{210}{300}\right) = 0.795\text{rad} = 45.6°$$

Power factor angle

Because the system is balanced we can work as it were a single-phase circuit.

$$500 = \sqrt{3} \cdot 18 \cdot I_1 \qquad I_1 = \frac{500}{\sqrt{3} \cdot 18} = 16.038\text{kA}$$

$$\vec{I}_1 = 16038 \cdot \varepsilon^{j \cdot 0.608}$$

RMS amperes

$$\vec{I}_1 = 16038(\cos(0.608) + j \cdot \sin(0.608))$$

$$\vec{I}_1 = 13164 + j \cdot 9161$$

RMS amperes

$$300 = \sqrt{3} \cdot 18 \cdot I_2 \qquad I_2 = \frac{300}{\sqrt{3} \cdot 18} = 9.623\text{kA} \qquad \vec{I}_2 = 9623 \cdot \varepsilon^{j \cdot 0.795}$$

RMS amperes

$$\vec{I}_2 = 9623(\cos(0.795) + j \cdot \sin(0.795)) \qquad \vec{I}_2 = 6739 + j \cdot 6870$$

RMS amperes

$$\frac{\Omega}{\Omega_b} = 1.2$$

Per unit value, machine based, of the G1 synchronous reactance.

$$\Omega_b = \frac{(\text{kV})^2}{\text{MVA}_b} = \frac{18^2}{500}$$

$$\Omega = 1.2 \cdot \Omega_b \qquad \text{See Eq. (1.42)}$$

The ohm values of the generators' synchronous reactances are:

$$\Omega_1 = \frac{1.2 \cdot 18^2}{500} = 0.778 \quad \text{ohm} \qquad \qquad \Omega_2 = \frac{0.95 \cdot 18^2}{300} = 1.026 \quad \text{ohm}$$

$$\frac{18000}{\sqrt{3}} + j \cdot 0 = 10392 \qquad \qquad \text{Line to neutral terminal voltage reference}$$

$$\overrightarrow{E_{g1}} = \overrightarrow{V_T} + \overrightarrow{I_1} \cdot j \cdot 0.778 = 10392 + (13164 + j \cdot 9161) \cdot j \cdot 0.778 = 3265 + j \cdot 10242$$

$$\left|3265 + j \cdot 10242\right| = 10750 \qquad \text{atan}\left(\frac{10242}{3265}\right) = 72.3185 \cdot {}^{c}$$

$$\overrightarrow{E_{g1}} = 10750 \cdot \varepsilon^{j \cdot 72°} \qquad \delta_1 = 72° \qquad \text{Line to neutral emf}$$

By definition the power angle is 72^{c} which is the angle between the phasor reference V_T and phasor E_{g1}. Below we follow another route to find δ_1 and δ_2.

From Eqns. (7.34 - 7.36) we obtain

$$P_{max1} = \frac{10750 \cdot 10392}{0.778} = 143.5 \quad \frac{MW}{per - phase}$$

$$143.5 \cdot 3 = 430.5 \qquad \text{3-Phase power in MW}$$

$$430.5 \cdot \sin\left(\delta_1\right) = 410 \qquad \qquad \delta_1 = \text{asin}\left(\frac{410}{430.5}\right) = 72°$$

$$\overrightarrow{E_{g2}} = 10392 + \overrightarrow{I_2} \cdot j \cdot 1.026 = 10392 + (6739 + j \cdot 6870) \cdot j \cdot 1.026 = 3343 + j \cdot 6914$$

$$\left|3343 + j \cdot 6914\right| = 7680 \qquad \qquad \text{atan}\left(\frac{6914}{3343}\right) = 64 \cdot {}^{c}$$

$$\overrightarrow{E_{g2}} = 7680 \cdot \varepsilon^{j \cdot 64°}$$

Line to neutral emf $\qquad \delta_2 = 64°$

$$P_{max2} = \frac{7680 \cdot 10392}{1.026} = 77.8 \quad \frac{MW}{phase}$$

$$77.8 \cdot 3 = 233.4 \ MW \quad \text{Three-phase max. power}$$

$$233.4 \cdot \sin(\delta_2) = 210 \qquad \delta_2 = asin\left(\frac{210}{233.4}\right) = 64°$$

Figure 7-20 shows the phasor diagrams of example 7-2

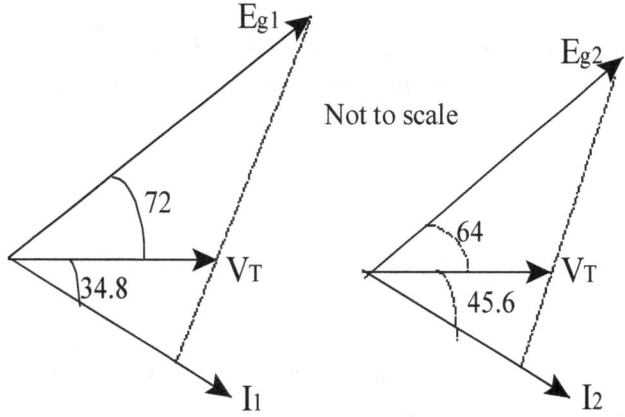

Figure 7-20 Phasor diagrams of the two generators of example 7-2

Now we compute the coherency of G1 and G2 in example 7-2   Coherency formula:

$$\frac{P_{sy1}}{M_1} = \frac{P_{sy2}}{M_2} \tag{7.46}$$

The synchronizing power coefficient Psy was defined in Eq. (7.12) and M is the inertial constant *angular momentum at synchronous speed.*

$$P_{sy1} = P_{max1} \cdot \cos(\delta_1) \ MW \qquad P_{sy2} = P_{max} \cdot \cos(\delta_2) \ MW$$

$$P_{sy1} = 430.5 \cdot \cos(72°) = 133 \ MW$$

$$P_{sy2} = 233.1 \cdot \cos(64°) = 102 \ \text{MW}$$

$$M = \frac{G \cdot H}{180 \cdot f} \quad \frac{\text{MW} \cdot \text{sec}^2}{\text{degree}}$$

$$M_1 = \frac{500 \cdot 6}{180 \cdot 60} = 0.278 \qquad\qquad M_2 = \frac{300 \cdot 4}{180 \cdot 60} = 0.111$$

$$\frac{P_{sy}}{M} = \frac{\text{MW}}{\frac{\text{MW} \cdot \text{sec}^2}{\text{degree}}} = \frac{\text{degree}}{\text{sec}^2} = \frac{\Delta\omega}{\text{sec}} \qquad \textbf{rotor acceleration} \tag{7.47}$$

The coherency formula see Eq. (7.46) means that in case of a disturbance the rotor acceleration of both machine must be equal or at least almost equal. Substituting in Eq. (7.46) we obtain:

$$\frac{133}{0.278} = 478 \qquad\qquad \frac{102}{0.111} = 919$$

$$478 < 919 \qquad \textbf{No coherent}$$

The rotor acceleration of the 300 MW generator is twice the rotor acceleratoion of the 500 MW generator. These generators should not be connected in parallel.

CHAPTER 8

Generator Ground Protection

8.1 AC Generators neutral grounding.

In general, it is a basic necessity to connect to ground the neutral of three-phase AC generators. This would not only benefit the load and the network to which the generator is connected, but also would allow to operate the generator in parallel with other power systems. In fact, the ground potential provides a common reference for any electrical power system in the planet earth. Otherwise you could not interconnect them. There are others not so obvious benefits like detection of a ground fault in one of the phases. The brief discussion that follows is tailored to large AC synchronous generators.

.8.2 Generator neutral grounding methods

There are many generator grounding methods. See IEEE standards C37.101-2006 and C62.92-2000. Listed below are the best known.

1 - Solidly grounded

2 - Effectively grounded

3 - Low resistance grounded

4 - High resistance grounded

5 - Resonant grounded

6 - Ungrounded

In solidly ground systems the neutral of the generator are directly connected to ground. This type of grounding provide a steady voltage reference during normal operation, but in case of a short circuit to ground in any coil of the generator stator windings the protection they offer is poor. However, solidly grounding the generator's neutral, limits the voltage to ground in any of the phases during normal generator operation and avoid the randomly change in values of the three-phase potentials. Besides, solidly grounding generator's neutral helps to prevent excessive line-neutral voltages on the phase windings due to line surges, lightning, or contact with nearby higher voltage lines. The definition of the term effectively grounded is as follows: grounded through an adequate low impedance, inherent in the interconnecting means or intentionally added. Low resistance grounded neutral, in its most simple form, consist of a resistor directly connected to ground. The low resistance value of the grounding resistor is with respect to the magnitude of the generator's transient reactance. The reader should keep in mind that in case of a ground fault in the power system--the lower is the resistance of the path of the short circuit current including the grounding resistor--the higher would be the short circuit current, and the possible damages inflicted to the generator's stator windings and in general to the power system. Furthermore, in some cases in which the ground-fault occurs in one of the stator windings, extensive stator damage could result due to long lasting short circuit current that continues to flow even with the main and DC field breakers opened (the spinning rotor keeps inducing emf in the windings). One practical solution is to switch to a high-resistance grounded system when a ground fault occurs in the generator stator windings. *Another solution would be to include in the low resistance grounding system a neutral breaker.* It is important to realize that the grounding resistor size also depends of the expected current unbalance during normal operation.

Inductive reactance grounded neutral method helps to decrease the rate of rise of the short circuit current when a fault to ground occurs, and also is very effective in decreasing the magnitude of the third and higher harmonic components of the short circuit current, because the value of the reactive voltage drop is proportional to the frequency of the current. Actually, the protection of generators against ground faults gets more complicated than the direct connection of a resistor or reactance to ground. A typical generator's grounding and generator stator windings protection against ground faults is illustrated in Fig. 8-1. This high-resistance scheme is effective against ground faults occurring at the generator windings except when the fault occurs near the neutral, at the beginning of one of the windings. Typically, the first coil is not protected, for this reason the 59G should be set low so that it is sensitive to faults occurring as close as possible to the generator neutral. However, this might result in unexpected coordination problems. One possible solution is to delay the 59G tripping to provide coordination with other relays elsewhere in the system. In the case of a ground fault in one of the generator stator windings, the voltage imposed in the 59G relay depends on the exact location of the fault. In fact the voltage across the relay becomes smaller as the ground fault location approaches the generator neutral. For a fault at any of the generator output terminals, the voltage across the primary of depicted grounding transformer would be 100% of the generated line to neutral voltage. However, if the ground fault occurs in the first or second coil of any of the stator windings the voltage across the transformer primary would be in the order of 5% or less (depends on the number of coils per phase) and typically, in this case, the 59G relay would not be activated. The reader must realize that the 59G relay cannot be set so low, that the expected load current unbalance would result in its tripping and take the generator out of service. So to avoid this occurrence the 59G digital relay should be tuned to the voltage fundamental frequency (60Hz) and should be insensitive to the third-harmonic (180Hz) voltage components that are occasional present in the generator neutral. Furthermore, the 59G relay must be set not to trip when a ground fault occurs elsewhere in the power system. Those types of faults must be cleared by other, of the many, relays installed in the power system. For instance the F4 ground fault in the transmission line should not trip the 59G relay.

When selecting surge protectors, the maximum over-voltages considered should not exceed double amplitude of the generator line-to-ground normal operating voltage. Furthermore, the generator grounding system should exhibit low resistance during normal operation, and high resistance to ground when a ground fault occurs in the generator armature. And the grounding method described in example 8-1 do just that.

Example 8-1 Compute the requirements for the grounding transformer and resistor of the generator high-resistance grounding depicted in Fig. 8-1. Assume the following data:

Generator rating = 1000 MVA. Line to line rated voltage = 24 kV

Step-up transformer = 1000 MVA, delta-grounded wye

Single-phase distribution transformer, 24 kV / 240 V

The goal is to protect the generator armature windings when a short circuit to ground occurs in one of the armature windings.

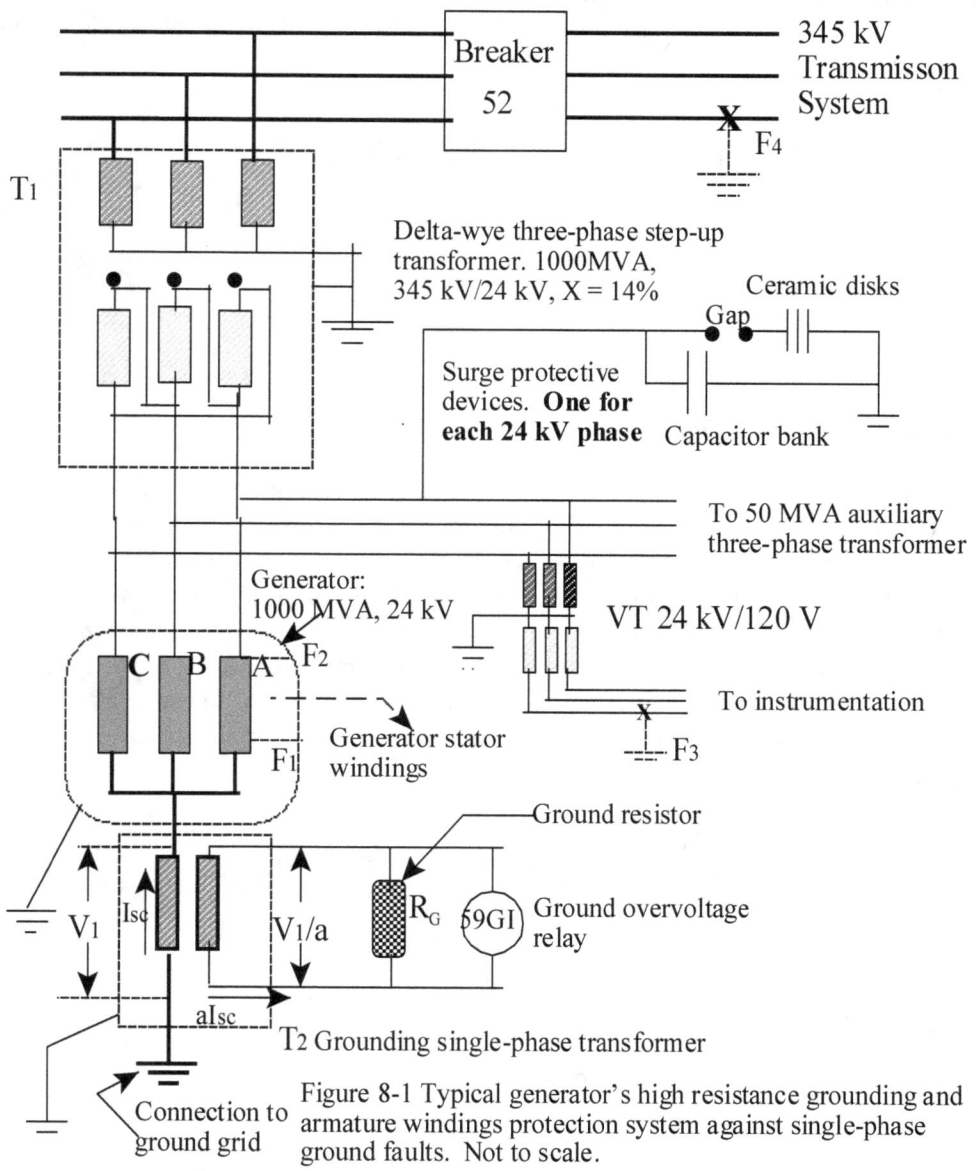

Figure 8-1 Typical generator's high resistance grounding and armature windings protection system against single-phase ground faults. Not to scale.

The basic equation that describes capacitor steady state operation is

$$q = C v_c \tag{8.1}$$

C = coupling capacitance, in farads   vc = voltage across plates, in volts

q =   charge on positive plates, coulombs

The charge on the positive plate is the result of a current circulating in the circuit connecting the plates. Eq. (8.1) expresses that the charge is proportional to the voltage between the plates. And we know that the charging current is equal to the rate of change of the charge on the plates. Differentiating Eq. (8.1) we get Eq. (8.2):

$$i = \frac{d}{dt}q = \frac{d}{dt}(C\,v_c) = C\left(\frac{d}{dt}v_c\right) \qquad\qquad i = C\left(\frac{d}{dt}v_c\right) \tag{8.2}$$

$$W = C\left(v_c\right)^2 \tag{8.3}$$

W = energy, in joules   C = capacitance, in farads

v_c = voltage across plates, in volts

Equation (8.2) shows that the discharge current in capacitor is proportional to the rate of change of the voltage across the capacitor. The energy stored in the electrical field of a capacitor is provided by Eq. (8.3). This energy is continuously interchanged between the capacitor's electrical field and the circuit inductive magnetic field. For this interchange to take place, it is necessary to have a circulating current. If the driving voltage were removed, the interchange would continue until all the energy is consumed in the resistance of the circuit. In case of a phase to ground short circuit there is not driving voltage across the coupling capacitance and the current **i** is a time decaying direct current. In this case, we obtain the following equations

$$R\,i + v_c = 0 \qquad\qquad RC\left(\frac{d}{dt}v_c\right) + v_c = 0 \qquad\qquad RC\left(\frac{d}{dt}v_c\right) = -v_c$$

$$RC\,dv_c = -v_c\,dt \qquad\qquad \frac{dv_c}{v_c} = \frac{-dt}{RC}$$

$$\int \frac{1}{v_c}\,dv_c \rightarrow \qquad\qquad \int \frac{-1}{RC}\,dt \rightarrow$$

$$\ln\left(v_c\right) = -\frac{t}{CR} + K' \qquad\qquad \text{The logarithm of a product is the sum of the logarithms}$$

$$v_c = K\varepsilon^{-\frac{t}{CR}} \qquad K = v_0 \qquad \text{Integration constant is equal to the value of } v_c \text{ at t =0}$$

$$v_c = v_0 \varepsilon^{\frac{-t}{T}} \tag{8.4}$$

V0 = value of the voltage across the capacitor at the instant of short circuit, in volts

T = time constant of the discharging circuit, in seconds

$$T := R\,C \tag{8.5}$$

Where R = Resistance of discharging circuit, in ohms

C = Capacitance, in farads

Applying Eqns. (8.2) and (8.4) we obtain

$$i = C\,v_0\,\varepsilon^{\frac{-t}{T}}\,\frac{-1}{T} = \frac{-v_0}{R}\varepsilon^{\frac{-t}{T}} \qquad \text{ampere} \tag{8.6}$$

The initial discharge current, t = 0 $\qquad i_0 = \dfrac{-v_0}{R}$ $\qquad\qquad$ (8.7)

Equation (8.6) is a decaying direct current. In the case depicted in Fig 8.1 the maximum value of a phase discharge current occurs when a short circuit from that phase to ground occurs at the peak value of the applied sinusoidal phase voltage or 19595 volts. In a three-phase power system the peak values of the sinusoidal voltages applied to the conductors are separated in time by 1200 electrical degrees. The reader should be aware that the capacitances we are working with are coupling capacitances, in fact each phase conductor is coupled through the electrical field to ground and to the other two phase conductors. And the value of the charge accumulated in the surface of each phase conductor is proportional to the instantaneous magnitude of the conductor's voltage, see Eq. (8.1). Furthermore, the coupling capacitances between the other two phases and ground also discharge. Actually, the line-to-ground fault current is fed by all affected phase to ground coupling capacitances in the generator installation. As listed below.

Equipment phase capacitance to ground:

Generator (**Cg**)	0.250 μF
Surge protector (Csp)	0.130 μF
Generator step-up transformer (Csut)	0.0038 μF
Auxiliary Transformer (Cauxt)	0.0011 μF

ISO-phase Busses (Cbusses)   0.006 μ F

All the capacitance listed above are between each of the phases of the 24kV level and ground and because they are in parallel they must be added to obtain the total. Actually, in case of a line to ground fault the coupling capacitances to ground of all three phases discharge.

$$C_T = 0.250 + 0.130 + 0.0038 + 0.0011 + 0.006 = 0.3909 \quad \frac{\mu F}{phase}$$

The capacitive reactance per phase is

$$X_c = \frac{1}{j\omega C_T} = \frac{1}{j2\pi\, 60 \times 0.3909 \times 10^{-6}} = -6786j \quad \frac{ohms}{phase}$$

The total capacitive reactance-to-ground (Xcg) of the three-phases is equal to the parallel combination of the three capacitive reactance per-phase to ground.

$$X_{cg} = \frac{-j\,6786}{3} = -j\,2262 \quad ohms \qquad \left|X_{cg}\right| = 2262 \quad ohms$$

The per-unit capacitive reactance per phase is

$$Z_b = \frac{\left(kV_b\right)^2}{MVA_b} = \frac{24^2}{1000} = 0.5760 \qquad \textbf{Base line to neutral ohms}$$

-6786j/0.5760 = -11781.25j   **Per-unit capacitive reactance per phase**

-2262j/0.5760 = -3927j   **Total per-unit capacitive reactance**

The single-phase distribution transformer depicted on Fig 8-1 is used to ground the generator neutral. And it has a primary to secondary voltage ratio of 24000/240 volts. Neglecting the distribution transformer exciting current its primary ampere-turns are equal to the secondary ampere-turns. Symbolically

$$N_1 I_1 = N_2 I_2$$

$$a = \frac{N_1}{N_2} = \frac{\left|V_1\right|}{\left|V_2\right|} = \frac{24000}{240} = 100 \qquad a := 100$$

On Fig. 8-1 the resistance RG installed in the secondary of the distribution transformer must reflect in the primary with a value that could be *selected* as equal or **larger** than |Xcg|.

$$a^2 R_G = 2400 \qquad R_G = \frac{2400}{10^4} = 0.24$$

Value in ohms of grounding resistance in the secondary

$0.24/0.5760 = 0.4167$ $R_G = 0.4167$ **Per unit grounding resistance**

The time constant of the capacitance discharge current per phase, see Eq. (8.5), is:

$$T = 2400 \times 0.3909 \times 10^{-6} = 938.16 \times 10^{-6} \text{ sec}$$

All phases are assumed to discharge simultaneously.

The initial discharge primary current, see Eq. (8.7), is

$$i_0 = \frac{19596}{2400} = 8.165 \text{ amperes}$$

The reflected initial discharge current in the secondary is

$$a\, i_0 = 816.5 \text{ amperes}$$

The goal is to protect the armature windings when a short circuit to ground occurs in one of the armature windings near the neutral of the generator. Let us consider the case in which the voltage applied to the primary of the grounding transformer is 2.5% of full line to neutral voltage as shown in Fig. 8-1

$$0.025 \frac{24000}{\sqrt{3}} = 346 \qquad \text{Line to neutral rms primary volts} \qquad V_1 = 346$$

$$\frac{346}{a} = 3.46 \qquad \text{Secondary rms volts}$$

$$V_{2min} = 3.46 \text{ volts}$$

This is the minimum rms secondary voltage that should activate the 59G relay.

$$I_{2min} = \frac{3.46}{0.24} = 14.42 \qquad \text{Minimum secondary rms amperes that would activate 59G.}$$

This is the rms current in the secondary of the grounding transformer when a short circuit to ground in one of the armature windings applies 2.5% of the generator line to neutral voltage to the primary of the grounding transformer. **Computing now the case when the winding short circuit to ground occurs in one of the generator output terminals. See Fig. 8-2**

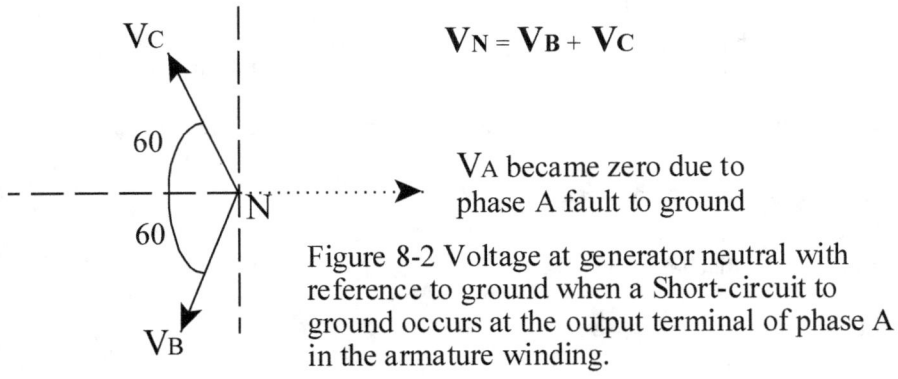

$$\mathbf{V_N = V_B + V_C}$$

V_A became zero due to phase A fault to ground

Figure 8-2 Voltage at generator neutral with reference to ground when a Short-circuit to ground occurs at the output terminal of phase A in the armature winding.

The voltage at N (the generator neutral point) with reference to ground during normal operating condition is zero. Symbolically.

$$\overrightarrow{V_A} + \overrightarrow{V_B} + \overrightarrow{V_C} = 0$$

Generator line to neutral rated output voltage

$$V_{LN} = \frac{24000}{\sqrt{3}} = 13856$$

Voltage at N, when a short circuit to ground occurs at the output terminal of phase A.

$$\overrightarrow{\left(V_B + V_C\right)} = [13856(\cos(240°) + j\sin(240°)) + 13856(\cos(120°) + j\sin(120°))]$$

$$\overrightarrow{V_N} = \overrightarrow{\left(V_B + V_C\right)} = -13856 \qquad \left|-13856\right| = 13856$$

Let us now assume that the generator output voltage phasors had rotated 90 degrees and the phasors B and C are now as follows

$$\overrightarrow{V_B} = 13856(\cos(330°) + j\sin(330°)) = 12000 - 6928j$$

$$\overrightarrow{V_C} = 13856(\cos(210°) + j\sin(210°)) = -12000 - 6928j$$

$$\overrightarrow{V_N} = \overrightarrow{\left(V_B + V_C\right)} = -13856j \qquad \left|-13856j\right| = 13856$$

In Fig. 8-2 the rms voltage across the grounding transformer primary is: -13856 volts.

When a short circuit to ground occurs at one of the generator output terminals the *instantaneous* voltage at N, the generator neutral point, is given by Eq. (8.8). The reader should understand than the line to neutral voltage phasors angular position at the instant of short circuit occurrence could be any angle, but the separation between phasors is always 120 degrees. The short circuit to ground could occur when the end of the upper coil of a phase winding in the generator armature touches the generator metal enclosure which is connected to the ground mat. When the polarity of the neutral point, N, is negative the path of the AC short circuit current is from the ground mat, through the primary of grounding transformer, to N, through the faulted phase winding, to the ground mat. When the polarity of the neutral point is positive the short circuit current path is from N, through the primary of the grounding transformer, to the ground mat, to the generator enclosure, through the faulted winding, to N. See Fig. 8-1. In Fig. 8-2 phase A is the faulted phase and V$_A$ is equal to zero. During this fault condition the *neutral potential* with reference to ground becomes equal to the addition of phasors B and C. See Fig. 8-2

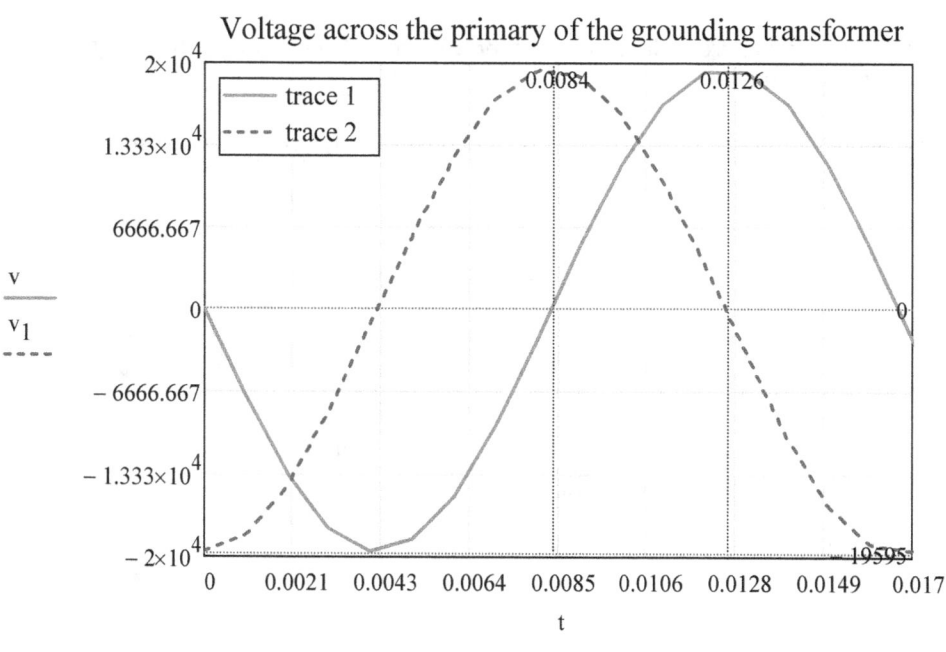

Figure 8-3 shows the instantaneous voltage at the generator neutral point when a short circuit to ground occurs at one of the generator output terminals.

$$v = 19595 \sin\left(377t + \theta_B\right) + 19595 \sin\left(377t + \theta_C\right)$$

$$\sqrt{2}\,13856 = 19595 \quad \text{volts} \qquad \text{Sinusoidal peak value}$$

$t := 0, 0.001 \ldots 0.018$

$v = 19595 \sin(377t + \theta) + 19595 \sin[377t + (\theta + 120°)]$ (8.8)

$v := 19595 \sin(377t + 120°) + 19595 \sin(377t + 240°)$

$v_1 := 19595 \sin(377t + 210°) + 19595 \sin(377t + 330°)$ (8.9)

The rms value of the sinusoidal voltage depicted in Fig. 8-3 is:

$$V_{rms} = \frac{19595}{\sqrt{2}} = 13856$$

$V_1 = 13856$ volts   RMS voltage applied to the distribution transformer primary.

$$V_2 = \frac{13856}{100} = 138.56 \qquad \text{rms volts}$$

Secondary rms voltage at the grounding transformer due to short circuit to ground at the output terminal of any of the generator armature phase winding.

The ohm value of the _secondary reactance_ of the neutral grounding transformer, see Table 8-3, is computed as follows:

$$\frac{\Omega}{\Omega_{base}} 0.049 \qquad\qquad \Omega_b = \frac{kV^2}{MVA_b} = \frac{0.240^2}{0.020} = 2.88$$

$\Omega = 0.049 \times 2.88 = 0.141$ ohms

The secondary circuit impedance including the grounding resistance, the distribution transformer and neglecting the 59G relay impedance is

$Z_{T2} = 0.24 + 0.141j$

$|0.24+0.141j| = 0.2784$ ohms

$$I_2 = \frac{138.56}{0.2784} = 497.7 \qquad \text{rms secondary amperes}$$

$$I_1 = \frac{497.7}{a} = \frac{497.7}{100} = 4.98 \qquad \text{rms primary amperes}$$

The maximum power dissipated in the grounding resistor is:

$$P_R = 0.24 \times 497.7^2 = 59449 \quad \text{watts}$$ Resistor power rating = 60 kW

The IEEE Std 32 provides, see Table 8-1, the following data.

On Time	Temperature Rise in Degree C
Ten Seconds	760
One Minute	760
Ten Minutes	610
Extended Time	610
Continously	385

$40 + 760 = 800$

Table 8-1 IEEE Std 32 OnTime Rating and Permissible Temperature Rise for Neutral Grounding Resistors

The 60 kW resistor should be sized to operate for one minute at 140 volts, 500 amperes. The maximum voltage sensed by the over voltage relay, 59G, is 139 rms volts. And it should be activated when the secondary voltage is equal or higher than 3.46 rms volts. The grounding transformer KVA size is calculated as follows;

$$VA = V_2 I_2 = 138.56 \times 497.7 = 68961 \quad \text{Transformer KVA = 70} \qquad (8.10)$$

However, IEEE Std's C62.92.2-1989 and C37.101-2006 recommend to compute the transformer thermal rating using the secondary full rated voltage or 240 volts as follows:

$$240 \frac{240}{0.2784} = 206897 \, \text{volt} - \text{amperes} \qquad \frac{206897}{1000} = 207 \quad KVA$$

Recommended rating of the grounding transformer KVA = 207

Table 8-2 provided by the IEEE Standard C62.92.2 1989 that list the permissible grounding transformer overload factors for short periods of time

Duration of overload	Overload factor
10 seconds	10.5
1 minute	4.7
10 minutes	2.6
30 minutes	1.9
2 hours	1.4

Table 8-2 IEEE Std C62.92.2-1989 Permissible Short Time Overload Factors for Distribution Transformers Used for Neutral Grounding

Using the overload factors provided on Table 8-2 for grounding distribution transformer we obtain the following KVA ratings

$$\frac{70}{4.7} = 14.894$$ one minute rating   Computed using Eq. (8.10) and one minute overload factor.

$$\frac{207}{4.7} = 44.043$$ one minute rating   Computed applying IEEE Std C-37.101-2006 and one minute overload factor.

The one minute rating was used for the ground resistor, so we use the same for the grounding transformer and select a 20 KVA transformer which is larger than the indicated 14.894

8.3 Comments about the high-resistance grounding method using a distribution transformer.

1. Select the single-phase distribution transformer with a primary voltage rating equal or larger than the generator line to neutral voltage. And a transformer ratio larger than ten. The larger is the transformer ratio the smaller would be the secondary voltage and the ohm value and size of the grounding resistor installed in the secondary. And the larger would be the ohm value of the reflected primary resistance.
2. The larger is the reflected primary resistance the smaller would be the short circuit current circulating through the generator armature windings during ground-faults. Which will decrease the potential damage to the generator by reducing mechanical stresses and generated heat. IEEE, Std. C37-101-2006 recommends a primary reflected value of 2400 Ω or less for the grounding resistor.
3. The voltage drop in the distribution transformer primary reduces the over voltage transients incoming from the power system. These transients could have large DC

components. Arcing ground faults and their destructive effects are practically eliminated.

4. The surge protective devices must be selected taking in account the over voltage's generated during short-circuit to ground.

5. The high-resistance grounding scheme is less expensive than a directly connected resistor. Which must be big due to the heat it generates.

8.4 Short circuit current computation using symmetrical components.

In Example 8-1 we calculated the size of the ground resistor and the ratings of the grounding single-phase distribution transformer. In this section we use symmetrical components to compute the ground fault current in the circuit depicted in Fig. 8-1. The following material about sequence networks will help the reader to understand the reactance values used in short circuit computations or inserted in Tables.

- On a perfectly balanced system, that is, with equal line impedances and no mutual impedances. Positive sequence currents produce only positive sequence voltage drops, negative sequence currents produce only negative sequence voltage drops, and zero sequence currents produce only zero sequence voltage drops. Actually, there is not interaction between the phase sequence networks, and therefore they can be analyzed separately. See section 3.4.

- The impedance of non-rotating, symmetrical linear elements are independent of the phase sequence assuming that the voltage applied to them are balanced. This is the reason why transmission lines, have identical positive and negative sequence impedances. However the impedance of a transmission line to zero sequence currents is much larger than the impedance they present to positive or negative currents, because the zero sequence current path is different and the magnetic field created by the three zero sequence currents are in phase. In general the zero sequence inductive reactance of overhead transmission is around 2 to 3.5 times larger than its positive sequence reactance.

- Transformers are symmetrical non rotating static apparatus and as such their impedance are independent of the phase sequence of the voltages applied to them, assuming the applied voltage are balanced. Therefore, for any type of power transformers, it is assumed that the series impedances to all symmetrical components sequences are equal. Furthermore, the value of the impedance is considered to be equal to the value of the reactance.

- For perfectly balanced three-phase networks, the three sequence networks are separated with no mutual coupling between them and the NPS and ZPS networks cannot contain any current or voltage. However, if a fault or any other unbalanced producing event occurs in a balanced three-phase network, then all three phase sequence networks can contain current and voltage and all of them will share the common point of unbalance. As a result the three phase sequence networks must be connected at the point of unbalance or where the fault occurs.

- Table 4-2 provides the range of typical line to neutral reactance values for most types of synchronous generators. In this table four different values are provided for the positive sequences reactance. But, for short circuit computations the *subtransient reactance* is the one to be used and is the one listed on Table 8-3. In general, the positive and negative sequence reactances are equal. The zero sequence reactance is the smallest of the three, because the resultant magnetic flux in the air gap, produced by the three zero sequence currents, is zero except for the flux produced by the end turns of the armature windings and for the leakage flux.

Some important data necessary to compute the short-circuit currents in the system illustrated in Fig.8-1 is given on Table 8-3

Apparatus	T_1	T_2	G
MVA	1000	0.020	1000
kV	24	24	24
Ratio	14	100	
X	0.14 pu	0.049 pu	1.45 pu
Pos. Sequence reactance	0.14 pu	0.049 pu	0.15 pu
Neg. Sequence reactance	0.14 pu	0.049 pu	0.15 pu
Zero Sequence reactance	0.14 pu	0.049 pu	0.05 pu

Table 8-3 Apparatus factory ratings for Fig. 8-1

Let us select 1000 MVA and 24 kV as the bases for the entire power system. The new per unit values of the generator and transformers are calculated using Eq. (1.45) and provided in Table 8-4. The new per unit values of T2 are listed in Table 8-4

$$X_{T2} = 0.049\frac{1000}{0.020} = 2450 \quad pu$$

Per unit values of T1 and G remain the same

Apparatus	T$_1$	T$_2$	G
MVA	1	0.000020	1
kV	1	1	1
Ratio	14	100	
X	0.14 pu	2450 pu	1.45 pu
Pos. Sequence reactance	0.14 pu	2450 pu	0.15 pu
Neg. Sequence reactance	0.14 pu	2450 pu	0.15 pu
Zero Sequence reactance	0.14 pu	2450 pu	0.05 pu

Table 8-4 Apparatus per unit values based on 1000 MVA and 24 kV

The following assumptions are usually made to do fault analysis using symmetrical components

- The power system is balanced before the short-circuit occurrence
- The resistance components of transformers and generators are neglected
- Transformers shunt admittances and transmission line capacitances are neglected
- The mechanical power input to generators remains constant
- The classical model of generators is used. The emfs and subtransient reactances remain constant
- Normal operating load is neglected

The conditions imposed by the short-circuit to ground on phase A of Fig. 8-1 are:

$$\overrightarrow{I_B} = 0 \qquad \overrightarrow{I_C} = 0 \qquad \overrightarrow{V_A} = 0 \tag{8.11}$$

From section 5.1, Eq. (5.8) we have:

$$\overrightarrow{I_{A1}} = \frac{\overrightarrow{V_{A1}}}{\left(Z_0 + Z_1 + Z_2\right)} \tag{8.12}$$

We know from Eq. (5.4), section 5. 1 $\qquad \overrightarrow{I_{A1}} = \overrightarrow{I_{A2}} = \overrightarrow{I_{A0}} = \frac{\overrightarrow{I_A}}{3} \tag{8.13}$

IEEE Std C37.101-2006 recommends that the value of the ground resistor in the secondary of *T2 must be such that the reflected value in the primary is 2400 ohms or less*

$$a^2 R_G = 2400 \qquad \underset{\wedge\wedge\wedge}{a} := 100 \qquad R_G = \frac{2400}{10000} = 0.24 \qquad \text{Ohms}$$

$$Z_b = \frac{(kV_b)^2}{MVA_b} = \frac{24^2}{1000} = 0.5760 \qquad \text{Base, line to neutral ohms}$$

$$\frac{2400}{0.5760} = 4167 \qquad \text{Per unit primary reflected value of } R_G \qquad 3R_G = 12501$$

T2 zero sequence per unit reactance of is: 2450, see Table 8-4.

Capacitive reactance per phase is: $\quad X_c = \dfrac{1}{j377(0.3909)\,10^{-6}} = -6786j \quad$ ohms

In case of a fault to ground in one phase, all the coupling capacitances of the three phases discharge to ground. So the capacitive reactance in the zero sequence network (Xcg) is equal to the parallel combination of the capacitive reactance to ground of all three phases

$$X_{cg} = \frac{-6786j}{3} = -2262j$$

$$X_{co} = -2262j \quad \text{ohms} \qquad \text{Zero sequence shunt capacitive reactance to ground}$$

$$\frac{-2262j}{0.5760} = -3927j \qquad \text{Per unit shunt \underline{capacitive} reactance}$$

Usually the positive and negative sequence impedances of the system are very small when compared to system zero sequence capacitive reactance and the zero sequence impedance of the generator is also very small when compared to the ground resistor value. Nevertheless, for didactical purpose all the impedances are included in the fault computations. The zero-sequence impedance of a transmission line includes the impedance of the ground and ground wires. *So the return circuit-segment of the zero-sequence network has zero impedance, and it is the reference bus of the system.* And therefore, voltages measured with respect to the reference bus of the zero sequence network gives the voltage to ground. Again, the system reference bus which is the return circuit of the zero-sequence network has zero impedance. At this point the reader is strongly encouraged to review section 4.7. The reader should remember than the generator stator winding are stationary. So, we assumed that the zero sequence capacitive reactance is equal to the positive and negative capacitive reactances. Furthermore the generator zero sequence per unit reactance (0.05) was neglected.

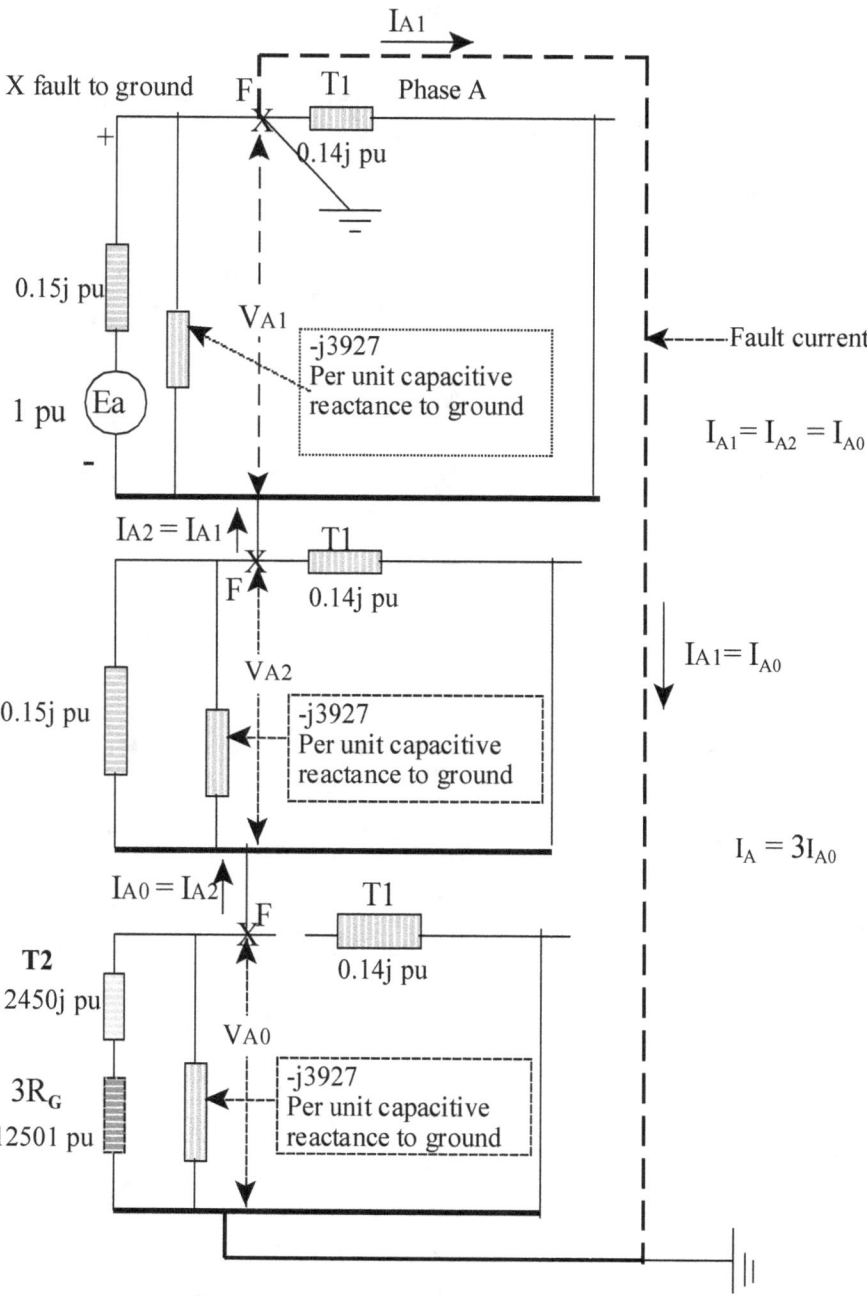

Figure 8-4 Sequence networks of phase A of the power system illustrated in Fig. 8-1

Equations (8.11 - 8.13) are the foundations to represent the line to ground fault by connecting the three sequence networks in series via the point at which the fault occurs as indicated on Fig. 8-4. The contribution of the 345 kV transmission line and the rest of the power system to the fault had been neglected.

For the positive and negative sequences networks neglecting the load

$$\frac{0.15j(-3927j)}{0.15j - 3927j} = 0.1500j \quad \text{pu impedance} \qquad \vec{Z}_1 = 0.15j \qquad \vec{Z}_2 = 0.15j$$

For the zero sequence network we have

$$\frac{(12501 + 2450j)(-3927j)}{(12501 + 2450j) - (3927j)} = 1217 - 3783j \quad \text{pu impedance} \qquad \vec{Z}_0 = 1217 - 3783j$$

From Eq. (5.4) we know that $\qquad \vec{I}_{A0} = \vec{I}_{A1} = \vec{I}_{A2} = \dfrac{\vec{I}_A}{3} \qquad\qquad \vec{I}_A = 3\vec{I}_{A0}$

Phase A is faulted to ground so $\qquad \vec{V}_A = 0$

From matrix Eq. (5.5) and knowing that $\quad \vec{I}_{A1} = \vec{I}_{A0}$ we obtain:

$$\vec{V}_{A0} = -\vec{Z}_0 \vec{I}_{A0}$$

$$\vec{V}_{A1} = \vec{V}_F - \vec{Z}_1 \vec{I}_{A1} \qquad\qquad \vec{V}_F \;\;\text{replaced}\;\; \vec{E}_a = 1$$

$$\vec{V}_{A2} = -\vec{Z}_2 \vec{I}_{A2}$$

$$\vec{V}_A = \vec{V}_{A0} + \vec{V}_{A1} + \vec{V}_{A2} = \vec{V}_F - \vec{Z}_0 \vec{I}_{A0} - \vec{Z}_1 \vec{I}_{A0} - \vec{Z}_2 \vec{I}_{A0}$$

$$\vec{V}_A = \vec{V}_F - \vec{I}_{A0}\left(\vec{Z}_0 + \vec{Z}_1 + \vec{Z}_2\right) = 0 \qquad\qquad \vec{I}_{A0} = \frac{\vec{V}_F}{\left(\vec{Z}_0 + \vec{Z}_1 + \vec{Z}_2\right)}$$

$$\overrightarrow{I_{A0}} = \frac{1}{1217 - 3783j + 0.15j + 0.15j} = 0.000077 + 0.00024j \qquad pu$$

$$\left|\overrightarrow{I_{A0}}\right| = \left|0.000077 + 0.00024j\right| = 0.000252 \qquad pu$$

From Eq. (1.41) we obtain the line base current

$$I_b = \frac{MVA_b}{\sqrt{3}\,kV_b} = \frac{1000}{\sqrt{3}\,24} = 24.056 \quad \textbf{kiloampere} \qquad 24056 \text{ amperes}$$

$$\overrightarrow{I_{A0}} = 24056(0.000077 + 0.00024j) = 1.852 + 5.773j \qquad amperes$$

$$\left|\overrightarrow{I_{A0}}\right| = \left|1.852 + 5.773j\right| = 6.063 \quad amperes$$

$$\overrightarrow{I_A} = 3(1.852 + 5.773j) = 5.556 + 17.319j \qquad atan\left(\frac{17.319}{5.556}\right) = 72.214°$$

$$\left|\overrightarrow{I_A}\right| = \left|5.556 + 17.319j\right| = 18.188 \quad amperes$$

The short circuit current leads the line to neutral voltage VF by $72.2°$

In the high resistance grounding method the reflection of the ground resistance into the primary of the grounding transformer is so high in comparison with the apparatus capacitances and impedances that make them negligible. Furthermore, you could do the computations only in the zero sequence network and obtain acceptable results. For instance:

$$\overrightarrow{I_A} = \frac{3 \times 1}{(1217 - 3783j)} = 0.0002 - 0.0007j \qquad pu$$

$$\overrightarrow{I_A} = 24056(0.0002 - 0.0007j) = 4.811 - 16.839j \qquad amperes$$

$$\left|\overrightarrow{I_A}\right| = \left|4.811 - 16.839j\right| = 17.513 \quad amperes \qquad atan\left(\frac{-16.839}{4.811}\right) = -74.055°$$

It is recommended that to select surge protectors the maximum over voltages considered should not exceed double amplitude of the generator line to ground operating voltage.

Furthermore, the generator grounding system should exhibit low resistance during normal operation, and high resistance to ground when a ground fault occurs in the generator stator.

CHAPTER 9

Modeling and Assumptions

9.1 Introduction

Synchronous generators are by far the most used type of generator in the power generation industry. They generate real and reactive power, and their output voltage is easy to control by changing the excitation or the rotor angular velocity. Furthermore, in large power systems containing many generators, they rotate synchronously. There are two types of synchronous generator round-rotor and salient pole. They consist of a stator, on which the three-phase armature windings are wound 120° apart from one another, and a rotor, on which the DC field winding is wound. When the driving turbine rotates the rotor, the rotating magnetic flux induces sinusoidal voltages in the stator windings. The magnitudes of the voltage induced in the stator windings (which are 120° out of time phase from one another) depend on the magnitude of the field DC current and their frequencies depend on the rotor angular velocity. Round rotors are used in high-speed machines usually driven by high-speed steam or gas turbines (combustion turbines). This is so because round-rotor generators can withstand better the high centrifugal forces associated with turbine-driven high-speed machines; also their rotors are smaller in diameter and easier to dynamically balance. Salient pole generators are used for lower-speed applications such as generators driven by water turbines. This type of machine needs several pairs of magnetic poles to generate power of 60 hertz (Hz). Fortunately, at the low speed at which they operate, the centrifugal forces are lower than those experienced by turbine-driven generators. Therefore the rotor diameter could be larger and salient poles can be used. In round-rotor generators, the air gap between the stator and rotor is uniform with constant reluctance. However, in salient pole machines, the air gap varies along the generator circumference with the smaller gap along the direct axis of the rotor field winding and the larger gap along the neutral axis of the field winding, which is commonly called the *quadrature axis*. Therefore, in salient pole generators the reluctance of the air gap is not constant. To facilitate the mathematical analysis of salient pole generators, the magnetomotive force (mmf) is expressed in terms of its components along the direct and quadrature axes. The electromotive force (emf) produced by the changing flux linkage of each component of the acting mmf is considered as generated in two separate magnetic circuits, each with constant but different values of reluctance. This approach yields a salient pole generator model that consists of two parts, one for the direct axis and the other for the quadrature axis. Electrical engineers and engineering professors alike are in the neglecting and assuming business. So when faced with a difficult problem— very complex and hard to clearly express in mathematical language, or the data required is hard to get or nonexistent— we simplify the problem by neglecting things that we believe will not change the outcome very much. Actually, we keep neglecting until we can arrive at a "solution." That is the approach we take to analyze the transient stability of a multi machine power system. We use the classical (and very simple) model to represent any generator connected to the network. In a power system operating in stable condition, the angular positions of the synchronous machines' rotors remain constant relative to one another when no disturbance occurs. *Small disturbance stability* occurs when a power system operating in a steady-state stable condition, following some small disturbance, returns to the same steady-state operating condition or very close to it. Furthermore, the power system is *transient stable* if, following a large or sudden disturbance, it reaches a different but acceptable steady-state operating condition.

9-2 Classical Model

The classical synchronous generator model is shown in Fig. 9-1. This simple model is by far the most used in stability studies, although it is not the best machine representation. In fact, the classical model is used even in the case of generators with salient pole rotors. In the classical model, all phasors, such as generated voltage and current, are usually expressed with reference to the generator terminal voltage, and the resistive portions of all the impedances are neglected. In the classical model, the generator reactance is easily combined by simple addition with the network reactance, such as the reactance of transformers and transmission lines. In generators with salient pole rotors, the direct axis transient reactance is different from the quadrature transient reactance. But the fact that the reactance of the transmission line connecting the generator with the load busses is usually larger than the transient values of the generator's direct and quadrature reactances validates the use of the classical model to represent machines with salient poles and no uniform flux linkages. Sometimes, in stability studies, salient pole generator models consist of two classical models, one for the direct axis transient reactance and the other for the quadrature transient reactance. Although the two magnetic circuits could be considered as separate from each other, the electric circuits of both models are interconnected in parallel and can be resolved into a single one. In reality, two diagrams complicate the computations.

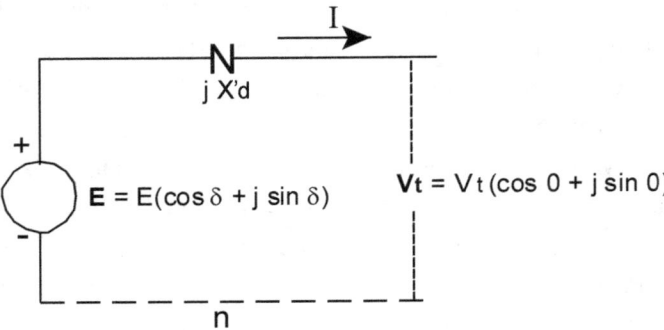

Figure 9-1 Classical generator model

When the load connected to the network suddenly changes or a fault occurs, the main field flux linkages and the generated emf are assumed constant for a short time of around 1 second, although the flux is decaying as determined by the time constant of the field winding circuit. Actually, the armature reaction (Lenz' law) plus the response of the excitation system trying to control the network disturbance tends to maintain the flux and the generated emf constants for at least the assumed 1 second. This period of time is large enough for stability studies concerned with the rotor's first swing. In these studies it is valid to consider the generated emf constant. The problem with the number of swings is that the classical model and the network representation itself are only good for linear parameters. But when a fault occurs, the magnitude of the current increases approximately from 10 to 16 times the normal generator rating, and this high value of the armature current could saturate the magnetic circuit. In addition, harmonics are introduced into the power system with the consequent voltage distortions. In particular, the even harmonics produce a DC offset that tends to saturates all magnetic circuits, such as transformers' cores, reactors, and the generator' s stator. So the system becomes very quickly nonlinear, and therefore the classic model and associated computation procedures are invalid after the first rotor swing or after 1 second if you are optimistic.

The classical model of Fig. 9-1 shows the emf **E** in series (behind) with the generator transient reactance X'd. This simple model provides a single electrical diagram to represent the generator, instead of two separate ones for the case of salient pole rotors, is based on the assumption that X'd = X'q or that the direct-axis transient reactance is equal to the quadrature transient reactance even during transient conditions. However, the results obtained using the single diagram are acceptable because usually the larger value of the reactance of the electrical network, the step-up transformer included, makes any difference between X'd and X'q insignificant. The *generated emf,* which is assumed constant for the duration of the first swing (or for 1second approximately), is defined as the value of **E** during the transient. The power angle δ is the angle between **E** and **Vt** where **Vt** is designated as the phasor reference. In this manner each generator would have its own individual phasor reference. So to avoid having a power system with several phasor references, the neutral of the power system, to which all the generators are connected, is selected as the common reference for the entire power system, and it is assumed that this common reference rotates at synchronous angular speed. In normal operating condition the generator synchronous reactance is constant and is designated as X_s, See Chapter 6.

$$X_s = X_{ar} + X_{l\epsilon} \qquad \text{steady state synchronous reactance} \qquad (9.1)$$

X_{ar} armature winding single-phase reactance

$X_{l\epsilon}$ armature winding single-phase leakage reactance

X_s overall single-phase reactance of the armature winding.

The leakage reactance is due to spreading of the magnetic flux outside the magnetic circuit. This fringing flux increases the magnetic flux density in specific places of the magnetic circuit, such as pole faces tips or the teeth in between slots. And the required correction is attributed to the leakage reactance. The terminal voltage in the generator classical model could be written as show in Eq. (9.2).

$$\vec{V_t} = \vec{E_g} - \vec{I_a} \cdot \overrightarrow{\left(R_{ar} + j \cdot X_s\right)} \qquad (9.2)$$

$$\vec{X_s} = \overrightarrow{\left(X_{ar} + X_{le}\right)} \qquad \text{is called the synchronous reactance.}$$

Neglecting the resistance of the armature winding, R_{ar}, we obtain

$$\vec{V_t} = \vec{E_g} - j \cdot \vec{I_a} \cdot \vec{X_s} \qquad (9.3)$$

But in transient condition the generator synchronous reactance is not constant and it is usually designated as X_d' and in subtransient conditions as X_d''. Which is supposed to be valid for both the direct axis reactance and the quadrature axis reactance. So Eq. (9.3) becomes (9.4).

$$\vec{V_t} = \vec{E_g} - j \cdot \vec{I_a} \cdot \overrightarrow{X_d'} \qquad \text{or} \qquad \vec{V_t} = \vec{E_g} - j \cdot \vec{I_a} \cdot \overrightarrow{X_d''} \qquad (9.4)$$

9.3 Equal-Area Criterion of Stability

Let us consider the simple electrical network illustrated in Fig. 9-2. A three-phase symmetrical short circuit occurs at the F1 point; after a short time the fault is cleared by the two nearest breakers to point F1. *Before* the fault, the net mechanical power input to the generator's shaft P_s equals the net generator output electric power P_e, and therefore the power accelerating the rotor is zero, $P_a = P_s - P_e = 0$ In this condition the generator operates at synchronous speed ω_s, and power angle δ_0. After the fault is cleared, the generator output electric power P_e is not zero because the generator keeps delivering a reduced amount of power through the not affected transmission line, and that is good. If the fault were no three-phase symmetrical, then the generator could also keep delivering power through the not affected phases of the faulted transmission line and improve even more the chances for stability, provided that the fault-clearing breakers open only the affected poles. The power unbalance *during* fault conditions (Ps > Pe) produces an accelerating torque in the shaft (the turbine governor has not reacted yet), and the rotor increases its angular speed beyond synchronism. The fault must be cleared before the increasing power angle reaches a critical value of δ_c, which is specific for every event. The power angle plot for the F1 fault is shown in Fig. 9-3, where area A_1 represents the rotor accelerating period during fault condition; this area is surrounded by P_s, δ_0, δ_c, and P_{du}. When the fault is cleared, the generator output power jumps to a point in P_{af} in the after-fault power angle sine curve and starts developing area A_2 which initiates the braking period. Here P_{af} represents the generator electric output power after the fault is cleared, and because it is higher than P_s, one hopes, there is enough "space" between the curve and the horizontal line representing P_s to contain an area A_2, at least equal to A_1. If this, is the case, the generator will slow down to synchronous speed and fall back into stable operation. Thus the generator would be able to keep delivering power to the load still connected to the power system. Simply stated, the stability criterion is: A_2 must be equal to A_1. However, if the fault were cleared beyond δ_c, the braking period would not be long enough ($A_2 < A_1$) to reduce the angular speed to synchronous operation.

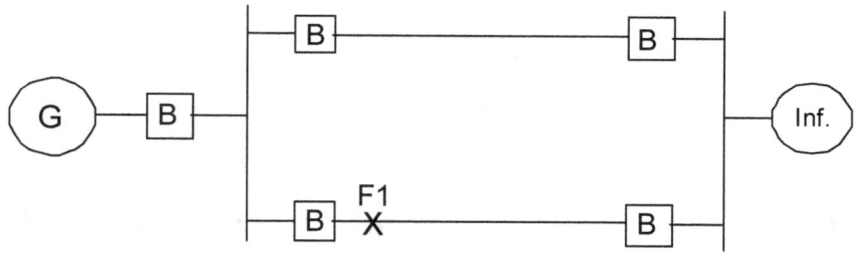

Figure 9-2 Network with a Three-phase fault at F1

The power angle diagram shown in Fig. 9-3 was plotted assuming the following data:

$P_s = 1.000$ Net mechanical power input to generator or shaft power

$P_{be} = 2.000 \cdot \sin(\delta)$ Generator electric output power before fault

$P_{du} = 0.5 \cdot \sin(\delta)$ Generator electrical output power during fault

$P_{af} = 1.5 \cdot \sin(\delta)$ Generator electrical output power after fault

The power angle δ_0 is provided by the first point of intersection of the horizontal line representing P_s with the P_{be} curve.

$$1.000 = 2.000 \cdot \sin(\delta_0) \qquad \delta_0 := \operatorname{asin}\left(\frac{1}{2}\right) \qquad \delta_0 = 0.524 \cdot \text{rad} \qquad \delta_0 = 30°$$

The power angle δ_{max} is provided by the second intersection of the horizontal line representing P_s with the P_{af} curve.

$$1.000 = 1.5 \cdot \sin(\delta_{max}) \qquad \delta_{max} := \pi - \operatorname{asin}\left(\frac{1}{1.5}\right)$$

$$\delta_{max} = 2.412 \cdot \text{rad} \qquad \delta_{max} = 138.19°$$

$$A_1 = 1 \cdot (\delta_c - \delta_0) - \int_{\delta_0}^{\delta_c} 0.5 \cdot \sin(\delta) \, d\delta$$

$$\cos(\delta_0) = 0.866$$

$$\int_{\delta_0}^{\delta_c} 0.5 \cdot \sin(\delta) \, d\delta = -0.5 \cdot \cos(\delta_c) + 0.5 \cdot \cos(\delta_0) = -0.5 \cdot \cos(\delta_c) + 0.5 \cdot 0.866$$

$$A_1 = \delta_c - 0.524 + 0.5 \cdot \cos(\delta_c) - 0.433 = \delta_c + 0.5 \cdot \cos(\delta_c) - 0.957$$

$$A_2 = \int_{\delta_c}^{\delta_{max}} 1.5 \cdot \sin(\delta) \, d\delta - 1 \cdot (\delta_{max} - \delta_c)$$

$$\cos(2.412) = -0.745446$$

$$\int_{\delta_c}^{\delta_{max}} 1.5 \cdot \sin(\delta) \, d\delta = -1.5 \cdot \cos(\delta_{max}) + 1.5 \cdot \cos(\delta_c) = 1.1182 + 1.5 \cdot \cos(\delta_c)$$

$$A_2 = 1.1182 + 1.5 \cdot \cos(\delta_c) - \delta_{max} + \delta_c \rightarrow A_2 = \delta_c + 1.5 \cdot \cos(\delta_c) - 1.2937$$

$$A_1 = A_2 \qquad \textbf{Stability criterion}$$

$$\delta_c + 0.5 \cdot \cos(\delta_c) - 0.957 = \delta_c + 1.5 \cdot \cos(\delta_c) - 1.2937 \quad 0.337 = \cos(\delta_c)$$

$$\delta_c := \text{acos}(0.337) \qquad \delta_c = 1.227 \cdot \text{rad} \qquad \delta_c = 70.31 \cdot^c \qquad \delta := 0, \frac{\pi}{12} .. \pi$$

$\delta_0 = 0.524 \text{rad}$ Initial power angle when the machine is operating synchronously and the mechanical power input is equal to the electric power output, neglecting losses.

$\delta_{max} = 2.412 \text{rad}$ Maximum power angle. Beyond this angle the input of mechanical power to the generator (Ps) is greater than the electric power generated by the generator after the fault has been cleared (Paf). Consequently the rotor angular speed, will increase beyond synchronous speed, and the generator will become unstable.

$\delta_c = 1.227\text{rad}$ Critical clearing angle. When a fault occurs, the power angle starts changing and keeps increasing. To avoid exceeding, the transient stability limit of the power angle δ_{max}, the fault must be cleared before the power angle reaches the critical angle δ_c.

The graphic interpretation of a power flow diagram as a function of the power angle is enough to predict the transient stability of a generator-infinity bus system without the need to solve the swing equation. The procedure is illustrated in Fig. 9-3. This very helpful procedure is called an equal-area criterion of stability. However, the solution of the swing equation is so easily accomplished using a numerical solver that the true value of the equal-area criterion is to utilize the power flow vs. power angle plot to find the values of δ_0, δ_c, δ_{max}.

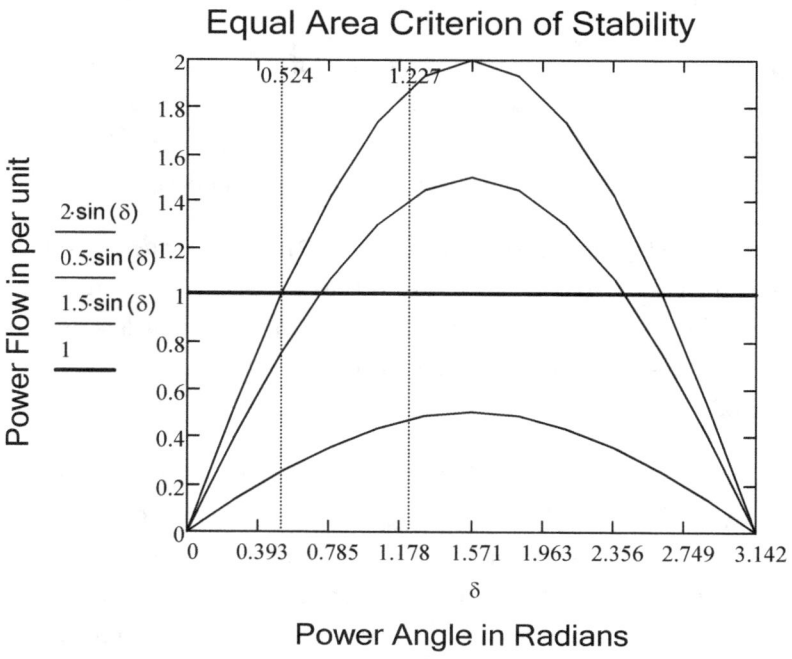

Figure 9-3 Power flow as function of δ , illustrating the equal-area criterion of stability.

9-4 Generator-Infinity Bus Network From a stability point of view, the most damaging types of short circuit faults, in decreasing order, are

1. Three-phase symmetrical short circuit.   2. Double line-to-ground short circuit

3. A line-to-line short circuit   4. A line-to-ground short circuit

The most common short circuit event is the single line to ground, and the most severe but least common is the three-phase symmetrical short circuit. And very rarely or never is the bolted short circuit occurrence, except for intentional sabotage. Because the rotor accelerating area A1 decreases when the generator keeps delivering power during the fault condition (the more power it delivers, the smaller is A1), it is very convenient to use parallel transmission lines. Furthermore, it is also convenient to install circuit breakers for protection or reclosing duty that are capable, in case of line-to-ground faults, of opening only the affected poles. The maximum power that could be transferred in a generator-infinity bus network without exceeding the stability limit is provided by Eq. (7.35), which is repeated below:

$$P_{max} = \frac{|E_g| \cdot |V_b|}{|X|}$$

Equation (7.35) indicates the methods that could be used to improve the transient stability of the classical generator-infinity-bus power system:

1. *Raise the system voltage.* Take into account that one half-wave of the voltage delivered cannot exceed the amount of volt seconds tolerated by connected step-up transformers. Otherwise the step-up transformer cores will saturate and if they are not tripped off the line, they will burn. In addition, the voltage delivered must not damage any connected load.

2. *Reduce the series reactance by using parallel transmission lines.* This both reduces X and increases the power angle of the critical stability limit. Also in the case of a fault in one of the transmission lines, the network will keep delivering power through the remaining line, with the net effect of increasing the system stability because A1 will be smaller than otherwise, and shorter the accelerating period.

3. *Reduce the series reactance by using series capacitor banks.* This will increase the stability limit, but capacitor use is risky. In the case of a short circuit in the network, the line current will contain a steady-state term and a transient term. In general, the transient term is oscillatory with a frequency that depends on the circuit parameters R, L, and C and therefore resembles the natural frequency of oscillation of the faulted circuit. Although this frequency is usually smaller than the system synchronous frequency (60 Hz), it could also be larger. Fortunately, this transient response is damped, and its time constant is usually smaller than 100 milliseconds (ms). The difference between the system frequency and the natural frequency of the faulted circuit is called the *frequency complement.* Actually, the transient term of the current circulating through the armature induces, at the complement frequency, currents and torques in the rotor that could cause a shaft failure or extreme vibration if the rotor natural frequency of oscillation is equal or close to the frequency complement.

4. Besides the conclusions derived from Eq. (7.35), it is essential to use high-speed circuit breakers. A fast fault interruption will decrease A_1. Furthermore, it is a plus to use breakers that open only the faulted phase. That will increase the power delivered during fault conditions and further reduce A_1.

9.5 Introduction to Stability of Multi machine Power Systems

An effort is always made to convert multi machine stability problems to the classical model of one machine feeding an infinity bus. One method lumps together coherent generators. The coherency equation, see Eq. (7.46), is the ratio of the generator synchronizing power coefficient over the angular momentum at synchronous speed. This equation provides the rotor acceleration in case of disturbance, see Eq.(7.47). And therefore generators with equal coherency will have the same rotor acceleration and will swing together responding to system disturbances. This model assumes that the fault affecting any of the generators, lumped or not, does not affect the other generators connected to the system. In general, this approach is not possible because the generators usually are interconnected by a mesh of short transmission lines, and a fault in any line actually affects all of them. The rotors of the generators so interconnected could swing asynchronously. "Meaning they swing to different rhythms." Actually the power grid could respond to a fault in many different ways, including these:

- The generator nearest to the fault sustains increasing synchronous oscillation before losing synchronism and is the only one to lose synchronism.
- The generator nearest to the fault is the first to lose synchronism and is soon followed by some of the other generators connected to the grid.
- Some generators near the fault sustain rotor swings without losing stability. However, one or more generators connected to the same power grid but far away from the fault, loses synchronism.
- It also could happen that the generator closer to the fault become unstable and loses synchronism without sustaining rotor swings. Meanwhile some generators at longer distance from the fault location sustain synchronous oscillations, but they eventually return to stable synchronous operation.

9.6 Modeling of Multi machine Power Systems Short circuits in the power

grid or other strong network disturbances could change the system operation from steady-state to transient. Under transient conditions, one, several, or all generators connected to the power grid could fall into an oscillatory mode of operation. The cause of the oscillatory operation of the entire system could be a single generator that under transient conditions fell in oscillatory mode of operation and whose oscillations and coherency transmitted through the interconnecting transmission lines affected the other generators and triggered their oscillatory operation as well. Their oscillations are usually composed of a 60 Hz fundamental, a low frequency harmonic close to the natural frequency of oscillation of the affected generators (1 to 4 Hz), and sometimes also high frequency harmonics. Often capacitors connected to the power grid are the source of these low or high frequency harmonics.

The swing curve of the machines undergoing simultaneous transient oscillations are determined by the 60-Hz fundamental and low-frequency components. Therefore the assumption is made that the network parameters should be computed for 60 Hz. Besides, modeling of multi machine power systems is very complicated and to decrease the amount of data and computations required in stability studies the assumptions listed below are frequently made:

1. The mechanical power input to each machine remains constant.
2. Damping power is negligible.
3. The classical model for a single machine could be used for each synchronous generator, that is, constant emf behind constant transient reactance. Both are assumed to remain constant through the transient.
4. The mechanical rotor angle of each generator is equal to δ, which is the phase angle of the generated emf phasor with respect to the generator terminal voltage.
5. All electrical loads could be represented by shunt passive impedances connected to ground with the values they have immediately before the disturbance event that triggered the oscillations. Classical stability models of multi machine power systems are those based in the above assumptions. Although these models are limited to first swing analysis, they are useful because they enhance our understanding of what could or did happen.
6. The equal-area criterion of stability is not applicable to multi machine power systems, and the three-phase symmetrical short circuit is not necessarily the worst fault that can occur. So all types of faults must be considered, especially double line-to-ground faults. Furthermore, instability studies require knowledge of the system conditions just before the fault occurs as well as the configuration of the network during and after the fault occurrence. Multi machine stability studies based on the classical model are not comprehensive enough because they do not consider the important factors listed below:
7. How the system performs 2 and 3 seconds after the fault. In this case, the nonlinearity of the system components will come into play, specially, the magnetic saturation of transformers, generators, and reactors.

8. The time response of the excitation system of each generator to a required sudden change in its output voltage.

9. How the generator and turbine shafts and their couplings respond to the twisting torques produced by possible low-frequency nonharmonic oscillations when they are almost equal to the torsional natural frequency of any of the shafts.

10. How fast the turbine valve controls respond to changes produced by the fault in the generator parameters

9.7 Power Flow in a Multi machine Network

The object of power flow studies is to determine the voltage magnitude and phase angle at each bus and the flow of real and reactive power in each line. A good single-line diagram of the power system is absolutely necessary to perform the analysis. The essential information that it must contain includes all connected equipment (generators, transformers, capacitors, etc.), ratings, and impedances; and the series impedances and shunt admittances of all transmission lines. From the single-line diagram the network admittance diagram is created. Generators and loads are considered outside the network, and they are not included in the network bus admittance matrix. To determine the network's power flow pattern, it is necessary to determine the admittance of all lines interconnecting the network's nodes (busses), and to specify at each bus either the net flow of power, real and reactive, into the network or the voltage magnitude and phase angle. At *load busses,* the most convenient selection is to specify net flow of power. At *generator busses* the most practical selection usually is to specify the magnitude and angle of the bus voltage. Rather than try to find a pure analytical solution to the problem, the standard method is to use an iterative process in which a set of estimated values are assigned (educated guess) to the unknown bus voltages. Then using the estimated set of bus voltage values and the real and reactive power specified, we calculate a new value for each bus voltage. In this way we obtain a new set of bus voltage values, and we are ready for another iteration of the procedure. The repetitive use of the procedure generates diminishing results, and it is stopped when the bus voltage changes are smaller than an established minimum value. When the voltage of load busses is changed, common practice is to maintain the reactive power flow constant. For generator busses the practice is to maintain the voltage magnitude constant. It is impossible to specify the net flow of real power in all the busses. So one generator bus, called the *swing bus*, is left without specifying the net real power flow. The generator or generators feeding the swing bus supply the difference between the total system net real power output plus losses and the net real power flow into the system at all the other busses. The power delivered to a load is negative input power to the system. Generators and interconnection with other power systems provide positive or negative power inputs into the network. A detailed application of the procedure is left for the example presented in Chapter 11

CHAPTER 10

System Stability Basic Knowledge

10.1 Brief Summary of Rotational Dynamics

Stability studies of rotating rigid bodies which are symmetrical about a fixed rotational axis requires the application of the mechanical formulas listed in Table 10-1 (at the end of Chapter 10), which also includes the corresponding formulas for similar concepts in linear motion.

Linear motion:

$$F = m \cdot \frac{d^2}{dt^2} S$$

Mass times linear acceleration

Rotational motion:

$$T = I \cdot \frac{d^2}{dt^2} \theta$$

Moment of inertia times angular acceleration (10.1)

Applying equation (10.1) to synchronous generators we obtain:

$$I \cdot \frac{d^2}{dt^2} \theta = T_a = T_s - T_e$$

(10.2)

I   Total moment of inertia of rotational masses in the generator and prime mover.

θ   Angular position of the rotor with respect to a fixed reference in mechanical radians.

t   Time in seconds

Ta   Net accelerating torque

Ts   Shaft torque delivered by prime mover minus the torque representing rotational losses

Te   Net electromagnetic torque produced by the generator

A synchronous generator delivers power to the electrical system and we define the shaft torque Ts and the electromagnetic torque Te as positive. Starting up the generator requires increasing the shaft torque to accelerate the generator rotor in the positive θ direction of rotation, generating a positive electromagnetic torque--which always oppose the driving shaft torque--and an electromotive force of the right polarity. When both torques are equal the acceleration torque Ta is zero and there is no positive or negative acceleration, and the rotor speed is then the synchronous speed. In this case, both the electromagnetic torque and the generated emf are constant.

A synchronous motor receives power from the electrical system, and consequently Ts and Te are inverted in sign. Te in this case represents the driving torque produced by the power delivered to the motor, and Ts represents the opposing counter torque of the mechanical load and rotational losses. During the starting period Te must be larger than Ts, but they are equal when the motor rotor reach synchronous speed. The torque T, the angular velocity ω, and the angular acceleration α are all vectors acting along the same axis. Therefore we can consider only their magnitudes and treat them algebraically.

The value of M at synchronous speed is called the *inertial constant,* which unfortunately is also the name assigned to H. In dealing with generators, it is better to express the stored rotational kinetic energy in megajoules or megawatt-seconds. The angular momentum M is:

$$M = I \cdot \omega \qquad \frac{Megajoule \cdot sec}{rad} \qquad \text{or} \qquad \frac{Megajoule \cdot sec}{degree}$$

Stored rotational energy: $\dfrac{I \cdot \omega^2}{2} = \dfrac{M \cdot \omega}{2}$ Megajoule or Megawatt·sec

The inertial constant H is the stored rotational energy of the generator at synchronous speed in megajoules per unit of generator rating in MVA.

H = Stored rotational energy in megajoules / generator MVA rating

1 megajoule = 1 megawatt-second, so H could also be expressed as:

H = Stored rotational energy in Megawatt-seconds / generator MVA rating

$$H = \frac{M \cdot \omega_s}{2 \cdot G} \qquad \text{G is the generator rating in MVA} \qquad (10.3)$$

GH = Stored rotational energy at synchronous speed in megajoules or megawatt-seconds.

$$G \cdot H = \frac{M \cdot \omega_s}{2} \quad \text{Megawatt-seconds} \tag{10.4}$$

For generators with one pair of poles and a frequency of 60 cycles per second, and expressing the angles in radians, we have

$$f := 6C \quad \frac{\text{cycle}}{\text{sec}} \qquad \omega_s := 2 \cdot \pi \cdot f \qquad \omega_s = 377 \quad \frac{\text{rad}}{\text{sec}}$$

$$\frac{\omega_s}{2} = \pi \cdot f = 188.5 \quad \frac{\text{rad}}{\text{sec}}$$

$$G \cdot H = M \cdot \pi \cdot f = M \cdot 188.5 \quad \text{Megajoule or Megawatt·sec} \tag{10.5}$$

$$M = \frac{G \cdot H}{\pi \cdot f} = \frac{G \cdot H}{188.5} = 0.0053 \cdot GH \quad \frac{\text{Megawatt·sec}^2}{\text{rad}} \tag{10.6}$$

For generators with one pair of poles and frequency of 60 cycles per second, expressing the angles in degrees, is

$$\omega_s := 360 \cdot f = 21600 \quad \frac{\text{degree}}{\text{sec}} \qquad \frac{\omega_s}{2} = 180 \cdot f = 10800 \quad \frac{\text{degree}}{\text{sec}}$$

Using equation (10.4) with electrical degrees, we obtain Eqns. (10.7) and (10.8)

$$M = \frac{G \cdot H}{180 \cdot f} = \frac{G \cdot H}{10800} = 0.000093 GH \quad \frac{\text{Megawatt} - \text{sec}^2}{\text{degree}} \tag{10.7}$$

The inertial constant H of a synchronous machine has a small range of values for each type of machine independently of his MVA and angular speed ratings. In fact, for all types of machines it ranges in value from 1 to 10 megawatt-second per MVA of the machine rating. Besides, H can be calculated provided that the weight and radius of gyration of all the rotating parts of the generator and prime mover are known. The manufacturers of synchronous generators usually provide enough data to compute H.

$$KE = \frac{I \cdot \omega_m^2}{2} \qquad \text{Rotational kinetic energy where } \omega m \text{ is the mechanical angular speed and I is}$$

the moment of inertia, is

For 3600 rpm machines we have:

Data: 3600 rpm $\qquad g = 32.17 \; \dfrac{ft}{sec^2}$

$$\omega_m = 2 \cdot \pi \cdot \frac{3600}{60} = 120 \cdot \pi = 377 \; \frac{rad}{sec} \qquad \frac{rad}{cycle} \cdot \frac{cycle}{sec}$$

$$I = \frac{W \cdot R^2}{32.17} \qquad \text{W in pounds and R in feet}$$

$$KE = \frac{1}{2} \cdot \frac{W \cdot R^2}{32.17} \cdot 377^2 \qquad foot - pounds \qquad (ft \cdot lb)$$

To convert foot-pound to megajoules, multiply by $\; 1.3564 \cdot 10^{-6} \quad$ <<<<

$$KE = \left(\frac{1}{2} \cdot \frac{W \cdot R^2}{32.17} \cdot 377^2 \right) \cdot \left(1.3564 \cdot 10^{-6} \right) \; megajoule$$

$$\left(\frac{1}{2} \cdot \frac{W \cdot R^2}{32.17} \cdot 377^2 \right) \cdot 1.3564 \cdot 10^{-6} \; simplify \rightarrow 2.9963 \times 10^{-3} \cdot R^2 \cdot W \qquad megajoule$$

H = stored rotational energy, megajoules / generator MVA rating

$$H = \frac{3 \times 10^{-3} \cdot R^2 \cdot W}{G} \; \frac{megajoules}{MVA} \qquad \textit{3600 rpm case} \qquad (10.8)$$

For 1800 rpm machines:

$$\omega_m = 2 \cdot \pi \cdot \frac{1800}{60} = 60 \cdot \pi = 188.5 \; \frac{rad}{sec} \qquad \text{Everything else the same}$$

$$\left(\frac{1}{2} \cdot \frac{W \cdot R^2}{32.17} \cdot 188.5^2 \right) \cdot 1.3564 \cdot 10^{-6} \; simplify \rightarrow 749.0821 \times 10^{-6} \cdot R^2 \cdot W \qquad megajoule$$

$$H = \frac{0.749 \cdot 10^{-3} \cdot R^2 \cdot W}{G} \; \frac{megajoules}{MVA} \qquad \textit{1800 rpm case} \qquad (10.9)$$

For 900 rpm machines:

$$\omega_m = 2 \cdot \pi \cdot \frac{900}{60} = 30 \cdot \pi = 94.2\zeta \; \frac{rad}{sec} \qquad \text{Everything else the same}$$

$$\left(\frac{1}{2} \cdot \frac{W \cdot R^2}{32.17} \cdot 94.25^2 \right) \cdot 1.3564 \, 10^{-6} \; simplify \rightarrow 187.271 \times 10^{-6} \cdot R^2 \cdot W \qquad megajoule$$

$$H = \frac{0.187 \cdot 10^{-3} \cdot R^2 \cdot W}{G} \; \frac{megajoule\zeta}{MVA} \qquad \textit{900 rpm case} \qquad (10.10)$$

10.2 Synchronizing Power Coefficient

Let us consider the single-phase version of a balanced three-phase electrical network which is illustrated in Fig. 10-1 and assume that is operating in steady-state condition at a power angle δ 0, where Eg is the generated line-to-neutral emf and Vb is the line-to-neutral voltage at the infinity bus. The reactance shown is the sum of the generator's synchronous reactance and the transmission line reactance. All the resistances have been neglected. The real power delivered by the generator is given by Eq. (10.11).

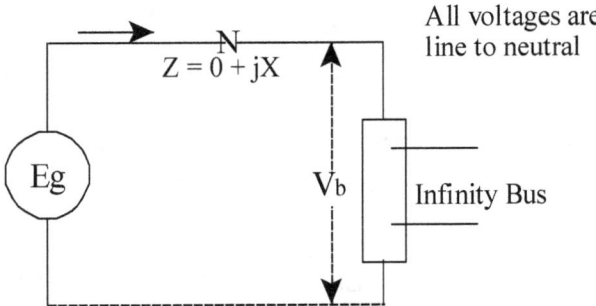

Figure 10-1   Simplified electrical network operating in steady
state condition

Single-phase real electrical power transferred between Generator and infinity bus

$$P_{e0} = \frac{\left| Eg \right| \cdot \left| V_b \right|}{\left| X \right|} \cdot \sin\left(\delta_0 \right)$$

$$(10.11)$$

Maximum single-phase real electrical power that the generator could deliver

$$P_{emax} = \frac{|E_g| \cdot |V_b|}{|X|}$$

(10.12)

$$P_{e0} = P_{emax} \cdot \sin(\delta_0)$$ Eq. (10.11) in concise format   (10.13)

The steady-state operating point of a generator must be such that it should not lose synchronism when small change occurs in its electrical power output. Let us consider that the shaft power remains constant when the output of electrical power changes. Assuming incremental changes in the parameters defining the initial operating point, we have

$$\delta = \delta_0 + \delta_\Delta \qquad P_e = P_{e0} + P_{e\Delta} \qquad P_s = constant$$

δ_0 is the initial operating point power angle

δ_Δ is the change in power angle

δ is the new power angle

P_{e0} electrical output power before change

$P_{e\Delta}$ change in the electrical output power

P_e new electrical output power after change

P_s shaft power

$$P_{e0} + P_{e\Delta} = P_{emax} \cdot \sin(\delta_0 + \delta_\Delta)$$

$$P_{e0} + P_{e\Delta} = P_{emax} \cdot \cos(\delta_0) \cdot \sin(\delta_\Delta) + P_{emax} \cdot \cos(\delta_\Delta) \cdot \sin(\delta_0)$$

If δ_Δ is small enough, we can write that $\sin(\delta_\Delta) = \delta_\Delta$ and $\cos(\delta_\Delta) = 1$

$$P_{e0} + P_{e\Delta} = P_{emax}(\delta_\Delta \cdot \cos\delta_0 + \sin\delta_0)$$

$$P_{e0} + P_{e\Delta} = \delta_\Delta \cdot P_{emax} \cdot \cos\delta_0 + P_{emax} \cdot \sin\delta_0$$

(10.14)

$P_{emax} \cdot \cos\delta_0$ *is defined as the synchronizing power coefficient.*

$$P_{sy} = P_{emax} \cdot \cos\delta_0 \quad \textbf{\textit{Synchronizing power coefficient}} \tag{10.15}$$

From Eqs. (10.13 to 10.15) we obtain

$$P_{e0} + P_{e\Delta} = \delta_\Delta \cdot P_{sy} + P_{e0} \qquad\qquad P_{e\Delta} = \delta_\Delta \cdot P_{sy} \tag{10.16}$$

The increment of the power angle times the synchronizing power coefficient provides the increment in electrical power delivered by the generator. The reader should understand that to obtain three-phase power he must multiply by 3 the values obtained applying Eqns. (10.12 and 10.13) or replace the line-to-neutral voltages in these equations with the line-to-line voltage of the 3-phase system. At the initial operating point the generator operates in steady-state condition, the power angle is δ_0 and the shaft power is equal to the generated electric power.

$$P_s = P_{e0} = P_{emax} \cdot \sin(\delta_0) \tag{10.17}$$

Subtracting Eq. (10.14) from Eq. (10.17), we obtain

$$P_s - \left(P_{e0} + P_{e\Delta}\right) = -P_{emax} \cdot \delta_\Delta \cdot \cos(\delta_0) \tag{10.18}$$

The very important rotor swing equation, see Eq. (10.29), is given below in a convenient format

$$P_s \text{ and } P_e \text{ are in per unit} \tag{10.19}$$

Expressing Eq. (10.19), in terms of the new value of the variables after the change, see Eq. (10.18) we obtain

$$\frac{2 \cdot H}{\omega_s} \cdot \frac{d^2}{dt^2}\left(\delta_0 + \delta_\Delta\right) + P_{emax} \cdot \delta_\Delta \cdot \cos(\delta_0) = 0 \tag{10.20}$$

$$\frac{2 \cdot H}{\omega_s} \cdot \frac{d^2}{dt^2}\delta = P_s - P_e$$

The synchronizing power coefficient, see Eq. (10.15), is the slope of the sine curve (differential of sine is the cosine) representing the electrical power, see Eqns. (10.11 and 10.13), transferred between generator and the load.

$$P_{sy} = \left(\frac{d}{d\delta} P_e\right)_{\delta_0} = P_{emax} \cdot \cos\left(\delta_0\right)$$

Evaluated for $\delta = \delta_0$

Since δ_0 is a constant, Eq. (10.20) can be expressed as

$$\frac{d^2}{dt^2}\delta_\Delta + \frac{\omega_s}{2 \cdot H} \cdot \delta_\Delta \cdot P_{sy} = 0$$

(10.21)

The system steady-state stability for small load changes is determined by Eq. (10.21), which is a linear second order differential equation whose solution depends on the sign of Psy or the sign of the slope of the power angle curve, see Eq. (10.22), at the operating point. If Psy is

positive, the increment of the power angle as a function of time δ_Δ (t) is oscillatory about δ_0, the original power angle, like an undamped swinging pendulum. However, the ever-present resistance in transmission lines, transformers, and in the generator's damper windings, which have been all neglected, introduce decay in the amplitude of the oscillations that eventually will stop the oscillations of the power angle. On the other hand, if Psy is negative, the power angle will increase without a limit after the occurrence of small change in the load. *Therefore the system would be steady-state stable only if the slope of Eq. (10.22) at the operating point were positive.* **This means that the generator must be operated with a power angle within the range of 0 to 90 degrees.**

$$P_e = P_{emax} \cdot \sin(\delta)$$

(10.22)

The condition for stable operation for small load changes is

$$\frac{d}{d\delta}\left(P_{emax} \cdot \sin(\delta)\right) > 0 \quad \delta = \delta_0$$

(10.23)

10.3 The Swing Equation

If the angular position of a generator rotor θ is measured with respect to a fixed reference on the stator, it would grows continuously with time. See Eq. (10.24). The important thing, however, is the angular position of the rotor with respect to a reference axis that rotates at synchronous speed. Symbolically,

$$\theta = \omega_s \cdot t + \delta$$

(10.24)

θ = angular position of the generator's rotor with respect to fixed reference, mechanical radians

ωs = generator rotor synchronous speed, mechanical radians per second

δ = rotor angular displacement from synchronous rotating reference axis, mechanical radians

Actually Eq. (10.24) is also valid if the angles are expressed in other consistent set of units. Eq. (10.25) *was obtained assuming ωs constant,* which is a fair assumption, and writing the derivative with respect to t.

$$\frac{d}{dt}\theta = \omega_s + \frac{d}{dt}\delta \qquad (10.25)$$

Equation (10.25) gives the rotor angular velocity, and it shows that is constant and equal to the generator synchronous speed ω_s only when

$$\frac{d}{dt}\delta = 0$$

So (d/dt) δ is the deviation from synchronism. When (d/dt) δ is constant, the generator rotor is running at a constant angular speed equal to the synchronous angular speed, ω_s, plus (d/dt) δ.

$$\frac{d^2}{dt^2}\theta = \frac{d^2}{dt^2}\delta \qquad (10.26)$$

Equation (10.26) provides the rotor angular acceleration, and it shows that is equal to the acceleration of the deviation. Furthermore it is valid in any consistent set of units, mechanical or electrical. From Eqns. (10.2) and (10.26) we obtain

$$I \cdot \frac{d^2}{dt^2}\delta = T_a = T_s - T_e \qquad (10.27)$$

$$\alpha = \frac{d^2}{dt^2}\delta \qquad \text{acceleration} \qquad \alpha \cdot I = T_a$$

From Table 10-1 we know that M = ω I and P = Tω. Multiplying both sides of Eq. (10.27) by ω, we obtain

$$\omega \cdot I \cdot \frac{d^2}{dt^2}\delta = P_a = P_s - P_e \tag{10.28}$$

$$M \cdot \frac{d^2}{dt^2}\delta = P_a = P_s - P_e$$

Swing equation (10.29)

M = rotor angular momentum at synchronous speed

δ = torque angle or rotor angular displacement from synchronous rotating reference

P_a = accelerating power

P_s = net mechanical power or shaft power

P_e = net electrical power generated

Another way of presenting the swing equation is derived below. From Eq. (10.3) we obtain

$$M = \frac{2 \cdot H}{\omega_s} \cdot G \qquad \frac{M}{G} = \frac{2 \cdot H}{\omega_s} = \frac{2 \cdot H}{2 \cdot \pi \cdot f} = \frac{H}{\pi \cdot f}$$

Where G is the generator rating in MVA. Substituting in Eq. (10.29) we obtain

$$\frac{2 \cdot H}{\omega_s} \cdot \frac{d^2}{dt^2}\delta = \frac{P_a}{G} = \frac{P_s - P_e}{G} \qquad \delta \text{ and } \omega_s \text{ should be expressed in a consistent set of units: radians or degrees}$$

$$\frac{2 \cdot H}{\omega_s} \cdot \frac{d^2}{dt^2}\delta = P_a = P_s - P_e \quad \textit{Here Pa, Ps and Pe are in per unit} \tag{10.30}$$

When δ is in *electrical radians,* Eq. (10.30) becomes

$$\frac{H}{\pi \cdot f} \cdot \frac{d^2}{dt^2}\delta = P_a = P_s - P_e$$

Swing equation in per unit (10.31)   When δ is in *electrical degrees,* Eq. (10.30) becomes

$$\frac{H}{180 \cdot f} \cdot \frac{d^2}{dt^2} \delta = P_a = P_s - P_e \quad \text{**Swing equation in per unit**} \quad (10.32)$$

In the case of a generator and infinity bus system, the swing equation becomes (10.33) by substituting Eq. **(7.34)** for P_e in Eq. (10.29).

$$M \cdot \frac{d^2}{dt^2} \delta = P_s - \frac{|E_g| \cdot |V_b|}{|X|} \cdot \sin(\delta) \quad (10.33)$$

According to Eq. (10.12) the maximum electrical power that could be transferred without exceeding the stability limit is

$$P_{max} = \frac{|E_g| \cdot |V_b|}{|X|}$$

So Eq. (10.33) becomes

$$M \cdot \frac{d^2}{dt^2} \delta = P_s - P_{max} \cdot \sin(\delta) \quad (10.34)$$

The swing equation is a second-order differential equation that could be solved by converting it to a system of two coupled first-order differential equations and applying any of the readily available differential equation numerical solvers. The author uses Mathcad 14. The solution is given in matrix (table) format and shows the power angle δ increasing with time, a clear indication that the system could be unstable for the disturbance considered. In that case, the disturbance (a fault or large, sudden load increase) must be cleared before the power angle reaches the critical value δ c to ensure that the system will remain stable after the fault is cleared.

10.4 Natural Frequency of Oscillation

A synchronous generator or motor in steady state operation runs at synchronous speed (commonly 377 radians per second). However some transient in the system, like for instance load changes or capacitors discharging introduces low frequency (sub-harmonic) oscillations in the electrical system which ride on top of the 60 Hz fundamental and consequently change the wave shape of the power. Fortunately these oscillations disappear quickly due to the damping effects of the electrical load, the synchronous machine itself, and the prime mover. The undamped oscillations are sinusoidal in wave shape and equation (10.35) provides its angular frequency.

$$f_n = \frac{1}{2 \cdot \pi} \cdot \sqrt{\frac{\omega_s \cdot P_{sy}}{2 \cdot H}} \quad \begin{array}{l} \text{natural frequency} \\ \text{of oscillation} \end{array} \quad \omega_n = 2 \cdot \pi \cdot f_n \qquad (10.35)$$

$$T_n = \frac{1}{f_n} \qquad \text{Period of the natural frequency of oscillation}$$

A machine natural frequency of oscillation is a function of the synchronizing power coefficient Psy and its inertia constant H that could not be negative but Psy could be, provided that cos δ_0 is negative or that the power angle is larger than 90^c and smaller than 270^c. If Psy is negative the natural frequency becomes complex. *In general, the H of different machines have different values and consequently different natural frequency of oscillation.* It could happen that a group of coherent machines oscillates together because they have the same natural frequency of oscillations and sometimes a group of coherent machines oscillates with respect to another group of coherent machines.

Table 10-1 Mechanical formulas used to analyze the transient stability of electrical networks

----

Quantity	Symbol	Formula	Unit

Angular displacement	θ		radians
Moment of inertia	I	$I = \int r^2\, dm$	$\dfrac{\text{Joules}^2}{\text{rad}^2}$
Angular velocity	ω	$\omega = \dfrac{d}{dt}\theta$	$\dfrac{\text{rad}}{\text{s}}$
Angular acceleration	α	$\alpha = \dfrac{d}{dt}\omega$	$\dfrac{\text{rad}}{\text{s}^2}$
Torque	T	$T = I \cdot \alpha$	$\dfrac{\text{J}}{\text{rad}}$
Angular momentum	M	$M = I \cdot \omega$	$\dfrac{\text{J} \cdot \text{s}}{\text{rad}}$
Work	W	$W = \int T\, d\theta$	J
Rotational kinetic energy	KE	$KE = \dfrac{I \cdot \omega^2}{2}$	J
Power	P	$P = T \cdot \omega$	Watts

10.5 Coherent Machines

Generators that swing together responding to system disturbances are called coherent machines. Coherent generators can be lumped together even if they have different speed ratings, provided that ω s and δ are expressed in a consistent set of units. When generators are lumped together to solve transient stability problems, the single equivalent generator must have a rating equal to the sum of the ratings of the individual generators. Also the angular momentum at synchronous speed of the equivalent machine, M, must be equal to the sum of the angular momentums at synchronous speed of the individual machines. Once the value of M is known for the equivalent machine, the inertial constant H of the equivalent machine could be calculated using equation (10.4). Furthermore, in a multi-machine system all the components must be expressed in a common MVA base including the inertial constant H. For instance:

Common base: 100 MVA

Generator 1 base: 80 MVA same than the generator 1 rating

Generator 1 inertial constant H1 = 5   Stored rotational energy in megajoules / 80

$$100 \cdot H_{common} = 80 \cdot 5 \qquad H_{common} = \frac{400}{100} = 4 \qquad \text{Megajoules per MVA}$$

In general the conversion   formula is: Common base x
Hcommon = Hg x generator base.   Or

Hcommon = Hg x generator base / common base   (10.36)

In stability studies H instead of M is the preferred constant, because it has a narrow range of values for each class of machine. For instance for 3600 rpm turbines driven generators the H-range is 4 - 7 Megajoules per MVA. Generator manufactures sometimes provides the WR2 that is the weight of the rotating parts in pounds times the square of the radius of gyration in feet, instead of H. To calculate H's follow the simple procedure given in Section 10.1, equations (10.8) - (10.10).

When applying the swing equation to several machines lumped together the following applies:

$$H = \sum_i H_i \qquad P_s = \sum_i P_{si} \qquad P_e = \sum_i P_{ei}$$

δ is the same for all the machines.

formula $\quad \dfrac{P_{sy}}{M} \quad$ (10.37) $\quad \dfrac{\text{Megawatts}}{\left(\dfrac{\text{Megawatts} \cdot \text{sec}^2}{\text{degree}} \right)} = \dfrac{\text{degree}}{\text{sec}^2}$

Meaning that coherent generators have the same rotor angular acceleration. So two generators are coherent if a common load disturbance produces the same rotor acceleration in both machines. Symbolically.

$$\frac{P_{sy1}}{M_1} = \frac{P_{sy2}}{M_2} \qquad\qquad (10.38)$$

However Eq. (10.38) is not comprehensive, because two or *more* machines could oscillate together, because they have the same *natural frequency of oscillation*, even when they are not supposed to do so according with Eq. (10.38).

CHAPTER 11

Transient Stability Example in a Four Generators Grid

11.1 Modeling of Multi-machine Electrical Power Grid.

Short circuits in the interconnecting network or other strong grid disturbances could change the power system operation from steady state to transient. Under transient conditions, one, several, or all generators connected to the power grid could fall into an oscillatory mode of operation. The cause of the oscillatory operation of the entire system could be a single generator that under transient conditions felt in oscillatory operation, and whose coherency transmitted through the interconnecting transmission lines affected the other generators and triggered their oscillatory operation as well. The transmitted oscillations, in general are composed of a 60 Hz fundamental and low frequency harmonics (1 to 4 Hz) close to the natural frequency of oscillation of the affected part of the system, and sometimes also contains high frequency harmonics. Often capacitors are the source of these low or high frequency harmonics.

The swing curves of the machines undergoing simultaneous transient oscillations are shaped by the 60 Hz and low frequency components. Therefore, the assumption is usually made that the network parameters should be computed for 60 Hz. Furthermore, to decrease the amount of data and computations required in multi-machine stability studies the assumptions listed below are frequently made:

- The mechanical power input to each machine remains constant
- Damping power is negligible
- The classical model for a single machine could be used for each synchronous generator, that is, constant emf behind constant transient reactance. Both are assumed to remain constant through the transient.
- The mechanical rotor angle of each generator is equal to the electrical power angle δ, which is the phase angle of the generator emf phasor with respect to the generator terminal voltage.
- All electrical loads could be represented by shunt passive impedances connected to ground with the values they have immediately before the disturbance event that trigger the oscillations.

Classical stability models of multi-machine power grids are those based in the above assumptions. Although these models are limited to first swing analysis, they are very useful because they improve our understanding of what could or did happen. The equal area criterion of stability is not applicable to multi-machine power grids, and the three-phases symmetrical short circuit is not necessary the worst fault that could occur. The reader should remember that symmetrical short circuit allows the single-phase analysis, which facilitates the solution of the problem. Nevertheless, all types of fault must be considered, especially double line- to-ground faults. Furthermore, instability studies require knowledge of the grid conditions just before the fault occurs. And it is necessary to determine the configuration of the network during and after the fault occurrence. In most protection schemes the fault is isolated by opening the two closest breaker. This practice might not be acceptable from a stability point of view. Multi-machine stability studies based in the classical model are not comprehensive enough because they do not consider the factor listed below.

1. How the network perform 2 and 3 seconds after the fall. In this case, the nonlinearity of the grid components will come into play, especially, the magnetic saturation of transformers, generators, and reactors.

2. The time response of the excitation source (DC rectifiers) of each generator to a sudden change of the generator output voltage.

3. How generators and turbines shafts and their couplings respond to the twisting torques produced by low frequency non-harmonics oscillations when they are almost equal to the torsional natural frequency of any of the shaft.

4. How fast the valves of the driving turbine coupled to each generator respond to the changes produced by the fault in the generator parameters.

11.2 Steps Required

A power grid is transient stable if it remains in synchronism when submitted to large disturbances. In this example the solution is carried out using time-domain methods of analysis, such as numerical integration techniques. In general, the solution of a multi-machine transient stability problem is a long and time consuming task that requires to make many assumptions and collect a great amount of hard to find data. The following steps are required.

1. Obtain or create a single line diagram of the entire grid
2. Collect all the necessary data
3. Obtain the most recent load flow study of the grid
4. Compute the power angles (initial power angles) of all generators connected to the grid during normal operation.
5. Compute M, the angular momentum at synchronous speed and Psy, the synchronizing power coefficient of all generators
6. Compute the natural frequency of oscillations of all generators
7. Determine which generators are coherent, and select the ones that should be lumped together.
8. Determine the grid connection configurations during and after fault conditions
9. Create the reactance diagram of each grid configuration. And select the configurations to be analyzed in the stability study.
10. Reduce all the grid configurations, using node elimination techniques (matrix algebra) or the classical star-mesh transformation formulas.
11. Use a numerical solver to determine the grid response to transient disturbances.

11.3 Numerical Solver Application to a Grid with Four Generators

The purpose of this section is to demonstrate the use of a numerical solver to obtain the response of a mini-grid to a symmetrical bolted three-phase short-circuit. Rather than following the steps required in Section 11.2 which are absolute necessary in a real case we will assume the result of steps 1 through 10 and jump to step number 11. In tables 11-1 to 11-4 the resistance of generators and transformers have been neglected.

Necessary formulas and concepts

M is the angular momentum at synchronous speed or inertial constant

$$M = 0.000093(G \cdot H) \ \frac{\text{Megawatt} \cdot \text{sec}^2}{\text{degree}}$$

Synchronizing power coefficient $\qquad P_{sy} = P_{emax} \cdot \cos(\delta_0) \qquad P_{emax} = \frac{|E_g| \cdot |V_t|}{X_s}$

Two machines are coherent if: $\qquad \dfrac{P_{sy1}}{M_1} = \dfrac{P_{sy2}}{M_2}$

Natural frequency of oscillations $\qquad f_n = \dfrac{1}{2\pi} \cdot \sqrt{\dfrac{\omega_s \cdot P_{sy}}{2 \cdot H}} \qquad \omega_n = 2\pi \cdot f_n$

Type	Steam	Steam	Combustion turbine	Combustion turbine
MVA rating	1200	1000	500	300
kV output	24	24	18	18
RPM	3600	3600	3600	3600
H MW-sec / MVA	8	7	5	4
X_d	1.2 per-unit	1.2 per unit	0.95 per-unit	0.95 per-unt
$X_d{}'$	0.16 per-unit	0.16 per-unit	0.12 per-unit	0.12 per-unit

Table 11-1 Generators Factory ratings

Generator	G1	G2	G3	G4
Type	Steam	Steam	Combustion turbine	Combustion turbine
MVA rating	1.2	1	0.5	0.3
kV output	0.048	0.048	0.036	0.036
RPM	3600	3600	3600	3600
H MW-sec / MVA	9.6	7	2.5	1.2
X_d	0.0023 pu	0.0028 pu	0.0025 pu	0.0041
X_d'	0.0003	0.0004	0.0003 pu	0.0005 pu
M	0.0011	0.0007	0.0001	0.00003
δ_0 Initial power angle	4.063^0	5.66^0	5.517^0	4.695^0
P_{sy}	10.001	8.038	4.135	2.556
fn Cycles / sec	2.23	2.347	2.81	3.189

Table 11-2 Generators Per-unit values based on 500 kV and 1000 MVA

$$H_2 = H_1 \cdot \frac{MVA_1}{MVA_2}$$ H in base 2 given H in base 1   (11.1)

$$Z_2 = Z_1 \cdot \frac{MVA_{b2}}{MVA_{b1}} \cdot \left(\frac{kV_{b1}}{kV_{b2}}\right)^2$$ Impedance in base 2 given impedance in base 1. (11.2)

$1.2 \cdot \dfrac{1000}{1200} \cdot \left(\dfrac{24}{500}\right)^2 = 0.0023$ $0.16 \cdot \dfrac{1000}{1200} \cdot \left(\dfrac{24}{500}\right)^2 = 0.0003$

$1.2 \cdot \left(\dfrac{24}{500}\right)^2 = 0.0028$ $0.16 \cdot \left(\dfrac{24}{500}\right)^2 = 0.0004$

$0.95 \cdot \dfrac{1000}{500} \cdot \left(\dfrac{18}{500}\right)^2 = 0.0025$ $0.12 \cdot \dfrac{1000}{500} \cdot \left(\dfrac{18}{500}\right)^2 = 0.0003$

$$0.95 \cdot \frac{1000}{300} \cdot \left(\frac{18}{500}\right)^2 = 0.0041 \qquad\qquad 0.12 \cdot \frac{1000}{300} \cdot \left(\frac{18}{500}\right)^2 = 0.0005$$

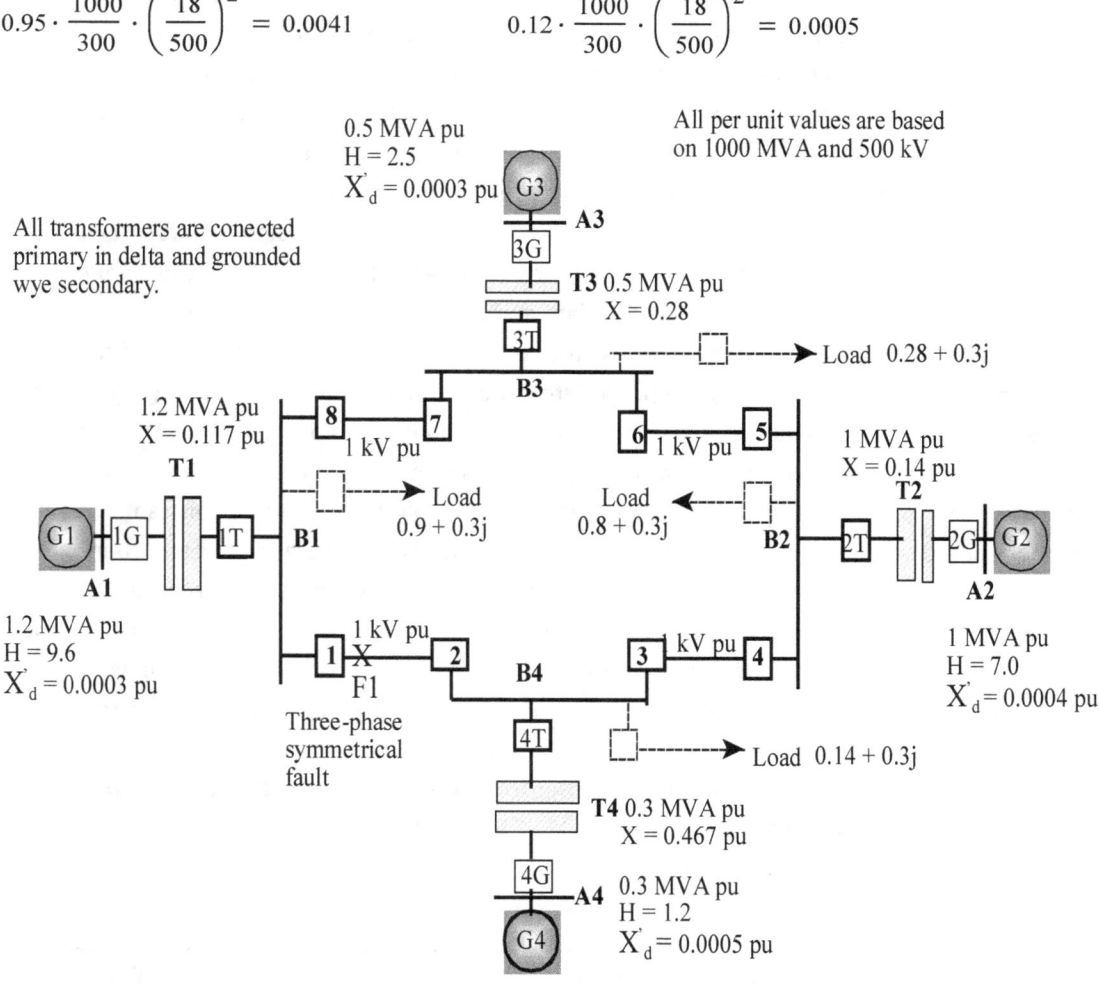

Figure 11-1 Single line diagram of a four generators mini-grid described in section 11.3. Not to scale

Transformers	T1	T2	T3	4
Type	Three-phase Oil/Air cooled	Three-phase Oil/Air cooled	Three-phase Oil/Air cooled	Three-phase Oil/Air cooled
MVA rating	1200	1000	500	300
kV output	500 kV	500 kV	500 kV	500 kV
X reactance	0.14 pu	0.14 pu	0.14 pu	0.154 pu
Connection	Delta-Wye	Delta-Wye	Delta-Wye	Delta-Wye

Table 11-3 Transformers Factory ratings. Reactances based in transformers rated MVA

To determine the grid's power flow pattern, it is necessary to determine the admittance of all the lines interconnecting the grid nodes (busses), and to specify at each bus either the net flow of power, real and reactive, into the grid or the voltage magnitude and phase angle. At *load busses* the most convenient selection is to specify net flow of power. At *generators busses* the most practical selection usually is to specify the magnitude and angle of the bus voltage. Rather than try to find a pure analytical solution to the problem, the standard method is to use an iterative process in which a set of estimated values are assigned (educated guess) to the unknown busses voltages. Then using the estimated set of bus voltage values and the real and reactive power specified, we calculate a new value for each bus voltage. In this way we obtain a new set of bus voltage values, and we are ready for another iteration of the procedure. The repetitive use of the procedure generates diminishing results, and it is stopped when the bus voltages changes are smaller than an established minimum value. When the voltage of load busses is changed, common practice is to maintain the reactive power flow constant. For generators the practice is to maintain the voltage magnitude constant It is impossible to specify the net flow of real power in all the busses. So one generator bus, called the *swing bus,* is left without specifying the net real power flow. The generator or generators feeding the swing bus supply the difference between the total system net real power output plus losses and the net real power flow into the system at all the other busses. *The power delivered to a load is negative input power to the system.* Generators and interconnections with other power systems provide positive or negative power inputs into the grid. A detailed application of the procedure is given in Chapter four of reference 1. The results of a power flow study for this example are provide in Table 11-6. Computations of per unit line impedances are based on 1000 MVA and 500 kV

Transformers	T1	T2	T3	T4
Type	Three-Phase Oil/Air cooled	Three-Phase Oil/Air cooled	Three-Phase Oil/Air cooled	Three-Phase Oil/Air cooled
MVA rating	1.2	1	0.5	0.3
kV output	1	1	1	1
X reactance	0.117	0.14	0.28	0.467

Table 11-4 Transformers Per-unit values based on 500 kV and 1000 MVA

Line	R	X	G	B
A1-- B1	--------------	0.1173j	--------------	-8.525j
A2 – B2	-------------	0.1404j	-------------	-7.123j
A3 – B3	--------------	0.2803j	-------------	-3.568j
A4 – B4	-------------	0.4675j	--------------	-2.139j
B1-- B3	0.08	0.50j	0.312	-1.95j
B2 – B3	0.10	0.48j	0.416	-1997j
B2--B4	0.12	0.55j	0.379	-1.736j
B4 – B1	0.14	0.60j	0.369	-1.581j

Base: 1000 MVA , 500 kV

Table 11-5 Line impedances and admittances in per unit

BUS	P	Q	V magnitude	NAME
B1			1.05	Swing bus / Generator G1
B2	0.8	0.2	1.05	G2 bus/Voltage magnitude constant
B3	0.4	0.2	1.05	G3 bus/Voltage magnitude constant
B4	0.21	0.2	1.05	G4 bus/Voltage magnitude constant
L1	- 0.9	- 0.3	1.00	Inductive load bus/constant Q
L2	- 0.8	- 0.3	1.00	Inductive load bus /constant Q
L3	- 0.28	- 0.3	1.00	Inductive load bus/ constant Q
L4	- 0.14	- 0.3	1.00	Inductive load bus/constant Q

Base: 1000 MVA, 500 kV

Table 11-6 Estimated per-unit real and reactive input power to the grid and bus voltage magnitude.

Generators supply power to the grid, loads receive power from the grid. In Table 11-6 *the three-phase power delivered to the loads appears as negative input power delivered to the grid, but it is positive input power delivered to the loads.* Because the three-phase power system is assumed symmetrical and balanced, the analysis is performed on a per-phase base; so the power delivered from phase to neutral is only one third of the total three-phase power. However, for per-unit computational purposes it is the same to use one one-third of the power and one-third of the three-phase MVA base as to use the total power and the total three-phase MVA base. From Eq. (11.3) we compute the magnitude and angle of the bus voltage with respect to the bus input current. Usually the neutral is selected as the phasor reference for the entire power system. *Table 11-6 provides the estimated values of the magnitude of the bus voltage at generator busses and the real and reactive power demanded at the load busses.* The Gauss-Seidel method of Grid's load flow analysis during normal operation is carry out by the repetitive application of the following equations.

$$V_k = \frac{1}{Y_{k.k}} \left[\frac{P_k - Q_k \cdot j}{\overline{V_k}} - \sum_{n=1}^{N} \left(Y_{k.n} \cdot V_n \right) \right]$$

(11.3)

$(n \neq k), N = \text{number of busses}$

Equation (11.3) applies only to load busses in which the real and reactive power are specified.

At generators busses where the voltage magnitude is specified the applicable equations are (11.4) and (11.5).

$$P_k - Q_k \cdot j = \overline{V_k} \cdot \sum_{n=1}^{N} \left(Y_{k.n} \cdot V_n \right) \qquad \text{n is allowed to equal k}$$

(11.4)

$$Q_k = -\text{Im}\left[\overline{V_k} \cdot \sum_{n=1}^{N} \left(Y_{k.n} \cdot V_n \right) \right]$$

(11.5)

At a particular generator bus Eq. (11.5) provides the reactive power and for the real component of the generator's power you should use the value estimated in Table 11-6. And then, use Equation (11.3) to compute the bus voltage. And correct it, if necessary. See chapter 4, reference 1. The mini-grid shown in Fig. 11-1 does not have any dedicated load bus, because its main purpose is to discuss its response to a transient event and how the oscillation of one generator could induce oscillations in the others. Assuming that the estimated values in Table 11-6 are essentially correct, because they should be the result of the power flow study required in Section 11.2, step 3.

The bus admittance matrix of the internal system depicted in Fig. 11-1 is:

$$Y_{bus} := \begin{pmatrix} 0.681 - 3.531j & 0 & -0.312 + 1.95j & -0.369 + 1.581j \\ 0 & 0.795 - 3.733j & -0.416 + 1.997j & -0.379 + 1.736j \\ -0.312 + 1.95j & -0.416 + 1.997j & 0.728 - 3.947j & 0 \\ -0.369 + 1.581j & -0.379 + 1.736j & 0 & 0.748 - 3.317j \end{pmatrix}$$

(11.6)

$$0.312 + 0.369 = 0.681 \qquad\qquad -1.95j - 1.581j = -3.531j$$

The bus admittance matrix is square, and the self admittances which are the diagonal elements are identified by repeated subscripts. *Each bus self admittance is equal to the sum of all the admittances terminating in that bus.* The rest of the matrix's admittances are the mutual admittances between busses, and each equals *the negative of the sum of all admittances connected directly between the busses identified by the double subscript.* There are four zeros in the matrix shown in Eq. (11.6) and of they are:

$$Y_{1.2}, Y_{2.1}, Y_{3.4}, Y_{4.3}$$

Balance of power computations

Generators supply power to grid and loads receive power from the grid.

Power input from generators to grid = Power out of the grid to load

Use power balance to *compute* the real and reactive powers that the swing bus (B1) provides to the load.

Real power:

$0.8 + 0.4 + 0.21 = 1.41$ $\qquad\qquad\qquad$ $-0.9 - 0.8 - 0.28 - 0.14 = -2.12$

$-2.12 + 1.41 = -0.71$ $\qquad\qquad$ $P_1 = 0.71$ $\quad$ Real power that G1 provides to the grid

Reactive power:

$0.2 + 0.2 + 0.2 = 0.6$ $\qquad\qquad\qquad$ $-0.3 - 0.3 - 0.3 - 0.3 = -1.2$

$-1.2 + 0.6 = -0.6$ $\qquad\qquad$ $Q_1 = 0.6$ $\quad$ Reactive power that G1 provides to the grid

11.4 Initial Power Angle Computations

The initial power angle is the angle between the generator emf and the generator bus voltage. The computations below are pre-transient and are based on complex power at the generator busses. Let us assume the following results from the load flow study required in step 3, section 11.2.

$$V_k := \begin{pmatrix} 0.804 + 0.675j \\ 1.019 + 0.254j \\ 0.944 + 0.46j \\ 0.755 + 0.729j \end{pmatrix} \quad \text{Load flow study results} \qquad\qquad (11.7)$$

$$k = 1, 2, 3, 4$$

G1-T1:

From Eq. (11.7) we obtain

$V_1 := 0.804 + 0.675$ $\qquad\qquad$ $\overline{V_1} = 0.804 - 0.675j$ $\qquad\qquad$ $\left| V_1 \right| = 1.05$

$$\arg(V_1) = 40.0152 \cdot {}^{\circ} \qquad \arg(\overline{V_1}) = -40.0152 \cdot {}^{\circ}$$

The complex power entering the system at bus B1, from balance of power computations, we get

$$S_1 = P_1 + Q_1 \cdot j \qquad S_1 := 0.71 + 0.6 \cdot j \qquad \overline{S_1} = 0.71 - 0.6j$$

$$|S_1| = |0.71 + 0.6 \cdot j| = 0.929 \qquad S_1 = V_1 \cdot \overline{I_1} \qquad \overline{S_1} = \overline{V_1} \cdot I_1$$

$$I_1 = 0.8855 - 0.0029j \qquad |I_1| = 0.8855 \qquad \arg(I_1) = -0.185 \cdot {}^{\circ}$$

I_1 Is practically in phase with V1 $\qquad\qquad \theta_1 := -0.185^{\circ}$

The pre-transient emf generated by G1 during steady state operation is

$$E_{g1} := \left[V_1 + (0.0003 + 0.117) \cdot j \cdot I_1\right]$$

$(0.0003 + 0.117) \cdot j$ Generator plus transformer per unit transient reactances

$$0.804 + 0.675j + (0.0003 + 0.117) \cdot j \cdot (0.8855 - 0.0029) = 0.804 + 0.779j$$

$$E_{g1} = 0.804 + 0.779j \qquad |E_{g1}| = 1.12 \qquad \arg(E_{g1}) = 44.078^{\circ}$$
   Referenced to neutral

The G1-T1 initial power angle is

$$\arg(E_{g1}) - \arg(V_1) = 4.063^{\circ} \qquad \delta_1 := 4.06^{\circ}$$
   G1 initial power angle

The emf generated by G1 leads bus B1 voltage by 4.06 degrees.

G2 - T2

From Eq. (11.7) we obtain

$$V_2 := 1.019 + 0.254 \qquad \overline{V_2} = 1.019 - 0.254j \qquad |V_2| = 1.05$$

$$\arg(V_2) = 13.9965 \cdot {}^{\circ} \qquad \arg(\overline{V_2}) = -13.9965 \cdot {}^{\circ}$$

The complex power entering the system at bus B2, from Table 11.6, is

$S_2 := 0.8 + 0.2$ $\left|S_2\right| = \left|0.8 + 0.2j\right| = 0.824\epsilon$ $S_2 = V_2 \cdot \overline{I_2}$

$\overline{I_2} = \dfrac{S_2}{\overline{V_2}} = 0.785 + 0.001$ $I_2 := 0.785 - 0.001$ $\arg(I_2) = -0.073 \cdot °$

$\theta_2 := -0.073^c$

$E_{g2} := \left[V_2 + (0.0004 + 0.14)j \cdot I_2\right]$ $E_{g2} = 1.019 + 0.364j$ $\left|E_{g2}\right| = 1.082$

$\arg(E_{g2}) = 19.6656 \cdot °$ $\delta_2 = 19.6656 - 13.9965 = 5.6691^c$

$\delta_2 := 5.67^c$ **G2 initial power angle**

The emf generated by G2 leads bus B2 voltage by 5.67^c

G3 - T3   From Eq. (11.7) we obtain

$V_3 := 0.944 + 0.46$ $\overline{V_3} = 0.944 - 0.46j$ $\left|V_3\right| = 1.05$ $\arg(V_3) = 25.979$

The complex power entering the system at bus B3, from Table 11-6 is

$S_3 := 0.4 + 0.2j$ $\left|S_3\right| = \left|0.4 + 0.2j\right| = 0.4472$

$S_3 = V_3 \cdot \overline{I_3}$ $\dfrac{S_3}{\overline{V_3}} = 0.426 + 0.004j$ $\overline{I_3} = 0.426 + 0.004j$

$I_3 := 0.426 - 0.004j$ $\left|I_3\right| = 0.426$ $\arg(I_3) = -0.538 \cdot °$ $\theta_3 := -0.538°$

$E_{g3} := V_3 + (0.0003 + 0.28)j \cdot I_3$ $E_{g3} = 0.945 + 0.579j$ $\left|E_{g3}\right| = 1.109$

$\arg(E_{g3}) = 31.496°$

$\delta_3 = 31.496 - 25.979 = 5.517°$ **G3 initial power angle**

The emf generated by G3 leads bus B3 voltage by 5.517°

G4 - T4

From Eq. (11.7) we obtain

$$V_4 := 0.755 + 0.729 \qquad |V_4| = 1.05 \qquad \arg(V_4) = 43.996^{\,c}$$

The complex power entering the system at bus B4, from Table 11-6, is

$$S_4 := 0.21 + 0.2 \qquad |S_4| = 0.29 \qquad S_4 = V_4 \cdot \overline{I_4} \qquad \overline{I_4} = 0.276 - 0.002$$

$$I_4 := 0.276 + 0.002 \qquad |I_4| = 0.276 \qquad \arg(I_4) = 0.415 \cdot {}^{\,c} \qquad \theta_4 := 0.415^{\,c}$$

$$E_{g4} := V_4 + (0.0005 + 0.467j) \cdot I_4 \qquad E_{g4} = 0.754 + 0.858j \qquad |E_{g4}| = 1.142$$

$$\arg(E_{g4}) = 48.691^{\,c} \qquad \delta_4 = 48.691 - 43.996 = 4.695^{c}$$

$$\delta_4 := 4.695^{c} \qquad \textbf{G.4 initial power angle.}$$

The emf generated by G4 leads bus B4 voltage by $\quad 4.695^{c}$

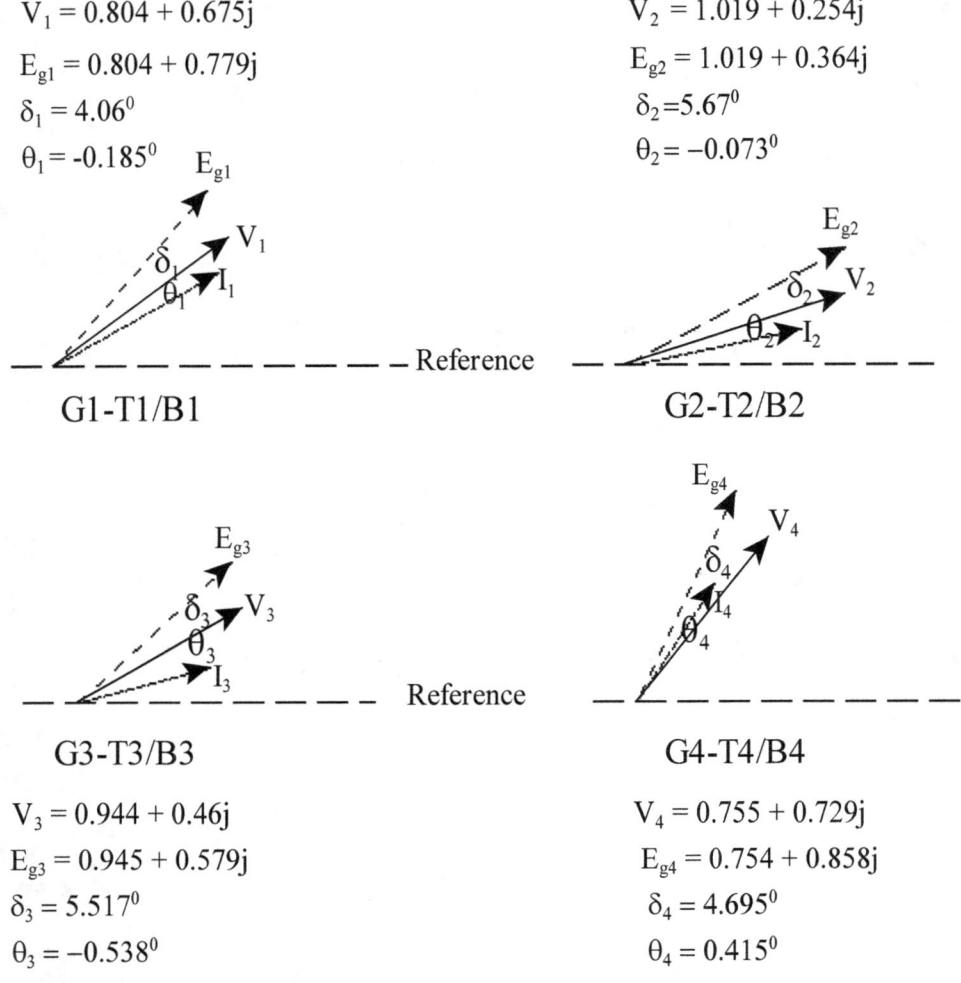

$V_1 = 0.804 + 0.675j$

$E_{g1} = 0.804 + 0.779j$

$\delta_1 = 4.06^0$

$\theta_1 = -0.185^0$

$V_2 = 1.019 + 0.254j$

$E_{g2} = 1.019 + 0.364j$

$\delta_2 = 5.67^0$

$\theta_2 = -0.073^0$

G1-T1/B1

G2-T2/B2

G3-T3/B3

G4-T4/B4

$V_3 = 0.944 + 0.46j$

$E_{g3} = 0.945 + 0.579j$

$\delta_3 = 5.517^0$

$\theta_3 = -0.538^0$

$V_4 = 0.755 + 0.729j$

$E_{g4} = 0.754 + 0.858j$

$\delta_4 = 4.695^0$

$\theta_4 = 0.415^0$

Figure 11-2 Pretransient gemerators phasor diagrams. Not to scale

11.5 Coherency Formula

Two generators coherency formula under steady state condition.

$$\frac{P_{sy1}}{M_1} = \frac{P_{sy2}}{M_2} \tag{11.8}$$

P_{sy} Is the synchronizing power coefficient, see Eq. (10.15)

M   Is the inertial constant or the angular momentum at synchronous speed. See Eq. (10.7)

$$P_{sy} = P_{emax} \cdot \cos\left(\delta_0\right)$$

The computations below are pre-transient and from Eq. (10.12) we obtain the maximum three-phase real electrical power that each generators could deliver. From section 11.4 we obtain the power angles δ 1, δ 2, δ 3, and δ .4

$$P_{emax1} = \frac{|E_{g1}| \cdot |V_1|}{|0.1173|} = \frac{1.12 \cdot 1.05}{0.1173} = 10.02(\quad \delta_1 := 4.06^\circ \quad \delta_1 = 0.071 \text{rad}$$

$$P_{emax2} = \frac{|E_{g2}| \cdot |V_2|}{|0.1404|} = \frac{1.08 \cdot 1.05}{0.1404} = 8.07^\cdot \quad \delta_2 := 5.67^\circ \quad \delta_2 = 0.099 \text{rad}$$

$$P_{emax3} = \frac{|E_{g3}| \cdot |V_3|}{|0.2803|} = \frac{1.109 \cdot 1.05}{0.2803} = 4.15 \measuredangle \quad \delta_3 := 5.517^\circ \quad \delta_3 = 0.096 \text{rad}$$

$$P_{emax4} = \frac{|E_{g4}| \cdot |V_4|}{|0.4675|} = \frac{1.142 \cdot 1.05}{0.4675} = 2.56! \quad \delta_4 := 4.695^\circ \quad \delta_4 = 0.082 \text{rad}$$

Synchronizing power coefficients

$$P_{sy1} := 10.026 \cdot \cos(4.06^\circ) = 10.001 \qquad P_{sy2} := 8.077 \cdot \cos(5.67^\circ) = 8.037$$

$$P_{sy3} := 4.154 \cdot \cos(5.517^\circ) = 4.1 \qquad P_{sy4} := 2.565 \cdot \cos(4.695^\circ) = 2.556$$

Each generator considered has one pair of poles and 3600 RPM, see Table 11-2. And its angular momentum is expressed by Eq. (10.7).

$$M = 0.000093 \cdot G \cdot H \qquad \frac{Megawatt - sec^2}{degree} \qquad \text{See Table 11-2}$$

$$M_1 := 0.000093 \cdot 1.2 \cdot 9.6 = 0.0011 \qquad M_2 := 0.000093 \cdot 1 \cdot 7 = 0.0007$$

$$M_3 := 0.000093 \cdot 0.5 \cdot 2.5 = 0.00012 \qquad M_4 := 0.000093 \cdot 0.3 \cdot 1.2 = 0.000033$$

$$\frac{P_{sy1}}{M_1} = 9091.\text{!} \qquad \frac{P_{sy3}}{M_3} = 34166.\text{'} \qquad \frac{P_{sy2}}{M_2} = 12346.4 \qquad \frac{P_{sy4}}{M_4} = 77454.\text{!}$$

There are not coherency among the four generators, because none of the possible pair comply with Eq. (11.8).

11-6 Generators Natural Frequencies of Oscillations

Equation (10.35) express the generators natural frequency of oscillation.

$$f_n = \frac{1}{2 \cdot \pi} \cdot \sqrt{\frac{\omega_s \cdot P_{sy}}{2 \cdot H}}$$ Natural frequency of oscillation

P_{sy} Is the synchronizing power coefficient

H Is the per unit stored rotational energy, megawatt-second / generator MVA rating

$$T_n = \frac{1}{f_n}$$ Period of the natural frequency of oscillation

$$\omega_n = 2 \cdot \pi \cdot f_n$$ Natural angular speed

$$\frac{1}{2 \cdot \pi} \cdot \sqrt{\frac{377 \cdot P_{sy1}}{2 \cdot 9.6}} = 2.230$$ cycle/sec $f_{n1} := 2.230$ base

$$\frac{1}{2\pi} \cdot \sqrt{\frac{377 \cdot P_{sy2}}{2 \cdot 7}} = 2.341$$ $f_{n2} := 2.341$ cycle / sec $\Delta_2 = 5\%$

$$\frac{1}{2\pi} \cdot \sqrt{\frac{377 \cdot P_{sy3}}{2 \cdot 2.5}} = 2.810$$ $f_{n3} := 2.810$ cycle/sec $\Delta_3 = 26\%$

$$\frac{1}{2\pi} \cdot \sqrt{\frac{377 \cdot P_{sy4}}{2 \cdot 1.2}} = 3.189$$ $f_{n4} := 3.189$ cycle/sec $\Delta_4 = 43\%$

Although G1 and G2 are not coherent still they *could* oscillate together because their natural frequencies of oscillations are only 5% apart.

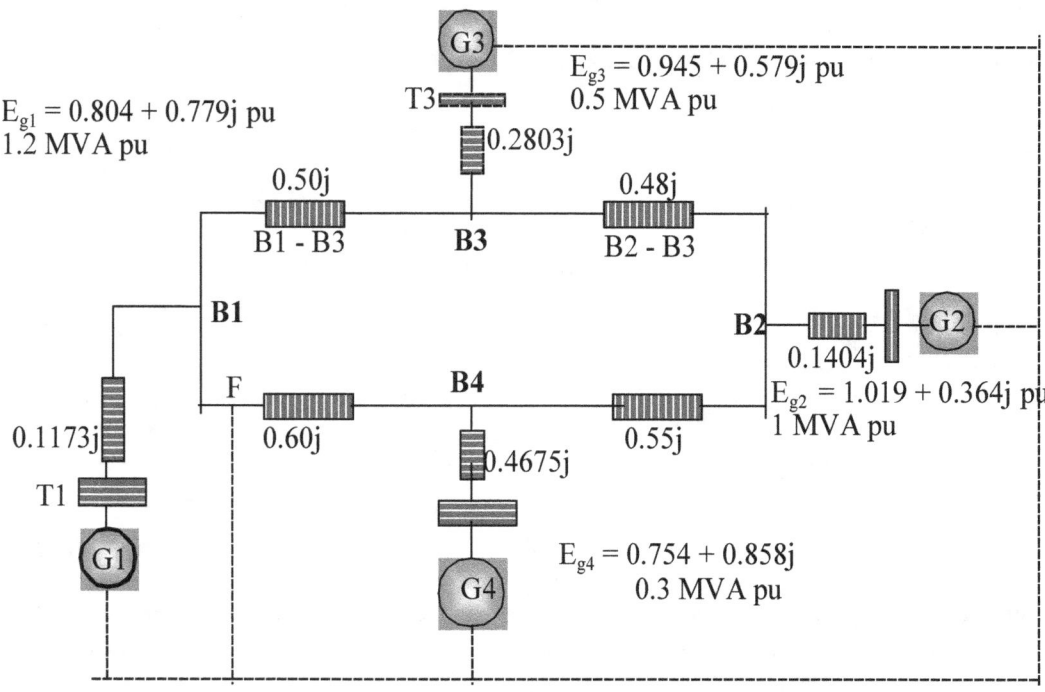

Figure 11-3 impedance diagram of power grid during fault

In Fig. 11-3 all generators send current to the fault and each of them uses a different path to reach the fault. Equations (11.8) provide the smaller impedance path for each generator. The reader should understand that the fault current contributed by each generator must return to its own neutral; it cannot return through others generator neutral. **During the three-phase symmetrical fault, the neutral of all generators are at the same potential than the neutral of the fault.** None of the nodes or busses in Fig. 11-3 can be eliminated, because each bus is connected to a generator. The flow of real power is unidirectional, you cannot have two different flows of real power flowing in opposite directions in the same branch at the same time.

G1 path impedance: 0.1173j

G2 path impedance:
$$0.1404j + \frac{(0.55j + 0.60j) \cdot (0.48j + 0.50j)}{(0.55j + 0.60j) + (0.48j + 0.50j)} = 0.67$$

G3 path impedance: $0.2803j + 0.50j = 0.78$

G4 path impedance: $0.4675j + 0.60j = 1.067$

The electrical real power transmitted by a generator is

$$P_e = \frac{\left| E_g \right| \cdot \left| V_t \right|}{\left| X \right|} \cdot \sin(\delta) \qquad (11.9)$$

The maximum real electrical power that a generator could transmit is

$$P_{emax} = \frac{\left| E_g \right| \cdot \left| V_t \right|}{\left| X \right|} \qquad (11.10)$$

The electrical real power could also be expressed as

$$P_e = P_{emax} \cdot \sin(\delta) \qquad (11.11)$$

Although the analysis is for a single-phase of a three-phase symmetrical power system, still the per-unit values obtained using the per-unit line-to-line voltages in Eqns. (11.9 - 11-11) are valid because the bases were changed accordingly. From section 11.4 we obtain the required per-unit voltages and from Eq. (11.8) the required reactance values.

$$P_{emax1} := \frac{1.12 \cdot 1.05}{0.1173} = 10.026 \quad \text{pu} \qquad (11.12)$$

$$P_{emax2} := \frac{1.082 \cdot 1.05}{0.67} = 1.696 \quad \text{pu} \qquad (11.13)$$

$$Pemax3 := \frac{1.109 \cdot 1.05}{0.78} = 1.493 \quad \text{pu} \qquad (11.14)$$

$$P_{emax4} := \frac{1.142 \cdot 1.05}{1.067} = 1.124 \quad \text{pu} \qquad (11.15)$$

Before the fault, each generator operates in steady state mode at constant power angle and no shaft acceleration. Therefore the pre-transient mechanical power delivered by the shaft must be equal to the generator real electrical power output plus the mechanical and electrical losses.

From section 11.3, balance of power computations, we obtain the real electrical power entering the grid at bus B1. For the other busses we use the values listed in Table 11-6. And the estimated per unit losses are 0.01 which are assumed constant and the same for all the machine. These losses have a damping effect in the transient.

$$P_1 := 0.71$$

$$P_{s1} = 0.71 + 0.01 = 0.72 \quad \text{Per unit shaft power in steady state condition} \quad (11.16)$$

$$\text{before fault.}$$

The shaft accelerating power during transient condition is provided by the swing equation.

$$\frac{H}{180 \cdot f} \cdot \frac{d^2}{dt^2}\delta = P_a = P_s - P_e \qquad \textbf{Per unit swing equation. See Eq. (10.32)}$$

We use δ without subindice to simplify plots; although each generator has its unique power angle

$$P_{e1} = \frac{|E_{g1}| \cdot |V_1|}{|X|} \cdot \sin(\delta)$$

Assuming that the emf of G1 and V1, (bus B1 voltage) remain constants at the pre-transient value we obtain the real electrical power provided by G1 <u>during</u> fault conditions

G1-T1:

$$P_{e1d} = \frac{1.12 \cdot 1.05}{0.1173} \cdot \sin(\delta) = 10.026 \cdot \sin(\delta) \qquad P_{s1} := 0.72$$

$$\text{See Eq. (11.16)}$$

$$P_{a1} = 0.72 - 10.026 \cdot \sin(\delta)$$

Assuming that the input mechanical power to the shaft remains constant during fault conditions. This is the shaft accelerating power of G1 during fault conditions.

The swing equation becomes $\qquad \dfrac{9.6}{\pi \cdot 60} \cdot \dfrac{d^2}{dt^2}\delta = 0.72 - 10.026 \cdot \sin(\delta) \quad \text{radians}$

$$0.051 \cdot \frac{d^2}{dt^2}\delta = 0.72 - 10.026 \cdot \sin(\delta) \qquad \frac{d^2}{dt^2}\delta = \frac{1}{0.051}(0.72 - 10.026 \cdot \sin(\delta))$$

$$\frac{d^2}{dt^2}\delta = 14.12 - 196.59 \cdot \sin(\delta) \qquad \textbf{G1 swing equation} \qquad (11.17)$$

G2-T2:

$$P_{e2d} = \frac{|E_{g2}| \cdot |V_2|}{|X|} \cdot \sin(\delta) = \frac{1.082 \cdot 1.05}{0.67} \cdot \sin(\delta) = 1.696 \cdot \sin(\delta)$$

$$P_{a2} = P_{s2} - 1.696 \cdot \sin(\delta) \qquad\qquad P_{s2} := 0.8 \qquad \text{See Table 11-6}$$

$$P_{a2} = 0.8 - 1.696 \cdot \sin(\delta) \qquad \text{Shaft accelerating power of G2}$$

The swing equation for this generator becomes

$$\frac{7}{\pi \cdot 60} \cdot \frac{d^2}{dt^2}\delta = 0.8 - 1.696 \cdot \sin(\delta) \qquad\qquad 0.0371 \cdot \frac{d^2}{dt^2}\delta = 0.8 - 1.696 \cdot \sin(\delta)$$

$$\frac{d^2}{dt^2}\delta = \frac{1}{0.0371} \cdot (0.8 - 1.696 \cdot \sin(\delta)) = 21.56 - 45.71 \cdot \sin(\delta)$$

$$\frac{d^2}{dt^2}\delta = 21.56 - 45.71 \cdot \sin(\delta) \qquad \textbf{G2 swing equation} \qquad (11.18)$$

G3-T3:

$$P_{e3d} = \frac{|E_{g3}| \cdot |V_3|}{|X|} \cdot \sin(\delta) = \frac{1.109 \cdot 1.05}{0.78} \cdot \sin(\delta) = 1.4929 \cdot \sin(\delta)$$

$$P_{a3} = P_{s3} - 1.429 \cdot \sin(\delta) \qquad\qquad P_{s3} := 0.4$$
$$\text{See Table 11-6}$$

$$P_{a3} = 0.4 - 1.4929 \cdot \sin(\delta) \qquad \text{G3 shaft accelerating power}$$

$$\frac{2.5}{\pi \cdot 60} \cdot \frac{d^2}{dt^2}\delta = 0.4 - 1.4929 \cdot \sin(\delta) \qquad\qquad 0.0133 \cdot \frac{d^2}{dt^2}\delta = 0.4 - 1.4929 \cdot \sin(\delta)$$

$$\frac{d^2}{dt^2}\delta = \frac{1}{0.0133} \cdot (0.4 - 1.4929 \cdot \sin(\delta)) = 30.08 - 112.25 \cdot \sin(\delta)$$

$$\frac{d^2}{dt^2}\delta = 30.08 - 112.25 \cdot \sin(\delta) \qquad \textbf{G3 swing equation} \qquad (11.19)$$

G4-T4:

$$P_{e4d} = \frac{1.142 \cdot 1.05}{1.067} \cdot \sin(\delta) = 1.1238 \cdot \sin(\delta)$$

$$P_{a4} = P_{s4} - 1.123\{ \qquad\qquad P_{s4} := 0.21 \qquad\qquad \text{See Table 11-6}$$

$$P_{a4} = 0.21 - 1.1238 \cdot \sin(\delta) \qquad \text{Shaft accelerating power of G4}$$

$$\frac{1.2}{\pi \cdot 60} \cdot \frac{d^2}{dt^2}\delta = 0.21 - 1.1238 \cdot \sin(\delta) \qquad\qquad 0.0064 \cdot \frac{d^2}{dt^2}\delta = 0.21 - 1.1238 \cdot \sin(\delta)$$

$$\frac{d^2}{dt^2}\delta = \frac{1}{0.0064}(0.21 - 1.1238 \cdot \sin(\delta)) = 32.81 - 175.59 \cdot \sin(\delta)$$

$$\frac{d^2}{dt^2}\delta = 32.81 - 175.59 \cdot \sin(\delta) \qquad \textbf{G4 swing equation}$$

$$(11.20)$$

11.7 Numerical Solution of the Swing Equation <u>During</u> Fault Conditions

G1- T1

$$\frac{d^2}{dt^2}\delta = 14.12 - 196.59 \cdot \sin(\delta) \qquad \text{From Eq. (11.17) we obtained the swing equation for G1}$$

This swing equation can be written as: $\dfrac{d}{dt}\left[\left(\dfrac{d}{dt}\delta\right)\right. = 14.12 - 196.59 \cdot \sin(\delta)\left.\right]$

Define two new functions δ0(t) and δ1(t):

$\delta_0(t) = \delta(t)$ Power angle $\delta_1(t) = \dfrac{d}{dt}\delta_0(t)$ Rate of change of the power angle

Now the original swing equation can be written as:

$$\frac{d}{dt}\delta_1(t) = 14.12 - 196.59 \cdot \sin\left(\delta_0(t)\right)$$

The new differential equation has two functions $\delta_0(t)$ and $\delta_1(t)$ instead of one, δ (t).

These two new functions are related by the following equation:

$$\delta_1(t) = \frac{d}{dt}\delta_0(t)$$

The original differential equation has been converted to a system of two differential equations, written with the derivatives on the left-hand side of the equal sign.

$$\frac{d}{dt}\delta_0(t) = \delta_1(t)$$

(11.21)

$$\frac{d}{dt}\delta_1(t) = 14.12 - 196.59 \cdot \sin\left(\delta_0(t)\right)$$ **Vector D element**   (11.22)

G2-T2

The original swing equation for the **G2** generator is provided by Eq. (11.18)

$$\frac{d^2}{dt^2}\delta(t) = 21.56 - 45.71\sin(\delta(t))$$ Define two new functions:

Power angle $\quad \delta_2(t) = \delta(t)$

Rate of change of power angle $\qquad \delta_3(t) = \dfrac{d}{dt}\delta_2(t)$

Now the original swing equation for G2 becomes:

$$\frac{d}{dt}\delta_3(t) = 21.56 - 45.71\sin\left(\delta_2(t)\right)$$

Similarly, the original swing equation for G2 is replaced by the following system of two differential equations.

$$\frac{d}{dt}\delta_2(t) = \delta_3(t)$$

(11.23)

$$\frac{d}{dt}\delta_3(t) = 21.56 - 45.71\sin\left(\delta_2(t)\right)$$ **Vector D element** (11.24)

G3-T3:

From Eq. (11.19) we get the swing equation for G3

$$\frac{d^2}{dt^2}\delta(t) = 30.08 - 112.25 \cdot \sin(\delta(t))$$ Define two new functions

$$\delta_4(t) = \delta(t) \qquad \delta_5(t) = \frac{d}{dt}\delta_4(t)$$ Power angle and rate of change of power angle

$$\frac{d}{dt}\delta_5(t) = 30.68 - 112.25 \cdot \sin\left(\delta_4(t)\right)$$

Similarly, the original swing equation for G3 is replaced by the following system of two differential equations.

$$\frac{d}{dt}\delta_4(t) = \delta_5(t) \tag{11.25}$$

$$\frac{d}{dt}\delta_5(t) = 30.08 - 112.25 \cdot \sin\left(\delta_4(t)\right) \qquad \textbf{Vector D element} \tag{11.26}$$

G4-T4: From Eq. (11.20) we get the swing equation for G4

$$\frac{d^2}{dt^2}\delta(t) = 32.81 - 175.59 \cdot \sin(\delta(t))$$

Power angle $\quad \delta_6(t) = \delta(t) \quad$ rate of change of power angle $\quad \delta_7(t) = \frac{d}{dt}\delta_6(t)$

The original swing equation becomes:

$$\frac{d}{dt}\delta_7(t) = 32.81 - 175.59 \cdot \sin\left(\delta_6(t)\right)$$

Similarly, the original swing equation for G4 is replaced by the following system of two differential equations.

$$\frac{d}{dt}\delta_6(t) = \delta_7(t) \tag{11.27}$$

$$\frac{d}{dt}\delta_7(t) = 32.81 - 175.59 \cdot \sin\left(\delta_6(t)\right) \qquad \textbf{Vector D element} \tag{11.28}$$

To use the numerical solver is necessary to define a derivative vector D containing the eight new functions of t. We use the Rkadapt solver to solve the system of second order differential equations.

Initial conditions:

$$\delta_0 := 0.071 \text{rad} \qquad \delta_1 := 0 \qquad 14.12 - 196.59 \cdot \sin(0.071) = 0.2$$

$$\delta_2 := 0.099 \text{rad} \qquad \delta_3 := 0 \qquad 21.56 - 45.71\sin(0.099) = 17.042$$

$$\delta_4 := 0.096 \text{rad} \qquad \delta_5 := 0 \qquad 30.08 - 112.25 \cdot \sin(0.096) = 19.3$$

$$\delta_6 := 0.082 \text{rad} \qquad \delta_7 := 0 \qquad 32.81 - 175.59 \cdot \sin(0.082) = 18.4$$

The elements of vector D are the rate of change of the new functions

$$D(t,\delta) := \begin{bmatrix} \delta_1 \\ 14.12 - 196.59 \cdot \sin\lfloor \delta_0(t) \rfloor \\ \delta_3 \\ 21.56 - 45.71\sin\lfloor \delta_2(t) \rfloor \\ \delta_5 \\ 30.08 - 112.25 \cdot \sin\lfloor \delta_4(t) \rfloor \\ \delta_7 \\ 32.81 - 175.59 \cdot \sin\lfloor \delta_6(t) \rfloor \end{bmatrix} \qquad ic := \begin{pmatrix} 0 \\ 0.2 \\ 0 \\ 17.042 \\ 0 \\ 19.3 \\ 0 \\ 18.4 \end{pmatrix}$$

$$t := 0 \qquad \qquad t2 := 0.1$$

$$\text{npoints} := 1000$$

$$Q := \text{Rkadapt}(ic, 0, 0.1, \text{npoints}, D)$$

Table 11-7 Power angles change and their acceleration during fault conditions

Q =

	0	1	2	3	4	5	6	7	8
500	0.05	0.0276	0.9022	0.8786	18.0868	1.0014	20.7112	0.9593	19.9015
501	0.0501	0.0277	0.9036	0.8804	18.0887	1.0035	20.7136	0.9613	19.904
502	0.0502	0.0278	0.905	0.8823	18.0907	1.0056	20.7161	0.9633	19.9064
503	0.0503	0.0279	0.9064	0.8841	18.0926	1.0077	20.7185	0.9653	19.9088
504	0.0504	0.028	0.9077	0.8859	18.0946	1.0097	20.7209	0.9672	19.9113
505	0.0505	0.0281	0.9091	0.8877	18.0966	1.0118	20.7234	0.9692	19.9137
506	0.0506	0.0282	0.9105	0.8895	18.0985	1.0139	20.7258	0.9712	19.9161
507	0.0507	0.0282	0.9119	0.8913	18.1005	1.0159	20.7282	0.9732	19.9185
508	0.0508	0.0283	0.9133	0.8931	18.1024	1.018	20.7307	0.9752	19.9209
509	0.0509	0.0284	0.9147	0.8949	18.1043	1.0201	20.7331	0.9772	19.9233
510	0.051	0.0285	0.916	0.8967	18.1063	1.0222	20.7355	0.9792	19.9257
511	0.0511	0.0286	0.9174	0.8985	18.1082	1.0242	20.7379	0.9812	19.9282
512	0.0512	0.0287	0.9188	0.9004	18.1102	1.0263	20.7404	0.9832	19.9305
513	0.0513	0.0288	0.9202	0.9022	18.1121	1.0284	20.7428	0.9852	19.9329
514	0.0514	0.0289	0.9216	0.904	18.1141	1.0305	20.7452	0.9872	19.9353
515	0.0515	0.029	0.923	0.9058	18.116	1.0325	20.7476	0.9892	...

$Q^{\langle 0 \rangle} = t$ Time in seconds

Power angles change $Q^{\langle 1 \rangle} = \delta_1$ $Q^{\langle 3 \rangle} = \delta_2$ $Q^{\langle 5 \rangle} = \delta_3$ $Q^{\langle 7 \rangle} = \delta_4$

Power angles acceleration $Q^{\langle 2 \rangle} = \frac{d}{dt}\delta_1$ $Q^{\langle 4 \rangle} = \frac{d}{dt}\delta_2$ $Q^{\langle 6 \rangle} = \frac{d}{dt}\delta_3$ $Q^{\langle 8 \rangle} = \frac{d}{dt}\delta_4$

$$M := \frac{G \cdot H}{\pi \cdot 60}$$ See Eq. (10.6)

From Table 11-2 we get the per unit values of G and H

$G_1 := 1.2$ $G_2 := 1$ $G_3 := 0.5$ $G_4 := 0.3$

$H_1 := 9.6$ $H_2 := 7$ $H_3 := 2.5$ $H_4 := 1.2$

The swing equation is provided by Eq. (10.29) $M \cdot \frac{d^2}{dt^2}\delta = P_a$

From Eq. (10.3) we obtain:

$$M = \frac{2 \cdot H}{\omega_s} \cdot G \qquad \frac{M}{G} = \frac{2 \cdot H}{\omega_s}$$

$$\frac{2 \cdot H}{\omega_s} \cdot \frac{d^2}{dt^2}\delta = \frac{P_a}{G} \qquad \omega_s = 2 \cdot \pi \cdot f \qquad \frac{2 \cdot H}{2 \cdot \pi \cdot f} \cdot \frac{d^2}{dt^2}\delta = \frac{P_a}{G}$$

$$\frac{H}{\pi \cdot f} \cdot \frac{d^2}{dt^2}\delta = P_a \qquad \textbf{Accelerating power Pa in per unit}$$

$$0.0053 \cdot H \cdot \frac{d^2}{dt^2}\delta = P_a \qquad \textbf{Per unit swing equation with δ expressed in radians} \qquad (11.29)$$

$$0.0053 \cdot H_1 \cdot Q^{\langle 2 \rangle} \qquad \textbf{Provides per unit accelerating power of G1} \qquad (11.30)$$

$$0.0053 \cdot H_2 \cdot Q^{\langle 4 \rangle} \qquad \textbf{Provides per unit accelerating power of G2} \qquad (11.31)$$

$$0.0053 \cdot H_3 \cdot Q^{\langle 6 \rangle} \qquad \textbf{Provides per unit accelerating power of G3} \qquad (11.32)$$

$$0.0053 \cdot H_4 \cdot Q^{\langle 8 \rangle} \qquad \textbf{Provides per unit accelerating power of G4} \qquad (11.33)$$

$$0.000093 \cdot H \cdot \frac{d^2}{dt^2}\delta = P_a \qquad \textbf{Per unit swing equation, δ is in degrees} \qquad (11.34)$$

The power angles are computed using the data in Table 11-7 as follows: Initial power angle plus the producto of time by the rate of change.

$$\delta1 := 0.071 + \overrightarrow{\left(Q^{\langle 0 \rangle} \cdot Q^{\langle 1 \rangle}\right)} \qquad 0.0053*9.6 = 0.0509 \qquad A_1 := 0.0509 \cdot Q^{\langle 2 \rangle}$$

$$\delta2 := 0.099 + \overrightarrow{\left(Q^{\langle 0 \rangle} \cdot Q^{\langle 3 \rangle}\right)} \qquad 0.0053 \cdot 7 = 0.0371 \qquad A_2 := 0.0371 \cdot Q^{\langle 4 \rangle}$$

Fig. 11.4 Power Angles and Accelerating Power

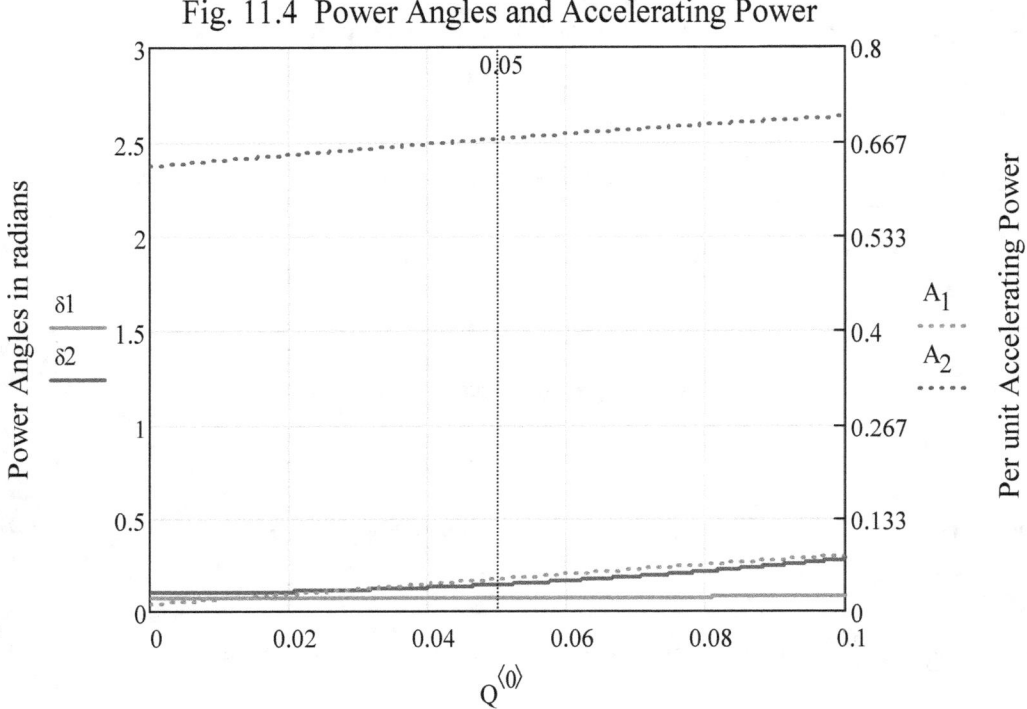

Time during fault in seconds

Figure 11-4 correspond to generators G1 and G2.

And from the plot it is obvious that the power angle of G1, at the bottom of the plot, is following a flat line and its accelerating power is increasing very slow. On the other hand, G2 is following a continuously growing curve and its accelerating power is large and increasing. The dotted curves represent the swing equations, which provide the accelerating powers in per unit. For steady-state operation the acceleration of the generator's shaft must be zero. Therefore, for stable operation the mechanical power delivered by the shaft minus the mechanical and electrical losses must be equal to electrical output power of the generator. In fact, both generators are unstable in short circuit condition. However, it does not matter because the fault would be cleared at 0.05 seconds.

$$\delta 3 := 0.096 + \overrightarrow{\left(Q^{\langle 0 \rangle} \cdot Q^{\langle 5 \rangle}\right)} \qquad 0.0053 \cdot 2.5 = 0.0133 \qquad A3 := 0.0133 \cdot Q^{\langle 6 \rangle}$$

$$\delta 4 := 0.082 + \overrightarrow{\left(Q^{\langle 0 \rangle} \cdot Q^{\langle 7 \rangle}\right)} \qquad 0.0053 \cdot 1.2 = 0.0064 \qquad A4 := 0.0064 \cdot Q^{\langle 8 \rangle}$$

Fig 11-5 Power Angles and Accelerating Power

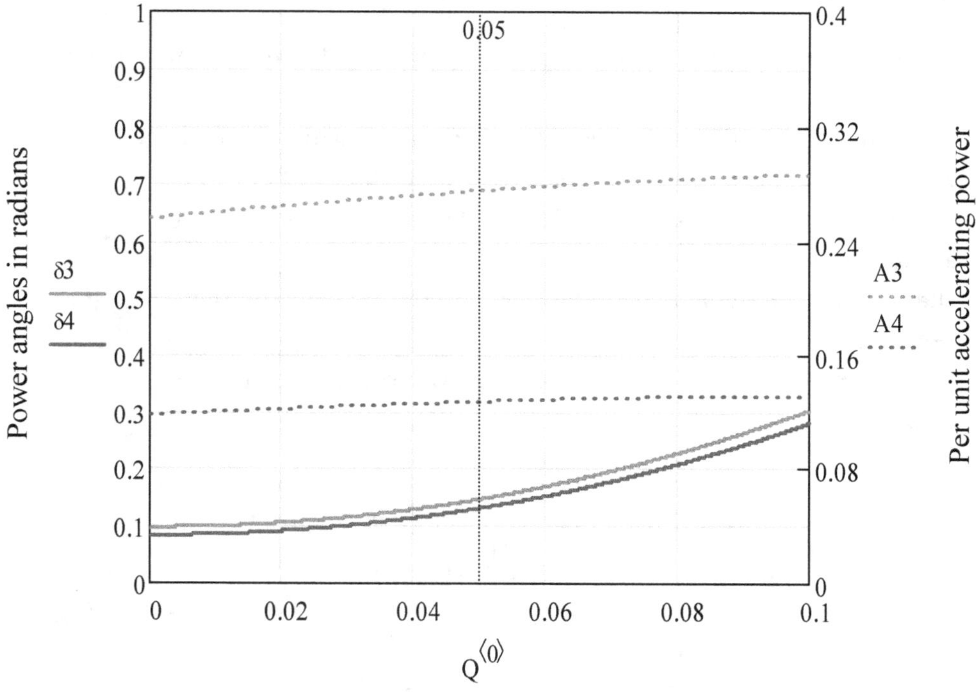

Time during fault in seconds

Figure 11-5 correspond to G3 and G4. It shows that the power angles of both generators are growing and that their accelerating powers are large and almost constant. In fact, G3 and G4 are unstable during the short circuit condition. However, it does not matter because the fault would be cleared at 0.05 seconds. *The best way to decrease the accelerating power is to increase the electrical power produced by the generator during and after fault conditions.*

$$P_a = P_s - \left(P_e + \text{losses}\right) = 0 \qquad \textbf{Required for stable operation} \qquad P_s = P_e + \text{losses}$$

$$P_e = P_{max} \cdot \sin(\delta) \qquad \text{Generator's power} \qquad P_{emax} = \frac{\left|E_g\right| \cdot \left|V_b\right|}{\left|X\right|} \qquad \text{To}$$

increase power without changing the magnitude of the generator's bus voltage the choices are:

1. Increase the magnitude of the generated emf.
2. Decrease the magnitude of the reactance between the generator and generator bus. This choice also has an impact on the magnitude of Vb.

The reader should be aware that before the breakers interrupt the fault, the following conditions applied:

- The neutrals of all the generators, and the symmetrical three-phase fault point are at the same potential.
- The analysis is based in real power flow. The swing equation is a real power flow differential equation and the real power flow is unidirectional.
- During short circuit condition the electrical load on each generator does not receive any real power.
- The current supplied by a generator sinks into the fault and return to the same generator.

11.8 Numerical solution of the swing equations after the fault has been clea

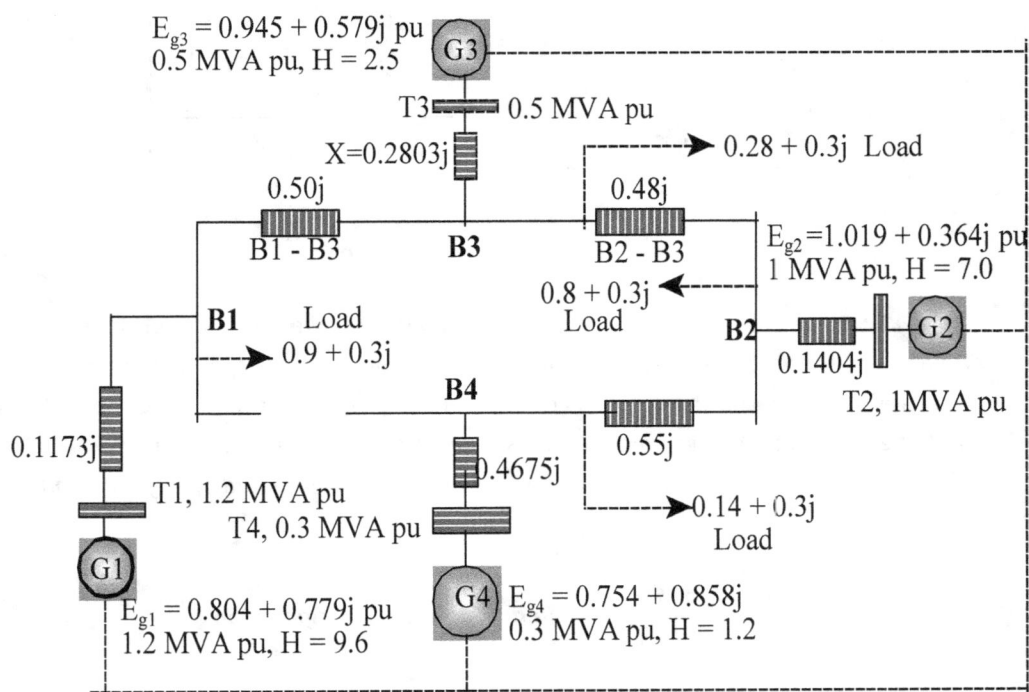

Figure 11-6 impedance diagram of power grid after fault

After the fault is cleared, the following conditions applied.

- The shaft mechanical power is assumed to remain equal to the pretransient value because the fault is cleared in 50 milliseconds, short enough time for the shaft's rpm to remain constant.
- The emfs generated by the generators remain constant at the pretransient values.
- The voltages at the generator's busses (B1, B2, B3, B4) return to the pretransient values. And the connected electrical loads are again receiving power.
- The opened mini-grid contains four voltages separated by the transmission lines impedances. *We assume that each generator only provides power to its own load.*
- It is a common practice to consider any connected load to the generator busses (B1, B2, B3, B4) as constant, and not to include them in the analysis.
- The generator's busses would maintain their voltages regardless of the load connected to them.

From the initial power angle computations, section 11.4, we obtained the voltage of each bus and the generators emfs. From Table 11-5 and figure 11-6 we obtained the generators plus transformers per unit reactances.

$$B = 1,2,3,4$$

$$V_B := \begin{pmatrix} 0.804 + 0.675j \\ 1.019 + 0.254j \\ 0.944 + 0.46j \\ 0.755 + 0.729j \end{pmatrix} \qquad X_B := \begin{pmatrix} 0.1173j \\ 0.1404j \\ 0.2803 \\ 0.4675 \end{pmatrix}$$

$$E_{gB} := \begin{pmatrix} 0.804 + 0.779j \\ 1.019 + 0.364j \\ 0.945 + 0.579j \\ 0.754 + 0.858j \end{pmatrix}$$

Table 11-7 provides the power angles of each generator after the fault is cleared (50 ms).

$$\delta_1 := 0.0276 \, \text{rad} \qquad \delta_2 := 0.8786 \, \text{rad} \qquad \delta_3 := 1.0014 \, \text{rad} \qquad \delta_4 := 0.9593 \, \text{rad}$$

From Table 11-2 we get the per unit generators inertial constants H.

$$H_1 := 9.6 \qquad H_2 := 7 \qquad H_3 := 2.5 \qquad H_4 := 1.2$$

Applying Eq. (10.12) we obtain the maximum power that each generator could deliver after clearing the fault.

$$P_{emax1a} = \frac{1.12 \cdot 1.05}{0.1173} = 10.026 \qquad P_{e1a} = 10.026 \cdot \sin(\delta1) \qquad \text{pu}$$

$$P_{emax2a} = \frac{1.082 \cdot 1.05}{0.1404} = 8.092 \qquad P_{e2a} = 8.092 \cdot \sin(\delta2) \qquad \text{pu}$$

$$P_{emax3a} = \frac{1.109 \cdot 1.05}{0.2803} = 4.154 \qquad P_{e3a} = 4.154 \cdot \sin(\delta3) \qquad \text{pu}$$

$$P_{emax4a} = \frac{1.142 \cdot 1.05}{0.4675} = 2.565 \qquad P_{e4a} = 2.565 \cdot \sin(\delta4) \qquad \text{pu}$$

The per unit shaft powers before, during and after the fault

$$P_{s1a} := 0.72 \quad P_{s2a} := 0.810 \quad P_{s3a} := 0.41 \quad P_{s4a} := 0.22 \qquad (11.33)$$

For the shaft power for G1 see Eq. (11-16) and for the others we used the values listed in Table 11-6. Immediately after the fault isolating breakers are open, there is a transient condition in which the swing equation is applicable. See Eq. (11.29)

$$0.0053 \cdot H \cdot \frac{d^2}{dt^2} \delta = P_a \quad \text{Accelerating power driving the power angle} \qquad (11.34)$$

G1-T1 $\qquad 0.0053 \cdot 9.6 \cdot \dfrac{d^2}{dt^2} \delta = 0.72 - 10.026 \cdot \sin(\delta) \qquad\qquad 0.0053 \cdot 9.6 = 0.051$

$$\frac{d^2}{dt^2} \delta = \frac{0.72 - 10.026 \cdot \sin(\delta)}{0.051} = 14.12 - 196.59 \cdot \sin(\delta) \qquad \textbf{Vector D element}$$

G2 - T2 $\qquad 0.0053 \cdot 7 \cdot \dfrac{d^2}{dt^2} \delta = 0.810 - 8.092 \cdot \sin(\delta) \qquad\qquad 0.0053 \cdot 7 = 0.037$

$$\frac{d^2}{dt^2}\delta = \frac{0.810 - 8.092 \cdot \sin(\delta)}{0.037} = 21.89 - 218.70 \cdot \sin(\delta)$$

Vector D element

G3-T3 $0.0053 \cdot 2.5 \cdot \dfrac{d^2}{dt^2}\delta = 0.408 - 4.154 \cdot \sin(\delta)$ $0.0053 \cdot 2.5 = 0.013$

$$\frac{d^2}{dt^2}\delta = \frac{0.408 - 4.154 \cdot \sin(\delta)}{0.013} = 31.38 - 319.54 \cdot \sin(\delta)$$

Vector D element

G4-T4 $0.0053 \cdot 1.2 \cdot \dfrac{d^2}{dt^2}\delta = 0.22 - 2.565 \cdot \sin(\delta)$ $0.0053 \cdot 1.2 = 0.006$

$$\frac{d^2}{dt^2}\delta = \frac{0.22 - 2.565 \cdot \sin(\delta)}{0.006} = 36.7 - 427.5 \cdot \sin(\delta)$$

Vector D element

The values listed on table 11-7 at 50 milliseconds, the time when the fault is cleared are

$\delta_1 = 0.0276$ $\dfrac{d}{dt}\delta_1 = 0.9022$

$\delta_2 = 0.8786$

$\dfrac{d}{dt}\delta_2 = 18.0865$

$\delta_3 = 1.0014$ $\dfrac{d}{dt}\delta_3 = 20.7112$

$\delta_4 = 0.9592$ $\dfrac{d}{dt}\delta_4 = 19.9015$

The reader should not confuse these angles with the initial power angles of section 11-4.

Time count starts at 50 milliseconds. <<<<

$$
ica := \begin{pmatrix} 0.0276 \\ 0.9022 \\ 0.8786 \\ 18.0868 \\ 1.0014 \\ 20.7112 \\ 0.9593 \\ 19.9015 \end{pmatrix}
\qquad
D(t,\delta) := \begin{bmatrix} \delta_1 \\ 14.12 - 196.59 \cdot \sin[\delta_0(t)] \\ \delta_3 \\ 21.89 - 218.70 \cdot \sin[\delta_2(t)] \\ \delta_5 \\ 31.38 - 319.54 \cdot \sin[\delta_4(t)] \\ \delta_7 \\ 36.7 - 427.5 \cdot \sin[\delta_6(t)] \end{bmatrix}
$$

$t1 := 0$ $\qquad$ $t2 := 1.00$ $\qquad$ $npoints := 1000$

$Q_a := Rkadapt(ica, t1, t2, npoints, D)$

Table 11-8 Power angles change and their acceleration after clearing the fault.

$Q_a =$

	0	1	2	3	4	5	6	7	8
0	0	0.028	0.902	0.879	18.087	1.001	20.711	0.959	19.901
1	0.001	0.029	0.916	0.897	18.109	1.022	20.742	0.979	19.938
2	0.002	0.029	0.93	0.915	18.13	1.043	20.773	0.999	19.974
3	0.003	0.03	0.945	0.933	18.152	1.064	20.804	1.019	20.01
4	0.004	0.031	0.959	0.951	18.173	1.084	20.834	1.039	20.045
5	0.005	0.032	0.973	0.969	18.194	1.105	20.864	1.059	20.08
6	0.006	0.033	0.987	0.988	18.214	1.126	20.893	1.079	20.114
7	0.007	0.034	1.001	1.006	18.235	1.147	20.922	1.099	20.147
8	0.008	0.035	1.015	1.024	18.255	1.168	20.951	1.12	20.181
9	0.009	0.036	1.029	1.042	18.275	1.189	20.979	1.14	20.213
10	0.01	0.037	1.043	1.061	18.295	1.21	21.007	1.16	20.245
11	0.011	0.038	1.057	1.079	18.314	1.231	21.034	1.18	20.277
12	0.012	0.039	1.071	1.097	18.333	1.252	21.061	1.201	20.307
13	0.013	0.041	1.085	1.115	18.352	1.273	21.087	1.221	20.338
14	0.014	0.042	1.099	1.134	18.371	1.294	21.113	1.241	20.367
15	0.015	0.043	1.113	1.152	18.389	1.315	21.138	1.262	...

Column headings of Table 11-8

Time in seconds $\quad Q^{\langle 0 \rangle} = t$

Power angles acceleration

Power angles change

$Q_a^{\langle 1 \rangle} = \delta_1$

$Q_a^{\langle 2 \rangle} = \dfrac{d}{dt} \delta_1$

$Q_a^{\langle 3 \rangle} = \delta_2$

$Q_a^{\langle 4 \rangle} = \dfrac{d}{dt} \delta_2$

$Q_a^{\langle 5 \rangle} = \delta_3$

$Q_a^{\langle 6 \rangle} = \dfrac{d}{dt} \delta_3$

$Q_a^{\langle 7 \rangle} = \delta_4$

$Q_a^{\langle 8 \rangle} = \dfrac{d}{dt} \delta_4$

Figure 11-7 and 11-8 are plots of the data shown in Table 11-8

$\delta 1a := \overrightarrow{\left(Q_a^{\langle 0 \rangle} \cdot Q_a^{\langle 1 \rangle} \right)}$

$0.0053 * 9.6 = 0.0509$

$A1a := 0.0509 \cdot Q_a^{\langle 2 \rangle}$

$\delta 2a := \overrightarrow{\left(Q_a^{\langle 0 \rangle} \cdot Q_a^{\langle 3 \rangle} \right)}$

$0.0053 * 7 = 0.0371$

$A2a := 0.0371 \cdot Q_a^{\langle 4 \rangle}$

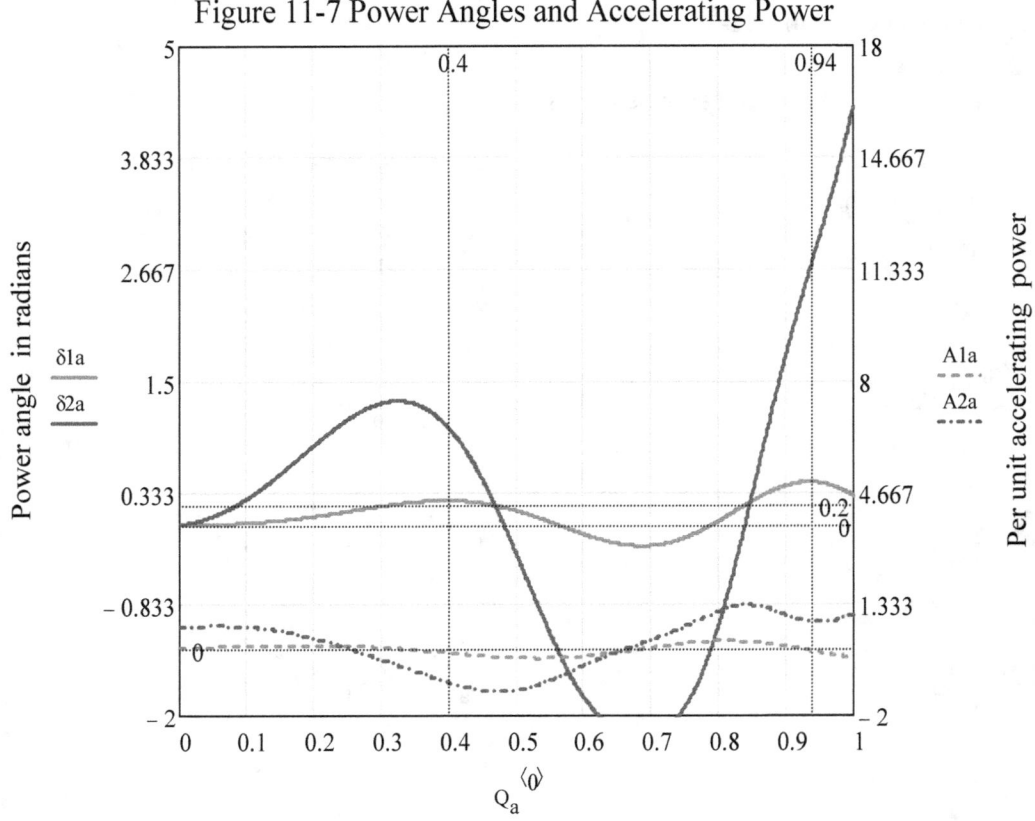

Figure 11-7 Power Angles and Accelerating Power

Time after fault in seconds

Figure 11-7 shows that the power angle of G2 is growing extremely fast. The machine will trip off the line. However the power angle of G1 is stabilizing at about 0.1 radians or 5.7 degrees, and is oscillating at 2 cycles per second, approximately. Furthermore, the amplitude of the oscillation of the accelerating power of the G1 power angle is very small and just over its zero line. So G1 could become stable. However, G2 goes from positive acceleration to negative acceleration and although the change is small, it is no good.

$$0.94 - 0.4 = 0.5 \qquad 1/0.5 = 2 \qquad \qquad \text{cycle/sec}$$

$$\delta3a := \overrightarrow{\left(Q_a{}^{\langle 0 \rangle} \cdot Q_a{}^{\langle 5 \rangle}\right)} \qquad 0.0053*2.5 = 0.0133 \qquad A3a := 0.0133 \cdot Q_a{}^{\langle 6 \rangle}$$

$$\delta4a := \overrightarrow{\left(Q_a{}^{\langle 0 \rangle} \cdot Q_a{}^{\langle 7 \rangle}\right)} \qquad 0.0053*1.2 = 0.0064 \qquad A4a := 0.0064 \cdot Q_a{}^{\langle 8 \rangle}$$

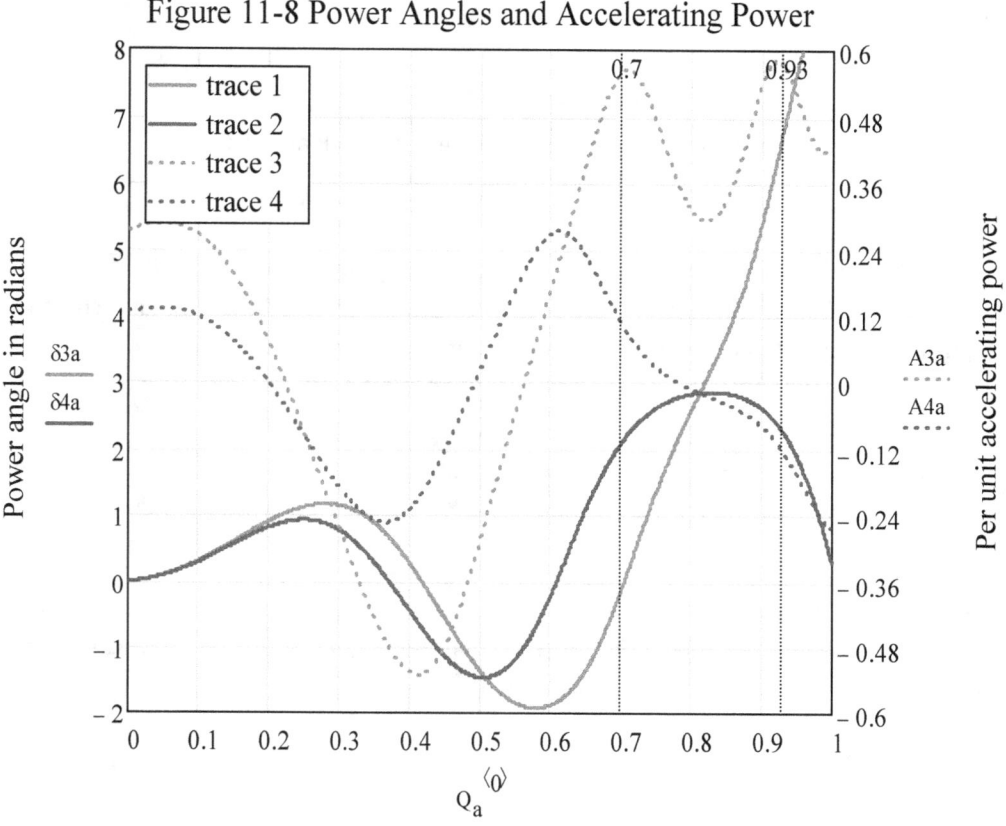

Figure 11-8 Power Angles and Accelerating Power

Time after fault in seconds

Figure 11-8 shows that the *accelerating powers* of the power angles oscillate with very large amplitude and that the power angles of G3 and G4 grow very fast after 0.5 sec. The frequency of oscillation of the accelerating power of G3 is around 4.3 cycles per second. And the accelerating power of G4 oscillates at 1.9 cycles per second. Their natural frequency of oscillations are 1.685 and 2.111 respectively. Both machines experience large sequentially negative and positive accelerations, and that is bad. They are definitely unstable.

G3: $\qquad 0.93 - 0.7 = 0.23 \qquad \dfrac{1}{0.23} = 4.3 \qquad \dfrac{\text{cycle}}{\text{sec}}$

G4:

$$0.62 - 0.08 = 0.54 \qquad \frac{1}{0.54} = 1.9 \qquad \frac{\text{cycle}}{\text{sec}}$$

The electrical power delivered by the generator after the fault is cleared is:

$$P = P_{max} \cdot \sin(\delta)$$

The shaft power should not be used to control the accelerating power, because the driver (turbine) rpm should be maintained within a narrow range. In fact, the best way to decrease the accelerating power is to increase the electrical power delivered by the generator.

$$\frac{H}{\pi \cdot f} \cdot \frac{d^2}{dt^2} \delta = P_a = P_s - P_e \qquad \textit{Swing equation}$$

In steady-state mode of operation, the *shaft acceleration is zero.* Therefore the pretransient mechanical power delivered by the shaft must be equal to the generator real electrical power output plus the electrical and mechanical losses. Symbolically:

$$P_s = P_{ess} + \text{Losses}$$

The symbolical expression of the shaft accelerating power is $\qquad P_a = P_s - P_e$

$$P_e = P_{max} \cdot \sin(\delta) \qquad \text{and} \qquad P_{max} = \frac{|E_g| \cdot |V_b|}{|X|}$$

The electrical power generated increases when the reactance between the generator and the corresponding buss decreases. So power computations at any of the B-busses location yields result larger than at any other location in the mini-grid, and therefore the accelerating power is smaller than at any other location in the mini-grid.

The choices to increase the electrical power are:

1. Increase the magnitude of the electrical voltage Eg.
2. Decrease the magnitude of the reactance X between the generator and the corresponding B buss.

Choice number two is the best and most practical.

Chapter 12

The Electromagnetic Pulse and the Smart Power Grid

12.1 Geomagnetic Storms

This natural kind of event happens unpredictable, although it is more probable during any of the sun 11 years' geomagnetic storms cycles. No all the occurring sun's magnetic storms affect the magnetic field shielding our planet. Except, in the case, that a giant burst in the sun corona ejecting a large stream of electrical charged particles produces a very strong magnetic field (moving charged particles produce magnetic and electric fields). And if this magnetic field happens to line up with earth own magnetic field, their interaction (charged particles interacting with the earth magnetic flux) makes the earth magnetic flux to move closer to the earth surface, where the magnetic reluctance is now smaller due to the flow of charged particles through the outer space surrounding earth. In a magnetic field the flux follows the path of least magnetic reluctance. The axis of the earth magnetic field pass through center of the earth and it is normally inclined 11 degrees with respect to the earth rotational axis. As a consequence of the geomagnetic storm, the earth's magnetic field becomes stronger and with a different inclination. The shift of the earth's magnetic flux generates voltages and induces circulating currents in closed metal circuits, like power transmission lines. The magnitude of the induced voltages and currents could reach high values with destructive results. Geomagnetic storms last as long as the charged particles keep coming from the sun's blast with great intensity and density. It could last hours. The flux lines representing the earth magnetic field must form closed loops, and they don't have a beginning or an end. The earth's magnetic flux lines go from the North pole which is located in the southern hemisphere to the South pole located in the northern hemisphere and then sinks into the earth center to complete the magnetic loop. The magnetic field extends through the entire planet, including the interior where it originates due to the rotation of the earth's molten iron core. The earth magnetic field extends for great distances into space. It could be detected in space for about sixteen earth radiuses in the direction of the sun.

Since the advent of electricity, back at the beginning of the 20th century, we have very few of this kind of event; in fact I can only remember the one that occurred in 1989 and affected the Northeast of the USA and part of Canada. So this event is rare and of limited scope. It did not affect the entire country and the electrical equipment destruction was very limited. So the power grid survived the event without major consequences. Nothing we can do will alter the relation between the sun and the planet earth. So, we need to deal with the geomagnetic storms accepting that anything we do, does not guarantee absolute protection.

12.2 Electromagnetic Interactions

Elementary particles have a dual aspect. Physicists use the particle aspect when discussing the interactions between them and the wave aspect when considering their propagation in space. The most recognized elementary particles are: electrons, protons, photons, positrons, and antiprotons. The photon (γ) is the particle of electromagnetic radiation and it has zero rest mass, zero rest energy, zero charge, spin 1 and infinity lifetime. To exist photon most move, it cannot exists at rest and it most travel at the speed of light.

Well known charged particles are: The electron (e-) which has a rest mass of **1**, a rest energy of **0.511** MeV/c2, **-1** electronic charge, spin **1/2**, and infinity lifetime. The positron (e+) has the same properties than the electron, except that the charge is **+1**.

The interaction between electromagnetic radiation and charged particles are always considered as the interaction between photons and charged particles and occurring at a single point in space and time. When an atomic fission weapon detonates, the nuclides inside the fission weapon exist in any one of the available quantized energy states from which they could decay with the emission of high energy photons. *A photon emitted from the nucleus of an atom in excited states is called a gamma ray.* Figure 12-1 depicts the basic electromagnetic interaction between a high energy photon and an electron using a Feynman diagram in space-time. The quantity of energy transferred to the electron depends of the angle of deflection. No energy is transferred when the angle is zero, and when the deflecting angle is equal to 180 degrees the maximum quantity of energy is transferred to the electron (Compton electron), in this case the scattered photon actually reverse its trajectory.

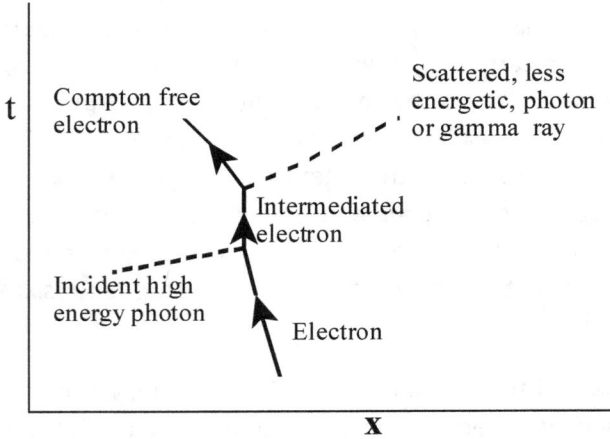

Figure 12-1 Feynman diagram of the basic electron-photon interaction.

Electrons are represented by solid line-arrows. Photons by dashed lines. A vertical line means particle at rest. A horizontal line means a particle moving at infinity speed. X provides the particle position in space and includes all three dimensions of classical space, t indicates the time.

Figure 12-2 provides a specific illustration of the Compton-effect triggered by an atomic high altitude detonation which creates an avalanche of high energy photons or gamma rays directed toward earth.

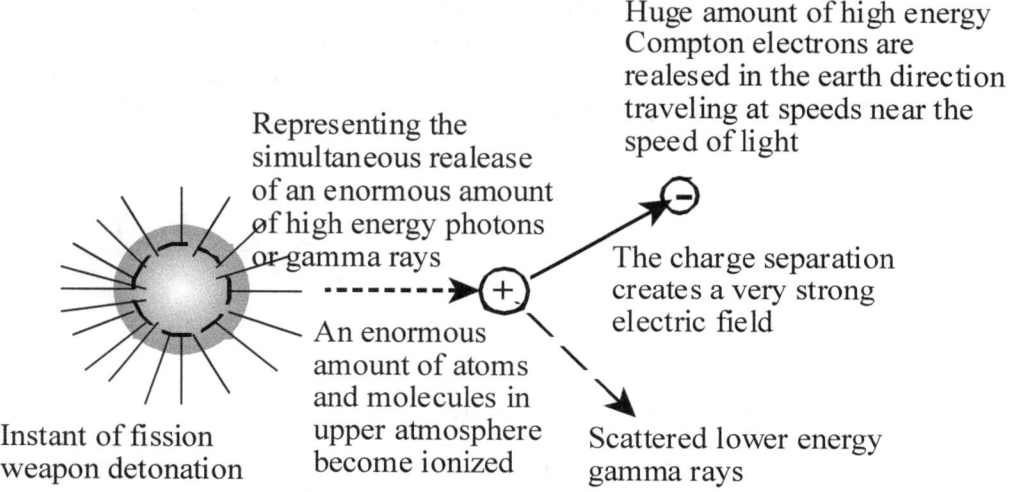

Figure 12-2 Illustration of the interactions occurring at the instant of detonation of a high altitude atomic weapon.

12.3 Elementary Quantum Electrodynamics and the Electromagnetic Force.

The theory of classical fields was developed back in the 19th century when the interactions between the electrical and magnetic fields was discovered by Faraday and mathematically described by Maxwell in his famous differential equations. In fact, our current technological development is largely based on our understanding of electromagnetic field. The following knowledge has been part of our civilization for many years.

- A changing magnetic field creates an electrical field, and the converse is also true.
- Electric and magnetic fields can exist in otherwise empty space and very remote from any material object.

- Electric and magnetic fields can coexist in the same space constituting the traveling electromagnetic field or electromagnetic radiation. Which is setup and maintained by a self-sustained action in which one field creates the other.

- Electromagnetic radiation, light included, consists of a self-propagating transverse oscillating wave that consists of electric and magnetic fields which are always in-phase but in two different planes which are at 90 degrees from each other.

The electromagnetic radiation (EMR) is emitted and absorbed by charged particles and propagates at the speed of light. It is composed of an electric field and a magnetic field which are in phase and oscillate perpendicular to each other, and also they are perpendicular to the wave propagating direction. The EMR nature is explain in classical physics as produced by accelerating charged particles. For instance, the electrical current produces EMR. However, quantum processes also emit EMR. Like when atomic nuclei are experienced gamma decay (gamma rays). The energy content in gamma ray photons is much higher than the energy of visible light photons. Photons are the basic ingredient of EMR and their energy content are proportional to the frequency of the radiation. This fact is symbolically expressed by the Planck's equation (12-1).

$$E = h \cdot f \quad \text{Joules per photon} \tag{12.1}$$

E is the energy content per photon

f is the photon frequency cycle per second

h is the Planck's constant, 6.62559 x 10-34 $\quad$ Joule $-$ seconds

Base in this solid foundation the concept of quantum fields was conceived and developed. In fact, the nature of the electromagnetic force is now described with great precision at the elementary particle level by the Quantum Electrodynamics (QED) theory. Full of the enthusiasm generated by the success of the QED in describing the electromagnetic force, scientist reached the conclusion that the same pattern of reasoning must be valid for all the other forces. However, trying to fit nature into the pattern has complicated the theory, and the picture is very imaginative but not very clear.

The QED theory adopted the pattern set by the Coulomb's law as the pattern to follow in describing the forces between elementary particles. As any engineering student knows, the symbolic expression of Coulomb's law is:

$$F = k \cdot \frac{Q_1 \cdot Q_2}{d^2} \qquad \text{Where k is a proportionality constant.} \tag{12.2}$$

The force (F) between two electric point charges (Q_1 and Q_2) is proportional to the product of their charges and inversely proportional to the square of the distance (d) between them. The force acts along the line joining the two point charges. If the two charges have the same sign, the force is repulsive; it would be attractive if the charges are of different signs.

Equation 12.2 can be transformed to express the strength of the electromagnetic interaction independently of the distance of separation and the units used to express this separation.

$$F \cdot d^2 = k \cdot \left(Q_1 \cdot Q_2 \right)$$

$$\frac{F \cdot d^2}{h \cdot c} = \frac{k}{h \cdot c} \cdot \left(Q_1 \cdot Q_2 \right) \qquad \text{If} \quad Q_1 = Q_2 = 1$$

$$\frac{k}{h \cdot c} = \frac{F \cdot d^2}{h \cdot c} \qquad\qquad (12.3)$$

Particles' electric charge are quantized. That is, any charged particle (with the exception of quarks) have an electric charge that is an integral multiple of the proton charge. Therefore, the interaction between two protons has the minimum possible value of the electro-magnetic interaction. So it is convenient to define the electric charge of a proton as the unit of positive charge. So the factor k/hc, which is a dimensionless number, see Eq. (12.3), provides an absolute measure of the strength of the electromagnetic interaction. The factor k/hc is called the *coupling constant* and for the electromagnetic interaction has a value of 0.0073.

$$\frac{k}{h \cdot c} = 0.0073 \qquad \text{Coupling constant for the electromagnetic interaction}$$

The QED theory, unlike the classic field theories, does not depend of the action at a distant concept; instead, the interaction between two particles is explained as the result of exchanging a virtual particle, which cannot be observed directly. The particle that conveys the electromagnetic force is a virtual photon, which contrary to a real photon is unobservable. Furthermore, the absorption of a virtual photon does not transmute the absorbing particle into another kind of particle.

A real photon is a particle of electromagnetic radiation that travels at the speed of light. It has energy but does not have any rest mass or charge. Real photons are easily observable, as light is made of real photons.

The electromagnetic force is explained in terms of the constant exchange of virtual photons between the interacting particles. However, the exchanged photons seems to violate the conservation laws of energy and momentum during the time interval between photon emission and photon absorption. The Quantum theory accepts theses violations provided that the excesses in energy and momentum are returned within the time and space constraints of the uncertainty principle. That is:

$$\Delta t = \frac{h}{\Delta E} \qquad \Delta x = \frac{h}{\Delta p}$$

h = Plank's constant, 6.62559x10-34 joule-sec

Δ E = magnitude of the energy violation

Δ p = magnitude of the momentum violation

Δ t = Time interval during which the violation is allow

Δ x = uncertainty in position

The position and velocity of a particle cannot both be known precisely at the same time, and therefore the particle energy is uncertain during a small period of time (Δ t).

The electromagnetic interaction consists of two events separated by a minute interval of time. In the first event, one particle loses energy and momentum, and the other particle has not gained them yet. Immediately after that, the second particle gains the energy and momentum lost by the first particle. For a brief but finite time interval, the conservation laws are violated. The Quantum theory interprets the interactions as follows: The first particle created or emitted something that could not be observed, and this something carried with it the missing energy and momentum and delivered it to the second particle. This something that could not be observed is called a virtual photon. To make the whole transaction legal, physicists imposed time and space limits within which this shuffle is tolerated.

The introduction of virtual particles into the scheme of things affected some very basic concepts, like:

- Virtual photons can spontaneously germinate out of empty space, survive the time allowed by the uncertainty principle, and disappear again.
- Electrically charged virtual particles can also be created in empty space provided that the charge conservation law is not violated. Therefore, they are created and destroyed in pairs of particles and antiparticles. For instance, an electron-positron pair can spontaneously sprout from a photon, virtual or real. Actually, the pair would be quickly annihilated and converted back into a photon.

According to the QED theory, the coupling constant of the electromagnetic interaction is not constant, but increases at very close range that is when the separation of the interacting charged particles is smaller than 0.000000000010 centimeters. The increase of the value of the coupling constant at very short range is explained as follows: The bare charge of interacting particles is really many times greater than the charge measured at long range (the value published), because any one of the charged particles is surrounded by a shield of virtual particles whose charge must be subtracted from the bare charge to obtain the measured charge. But at very close range the shield of virtual particles gradually loses its effectiveness, the coupling constant increases, and the strength of the electromagnetic interaction increases.

12.4 Electromagnetic Pulse

Electromagnetic pulses (EMP) are produced by atomic explosions *within the earth magnetic field.* The effects of the EMP are a function of many factors. One of the most important is the altitude of the detonation. The *electric pulse* produced by a high-altitude atomic fission detonation is designated as HEMP. The higher the altitude the larger would be the area affected by the atomic explosion. However, the severety of the effects caused by HEMP depends also of the intensity and orientation of the earth magnetic field at the particular location of the detonation. Besides, the non-uniform distribution of particles (molecules and atoms) in the upper atmosphere results in asymmetric charge distribution. Producing an asymmetric electrical field. **Furthermore, if the detonation is too high it would not encounter enough atoms and air molecules to be effective.** The test performed by the Soviet Union in 1962, with a relative small atomic weapon over a land mass yielded, in comparison with tests performed by the USA, very severe results. That event affected underground power lines. And it was discovered that there were problems with the ceramic insulators used on overhead power lines. *To fully understand this type of event it would require a large amount of (non-existing) data.*

The electric field pulse generated by a high altitude (around 400km) atomic fission detonation within the earth magnetic field is huge and extremely brief, it reaches its peak value in about 5 nanoseconds with and electric field intensity in the order of 50kV per meter. The electric field is produced by the interaction of gamma rays produced by the atomic explosion with atoms and molecules existing in the earth upper atmosphere. An enormous amount of practically simultaneous interactions produces a very intense electric field due to the charges separation illustrated in Fig. 12-2. And a large portion of the electrons liberated by the Compton Effect are directed toward earth. The incoming electrons polarize all the earth's dielectric materials within the very large area affected by the electrons liberated by the Compton Effect. See Fig. 12-3. The reader should remember that the earth magnetic field is created by earth's inner iron core and therefore it exists in the ground as well as in the atmosphere. As soon as they are created the enormous amount of incoming electrons are deflected by the earth's magnetic field which change their movement direction by 90 degrees, such that after interacting with the earth's magnetic field they move at right angle with respect to it. The loss-energy of the incoming electrons as a consequence of their lost in momentum due to their change in direction, is gained by the pulse's electric field. In a nut shell, *the EM pulse is created because the simultaneous interaction of an enormous amount of high energy electrons with the earth magnetic field.*

The generated pulse is perpendicular to the plane defined by the current (flow of electrons) phasor and the magnetic field phasor. The pulse has an odd shape; it rises very rapidly and then meanders for a few minutes. So, for convenience it is artificially divided in three sections: E1, E2, and E3. *In fact, the event consists in the three of them, and you cannot get only one of them.* The faster component is the E1, which produces an intense and extremely brief electric pulse that last one microsecond and reaches its peak value in about 5 nanoseconds inducing very high voltages in all metals within the area affected. *The E1 pulse is not like a customary pulse that propagates along transmission lines; On the contrary, it hits the entire target region simultaneously. The pulse is non-periodic and it cannot be analyzed by Fourier series. The E1 pulse is not part of a wave and consequently the wave length parameter does not exists.*

The rate of rise of the E1 pulse is so high, in the order of ten thousands volts per nanosecond, that power line surge protectors do not work; it goes through them or flash over them. In the high voltage side of the power system; the pulse disintegrates dielectric materials in insulators, transformers, breakers, and switches. In the low voltage side of the power system, the E1 pulse totally destroy everything including computers and communication equipment. In fact, old technology equipment based in vacuum tubes and electromagnetic relays are better than solid state devices for this kind of duty.

In summary: gamma rays consist of high energy photons emitted from the nuclei of atoms in excited states. An E1 pulse is produced when the gamma radiation from an atomic fission detonation interacts with atoms and molecules in the upper atmosphere and ionizes them by exciting and liberating electrons from their bound energy states. Many of these electrons travel toward earth close to the speed of light, and their interaction with earth's magnetic field change the direction of the electrons' flow, which now flow at right angle to the magnetic field, which simultaneously is moving closer to earth surface. This interaction produces a very intense and very brief electromagnetic pulse that covers most of the earth surface underneath the circular horizon. See Fig. 12.3

The E2 component is similar to a lightning pulse and it last from one microsecond to about one second after the atomic explosion. The protection required against the E2 component, by itself, would be similar to the protection used against lightning strikes. But, the E2 component follows immediately after the E1 component, encountering a damaged protection system and therefore it complements the destruction cause by E1 component.

The E3 pulse is similar to a geomagnetic storm. See section 12-1. It last from one second to hundreds seconds and is produced by the atomic explosion temporary distortion of the earth magnetic field. The E3 pulse induces additional voltages and currents in transmission lines and therefore it could damage power transformers by saturating theirs iron core.

12.5 Smart Power Grid.

 The power grids can be easily seen by anyone in the United States. It consists fundamentally of power generating plants, electrical transmission lines and electrical substations. And they have been around since the beginning of the 20th century and worked very fine, helping and contributing to the developing of this big country. In the other hand the smart-grid is a figment of the imagination of uneducated people trying to apply what they think they know. They want to expand the cell-phones and Disney Land technology to everything and by doing so, hope to participate in something important, the energy revolution. This group of well-meaning people had attracted a large number of desperate politicians looking for a cause and a bunch of "wannabe" entrepreneurs.

To begin with; photovoltaic cells, parabolic-mirrors sun collectors, wind powered electric generators, seaweed and cattle generated methane gas cannot and will not replace oil, gas, coal and hydro-power. And sometime in the 22nd century atomic-power will replace all of them. There is not any other way. In the meantime the green-energy gurus are building a monumental house of paper with no foundations and no resistance to any kind of intellectual attack.

In fact, the electrical power grid is composed of a giant conglomerate of generating power plants, high-voltage transmission lines, and electrical substations. In the United States there are three major electrical power grids: The Northeastern grid, the Western grid and the Texas grid.

Most of the proposed improvements to be incorporated in the "smart grid" takes place in the low voltage side of power distribution radial feeders and at the user's home and **not in actual high voltage power grid.** Below is a short listing of the proposed improvements some of them already in the planning state or implementation.

- Home demand management. Including controlling home and business air conditioners.

- Demand curtailment during peak time. Including automatically changing the set points of smart air conditioners in homes and business of an entire city. Or simple yet turning them off. Also turning off, at peak hours, all the smart-dish-washing machines. They hope that this reduction would make unnecessary the use of spinning generators reserve. *Here again the improvements proposed by the smart grid advocates means that some government agency will control the delivery of electrical power to the customers' home which nullify and void all customer rights to privacy and self-determination.*

- The first step to take total control would be to replace the induction watt-hour meters and all meter reading equipment with microprocessors based smart meters in the user home or business. Next they will introduce the smart electrical sockets which will become enforces of the government regulations.

- Instead of the 60 Hertz power they would supply variable frequency electrical power to home and business. They hope that this garbage will reduce the ON-time of consumers' electrical devices.

- The smart grid advocates would advertise that the above "improvements" will translate in higher efficiency, reliability and safety. WRONG!

The smart grid advocates should demand that the local power distribution feeders and branches be converted from overhead to underground; that change, for sure, will be enthusiastically welcome by the people.

12.6 Faraday Cage and the E1 Pulse

The Faraday cage or mesh is a box made of conducting material that can block the effects of an external electric field, static or non-static. But it cannot block static or slow changing magnetic fields, like the earth magnetic field. *However, the cage can block incoming electromagnetic radiation if the holes in the mesh are smaller than the wavelength of the magnetic radiation provided that the shielding conductors are of the proper thickness.*

Any incoming electromagnetic radiation is composed of electrical and magnetic waves; these waves move perpendicular to each other creating a self-propagating electromagnetic wave (the moving of one field creates the other). Both field components are eliminated or significantly attenuated by the Faraday cage. The electrical field is eliminated inside the cage because the external electric field changes the distribution of the electrical charges in the cage's conducting material in such a way (separates them) that it nullifies the electric field inside the cage. And the magnetic field is significantly attenuated by the charge movement within the walls (metal wires) of the enclosure. This charge movement (current) in the conducting material of the enclosure is of such direction that by the Lenz law, will oppose the cause (varying magnetic field) that produced the charge movement.

The High altitude E1 pulse produced by the detonation of a fission atomic weapon at high altitude within the upper atmosphere produces a cluster of very strong radial *electric fields* due to the electric charges separation occurring as a consequence of the interaction between the gamma rays released by the atomic explosion and atoms and molecules existing in the upper atmosphere, see Fig. 12-2. The enormous amount of high energy Compton-electrons simultaneously created within the impact cone of the atomic explosion, interacts with the magnetic field surrounding our planet and generates the E1 pulse. See Fig. 12-3   The E1 pulse does not comes isolated, actually is only a part of the total incoming HEMP that follows a high altitude atomic explosion. So, the hope is that the dielectric materials of the power lines insulators' will withstand the E1 pulse and the immediately following E2 and E3 pulses

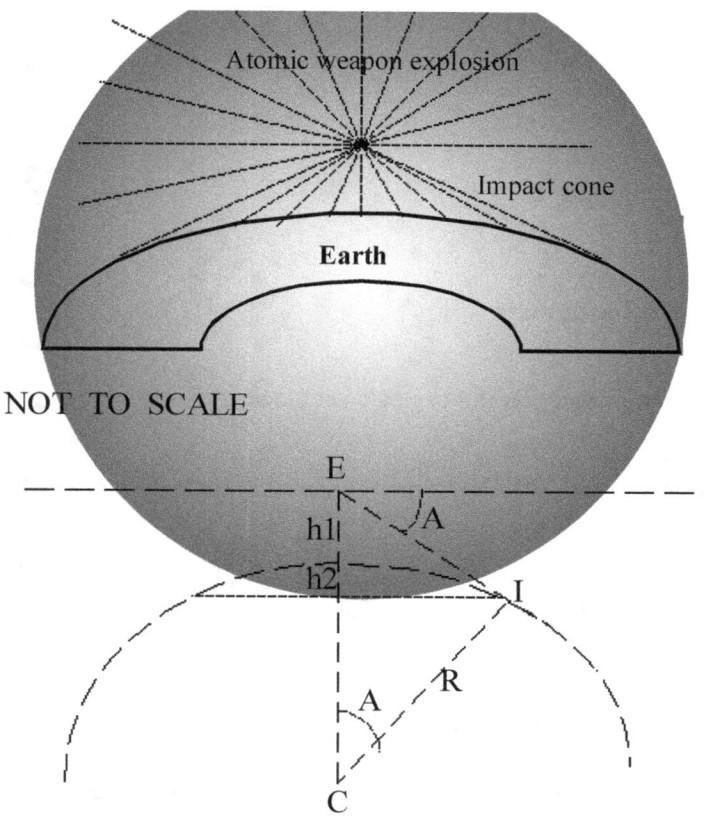

Figure 12-3 Illustration of the earth surface area affected
by the explosion of a high altitude atomic weapon

Assuming that earth is perfectly spherical and that the explosion is symmetrical, we have:

$R = 6378km$

$(R + h1)^2 = R^2 + (EI)^2$

$EI = \sqrt{(R + h1)^2 - R^2} = \sqrt{2 \cdot h1 \cdot R + h1^2}$

$R := 6378$

$h1 := 400 \quad km$

$\sqrt{2 \cdot h1 \cdot R + h1^2} = 2294 \qquad EI := 2294 \quad km$

$S = 2 \cdot \pi \cdot R \cdot h2 = \pi \cdot (EI)^2$ Formula: Surface area affected by the explosion

$$\pi \cdot 2294^2 = 16.532 \times 10^6 \quad \text{Square kilometers}$$

$$h2 = \frac{16.532 \times 10^6}{2 \cdot \pi \cdot R} = 378.299 \quad \text{kilometer}$$

The surface area affected by the E1 pulse is:

$$16.532 \times 10^6 \quad \text{Square kilometers that the explosion could cover}$$

$$0.3861 \cdot \left(15.532 \times 10^6\right) = 5.997 \times 10^6 \quad \text{Square miles that the explosion could cover}$$

USA: $\quad 9.372 \cdot 10^6 \quad$ Square kilometers

USA: $\quad 0.3861 \cdot \left(9.372 \cdot 10^6\right) = 3.619 \times 10^6 \quad$ Square miles

The reader should realize that the illustration in Fig. 12-3 gives the impression that the explosion coverage is smaller than really is.

12.7 Improving the Protection of the Smart Grid

Let us assume the following data:

Electric field intensity peak value = 50000 volts per meter

Electric field rate of rise = 10000 volts per nanosecond

$$\xi = 50 \frac{kV}{m}$$

The protection of any of the existing power grids against E1 electromagnetic pulse requires to improve the existing shielding as follows:

1. The number of overhead grounded wires running at the highest points of transmission line towers should be increased to no less than one per phase. The actual location of the shield wires depends of the tower design and the number of hot wires on the tower. If the phase wires are vertically installed, then the three ground wires must form a triangle around the three vertical phases. The cones of protection of the shield

wires must overlap not less than 50%. The shield wires must be sized to withstand the very high current flowing during electric charges re-arranging.

2. Substations must be completely surrounded by shield wires. The idea is to surround the substation equipment by a grounded Faraday cage. The roof shielding wires' horizontal separation must be 24 inches or less, where possible. The vertical sides must go all the way down to the ground grid. The vertical wire's separation should be equal to the separation on the roof, 24 inches or less. The substation fence could be part of the vertical sides. The E1 pulse is a single electric pulse that does not propagate and does not has a wave length. Therefore the size of the cage's holes is not critical, however, the size of the shielding wires is very critical.

3. The electrical connections from generators to step-up transformers and to substation busses must consist of bus-ducts with their thick metal enclosures grounded directly to the ground grid at several points. The substation-side-end of all bus ducts must be protected with surge suppressor, lightning arresters and air gaps.

4. In substations, incoming and outgoing transmission lines should fly-over the substation and end at a close-by dead-end structure, situated outside the substation. Equipment connecting conductors should have different surge impedance than the transmission line conductors and drop vertically into the substation maintaining all the necessary clearances to grounded structures and equipment and between phases. The actual connecting point must be at the output side of switches and breakers, with appropriates lightning arresters and air-gaps on each phase.

5. High voltage insulators in transmission lines and substations equipment must be designed to facilitate flash-overs at extremely high voltages.

6. Underground cables-runs must be made of single-phase cables, each inside a continuous metal shield. Each metal shield must be separately grounded at both ends.

7. The basic impulse level (BIL) rating of any insulator, equipment or device must be 150% of the IEEE recommended value for the specific duty.

8. Must-survive instrumentation should be located in safe-rooms. Surrounded by metallic walls and doors of the proper design.

9. The power generating building must be designed for complete protection of all generators, turbines, their auxiliary equipment and attending personnel.

10. The most important protection factor is a simple declaration by the President of the USA telling our potential enemies that a nuclear weapon attack on the United States will trigger a devastating H-bomb response coming from our submarines that are immune to the E1 electromagnetic pulse. *They don't need to ask me, our military forces already have the orders.*

11. A safer approach would be to move all transmission lines and substations to underground (more than 20 feet deep), which is a remote and very expensive solution. But, underground AC cables have significant capacitance which significant diminish their efficiency beyond 50 miles. If we go underground, it better be DC. Because DC cables can run for thousands of miles at high efficiency.

12.8 Recommended Design Improvements for the Smart Grid

- The difference in natural frequencies of oscillations, among all the generators connected to the grid should not be smaller than 7%.
- Do not allow the installation of coherent generators at different power generating stations connected to the same power grid. Coherent generators installed in the same power plant could be OK.
- Use large braking resistor banks to dissipate transients' energy; They should have a short-time rating in the megawatts size (100 MW or more) and should be automatically shunt-connected across the line only in case that the power angle is growing very fast or that the accelerating power of the power angle oscillate with large amplitude. The ON time rating of these braking resistors should be in the order of 1 second.
- Phasor measurement units (PMU), which directly measures the power angle between the buss voltages and the generated emfs should be placed in remote and important busses to watch the grid and automatically decide when to energize the braking resistors. The PMU unit must continuously scan all signals coming from remote locations.

Bibliography

Power Plant Stability, Capacitors, and Grounding

Numerical Solutions.

Orlando N. Acosta

McGraw-Hill Companies, Inc., 2012

Symmetrical Components for Power Systems Engineering

J. Lewis Blackburn

Marcel Dekker, Inc., 1993

Power Systems Modelling and Fault Analysis

Nasser D. Tleis

Elsevier Ltd., 2008

Elements of Power Systems Analysis, Third Edition

William D. Stevenson

McGraw-Hill, Inc., 1975

Electric Circuits

A First Course in Circuit Analysis for Electrical Engineers

By Members of the Staff of the Department of Electrical Engineering

Massachusetts Institute of Technology

John Wiley & Sons, Inc., 1940

Alternating-Current Circuits, 4th edition

Russell M. Kerchner and George F. Corcoran

John Wiley & Sons, Inc., 1960

314

Applied Protective Relaying

For Individual or Class Study

A New "Silent Sentinels" Publication

Westinghouse Electric Corporation

Relay-Instruments Division

Electric Machinery

The processes, Devices, and Systems of Electromechanical Energy Conversion.

Third Edition

A. E. Fitzgerald, Charles Kingsley, JR., Alexander Kusko

McGraw-Hill Book Company 1971

Principles of Electric Power Transmission

L. F. Woodruff

Second Edition

John Wiley & Sons, Inc. 1938

Power System Control and Stability

P. M. Anderson and A. A. Fouad

Iowa State University Press, 1977

Transmission and Distribution Electrical Engineering

Dr. C. R. Bayliss Ceng

Butterworth-Heinemann, 1996

Stability of Large Electric Power Systems

Richard T. Byerly, Edward W. Kimbark

IEEE Press, 1974

Power System Analysis

John J. Grainer, William D. Stevenson, Jr.

McGraw-Hill, Inc. 1994

High Voltage Engineering Fundamentals, 2nd Edition

E. Kuffel, W.S. Zaengl, J. Kuffel

Butterwort-Heinemann

Power System Dynamics, Stability and Control, 2nd Edition

Jan Machowski, Janusz W. Bialek, James R. Bumby

John Wiley & Sons, Ltd. 2008

Fundamentals of Engineering Numerical Analysis, 2nd Edition

Parviz Moin,

Cambridge University Press, 2010

IEEE Guide for Generator Ground Protection

IEEE Std C37.101-2006

IEEE Guide for the Application of Neutral

Grounding in Electrical Utility Systems—Part 1: Introduction.

IEEE Std C62.92.1-2000

IEEE Guide for the Application of Neutral

Grounding in Electrical Utility Systems—Part II-Grounding

of Synchronous Generator Systems.

IEEE Std C62.92.2-1989 (R2005)

www.ingramcontent.com/pod-product-compliance
Lightning Source LLC
Chambersburg PA
CBHW080759180526
45168CB00006B/2259